PRINCIPLES AND APPLICATIONS OF ION SCATTERING SPECTROMETRY

Wiley-Interscience Series on Mass Spectrometry

Series Editors

Dominic M. Desiderio
Departments of Neurology and Biochemistry
University of Tennessee Health Science Center

Nico M. M. Nibbering
University of Amsterdam

The aim of the series is to provide books written by experts in the various disciplines of mass spectrometry, including but not limited to basic and fundamental research, instrument and methodological developments, and applied research.

Books in the Series

PRINCIPLES AND APPLICATIONS OF ION SCATTERING SPECTROMETRY

Surface Chemical and Structural Analysis

J. Wayne Rabalais

Cullen Professor of Chemistry and Physics, University of Houston

WILEY-INTERSCIENCE

A JOHN WILEY & SONS, INC., PUBLICATION

Chemistry Library

CHEM
011894520

Copyright © 2003 by John Wiley & Sons, Inc. All rights reserved.

Published by John Wiley & Sons, Inc., Hoboken, New Jersey.
Published simultaneously in Canada.

No part of this publication may be reproduced, stored in a retrieval system, or transmitted in any form or
by any means, electronic, mechanical, photocopying, recording, scanning, or otherwise, except as
permitted under Section 107 or 108 of the 1976 United States Copyright Act, without either the prior
written permission of the Publisher, or authorization through payment of the appropriate per-copy
fee to the Copyright Clearance Center, Inc., 222 Rosewood Drive, Danvers, MA 01923, 978-750-8400,
fax 978-750-4470, or on the web at www.copyright.com. Requests to the Publisher for permission should
be addressed to the Permissions Department, John Wiley & Sons, Inc., 111 River Street, Hoboken,
NJ 07030, (201) 748-6011, fax (201) 748-6008, e-mail: permcoordinator@wiley.com.

Limit of Liability/Disclaimer of Warranty: While the publisher and author have used their best efforts in
preparing this book, they make no representations or warranties with respect to the accuracy or
completeness of the contents of this book and specifically disclaim any implied warranties of
merchantability or fitness for a particular purpose. No warranty may be created or extended by sales
representatives or written sales materials. The advice and strategies contained herein may not be suitable
for your situation. You should consult with a professional where appropriate. Neither the publisher nor
author shall be liable for any loss of profit or any other commercial damages, including but not limited
to special, incidental, consequential, or other damages.

For general information on our other products and services please contact our Customer Care Department
within the U.S. at 877-762-2974, outside the U.S. at 317-572-3993 or fax 317-572-4002.

Wiley also publishes its books in a variety of electronic formats. Some content that appears in print,
however, may not be available in electronic format.

Library of Congress Cataloging-in-Publication Data is available.

ISBN 0-471-20277-0

Printed in the United States of America.
10 9 8 7 6 5 4 3 2 1

QC
7022.7
S 3 R 3
2003
CHEM

To Becki and Lillian

CONTENTS

SERIES PREFACE

This series provides books written by experts in every area of mass spectrometry, including basic and fundamental research, instrument and methodological developments, and applied research. The books in this series will be of use not only to researchers who use mass spectrometry and wish to focus on one particular area, but also to teachers in the classroom and newcomers to the field of mass spectrometry. Mass spectrometry is being used in a variety of rapidly developing disciplines, and this series will provide an effective way to collect pertinent information in each area. Finally, the sum total of the research collected within this book series will be of interest to researchers in related areas such as chemistry, physics, biology, medicine, and nutrition.

PREFACE

Ion scattering and recoiling spectrometry has developed over the last 30 years into an important surface elemental and structural analysis technique. This is mainly due to its high sensitivity, its ability to detect all elements, and its capability of surface structure analysis. Such information is vital to understanding the physical and chemical properties and the reactivity of surfaces.

Scattering and recoiling contribute to our knowledge of surfaces through (1) elemental analysis, (2) structural analysis, and (3) analysis of electron exchange probabilities. For elemental analysis, it has high sensitivity to the outermost surface atomic layer, and it is capable of direct detection of all elements, including hydrogen. For structural analysis, it is capable of probing the positions of all elements in real space with an accuracy of ≤ 0.1 Å. It is sensitive to short-range order and provides a direct measure of interatomic distances in the first and near-subsurface layers as well as a measure of surface periodicity. For measurements of electron exchange probabilities, it is capable of measuring the scattered and recoiled neutrals and ions, thereby determining the scattered and recoiled ion fractions. The mechanism of charge transfer and the prediction of the charge composition of the flux of scattered, recoiled, and sputtered atoms in the interaction of low-energy ions with surfaces is a major unsolved problem in physics. Emphasis in this book has been placed on the experimental methods, the physical concepts, the time-of-flight method of ion scattering spectrometry, and the structural applications of the technique. The technique of ion scattering has now advanced to the level where it can have a significant impact in areas as diverse as thin film growth, catalysis, hydrogen embrittlement and penetration of materials, surface reaction dynamics, analysis of shallow interfaces, and analysis of plasma-surface interactions.

The book has been prepared with the aim of serving both a pedagogic need and a research need. The pedagogic function is particularly evident in Chapters 1 to 5; these chapters are written at the level of senior undergraduates or beginning graduate students in chemical physics. Many figures and illustrative diagrams are included to exemplify the discussions. The research function is particularly evident in Chapters 6 to 9; these chapters contain material that is at the brink of current research in the area. In order to keep a keen research edge on the book, specific references have been compiled at the ends of each chapter, and Chapter 10 is a bibliography of ion scattering publications. Many illustrative figures are provided throughout the text.

The text is organized as follows: Chapters 1, 2, and 3 are introductory, theoretical, and experimental; Chapters 4 and 5 provide general features and structural analysis; Chapter 6 contains the recent technique of scattering and recoiling imaging spectrometry; Chapter 7 provides examples of structure analysis; Chapter 8 addresses ion–surface charge exchange phenomena; Chapter 9 discusses hyperthermal ion–surface interactions. An attempt has been made throughout the text to amalgamate theory and experiment. The approach has been to use classical mechanical models to interpret experimental results, for I believe that the material can best be grasped and remembered in this manner. The emphasis is on time-of-flight techniques in ion scattering and recoiling spectrometry and their application to surface structure analysis. The detailed derivations of the classical scattering equations could be omitted in a first reading; however, it was felt that this treatment should be included for the sake of students or researchers who wish to explore extensions or modifications of the methods.

The writing of this manuscript is a logical, if not inevitable, consequence of my involvement with ion scattering spectrometry over the last 25 years. I am deeply indebted to the many persons who assisted in my efforts. These people include students, postdoctoral associates, visiting scientists, and other researchers throughout the world who I have interacted with at meetings and conferences; they are too numerous to mention all of them. I am grateful to my wife, Becki, and my granddaughter, Lillian, for their encouragement, patience, and assistance throughout the course of this work.

My research in this field and the preparation of the manuscript was prepared during my employment at the University of Houston. Much of the research efforts and the preparation of the manuscript were supported by the National Science Foundation, the R. A. Welch Foundation, and the Texas Advanced Research Program. I am grateful to the university and the granting agencies for their assistance.

I acknowledge my debt to the American Chemical Society, American Institute of Physics, American Physical Society, Elsevier Science Publishers, Wiley-VCH Verlag GmbH, Burle Industries, Inc, Korean Chemical Society, and Peabody Scientific for permission to reproduce figures and tables.

Critical comments by the readers will be greatly appreciated, for it is difficult to write a completely up-to-date book in such a rapidly advancing field and to list all of the publications in ion scattering in the Bibliography. It is my humble hope, however, that the book will prove useful to students and researchers of this fascinating and intriguing technique.

Houston, Texas   J. WAYNE RABALAIS
April 2002

1

INTRODUCTION

Collisions of energetic ions with the atoms of a solid surface can result in scattering of the primary ions and recoiling of the surface atoms. These scattered and recoiled ions and atoms have discrete kinetic energies that are determined by the nature of the collision. Analysis of these energies and their angular distributions can provide direct information on the identity of the surface atoms and their structural arrangement in the surface.

1.1. ION SCATTERING SPECTROMETRY

Ion scattering spectrometry consists of using a monoenergetic, mass-selected, collimated beam of ions in the low keV energy range to irradiate a surface. As a result of the interaction of these ions with an atom or several atoms of the target, some of the primary ions are reflected from the surface, and some of the target atoms can be recoiled in such a direction that they also leave the surface. Both the scattered primary and recoiled surface particles are atoms that may be in neutral, positive, or negative charge states due to electronic charge exchange processes with the surface itself. These scattered and recoiled atoms have discrete kinetic energies in the low keV range as a result of quasi-single collisions from the impinging ions. The ions and atoms that are scattered or recoiled at a well-defined scattering–recoiling angle are analyzed for their kinetic energies using an electrostatic or magnetic energy analyzer or for their velocities using time-of-flight techniques (TOF). The atoms and ions are detected by an electron multiplier detector with a small acceptance solid angle. The detector signal is plotted as a function of the analyzer pass energy or the TOF. When a large-area, gated, position-sensitive microchannel plate detector is used, time-resolved spatial distribution images of the scattered and recoiled atoms are obtained. Since the scattering process is mass-dispersive, the energies or TOFs of the scattered and recoiled atoms provide a mass spectrum of the constituent atoms in the surface.

The energy and TOF spectra and the images contain information about the elemental composition of the surface and the surface atomic structure, thereby making ion scattering spectrometry a surface compositional and structural analysis technique. This analysis is straightforward because the kinematics of energetic atomic collisions is accurately described by classical mechanics. Such scattering occurs as a result of the mutual coulomb repulsion between the colliding atomic cores, that is, the nucleus plus core electrons. The scattered atom loses some of its energy to the target atom. The latter, in turn, recoils into a forward direction. The final energies of the scattered and recoiled atoms and the directions of their trajectories are determined by the masses of the pair of atoms involved and the closeness of the collision. By analysis of these final energies and angular distributions of the scattered and recoiled atoms, the elemental composition and structure of the surface can be deciphered.

Atomic collisions in the keV energy range can result in transfer of energy to both translational and internal degrees of freedom of the target atoms. If the collision involves only transfer of translational energy, it is known as an *elastic collision*. This is sometimes called a *hard sphere* or *billiard ball* collision. If the collision involves transfer of both translational and internal energy, it is known as an *inelastic collision*. Transfer of energy to internal degrees of freedom results in electronic excitation of the target and/or the projectile atom. This can produce a variety of phenomena, such as secondary electron, Auger electron, and photon emission. The physics of the mechanism of electronic excitation in atomic collisions is currently an active area of investigation. Since these inelastic energy losses are typically less than 5% of the beam energy for atomic collisions in the low keV energy range, they can be ignored for most elemental and structural analyses.

1.2. IMPORTANCE OF SURFACES

When we observe the world around us, we see predominantly the surfaces of objects. Interactions of these objects with their surroundings are largely through their surfaces. A detailed understanding of the surface and its properties is of utmost importance for understanding its interactions with the environment and for developing materials with specific properties. Since the surface forms the boundary between materials and gases or liquids, it is the surface that reacts with the environment. Adsorbed species on surfaces occupy specific chemically active sites, and atoms of the surface itself often arrange themselves so as to have different symmetry or different interlayer spacings from that of the bulk; the former is known as *reconstruction* and the latter as *relaxation*. Knowledge of surface composition and atomic positions is important for many applications: (a) understanding the oxidation of metals resulting in rusting and tarnishing; (b) the reactive sites in catalysis; (c) the chemical conversion of gaseous molecules by a catalyst; (d) friction; (e) adhesion of paint and glue to surfaces; (f) defining atomic templates for epitaxial film growth; and (g) fabricating well-defined interfaces between different materials. From a heuristic viewpoint, we are interested in knowing if atomic sites and bond lengths in surfaces are as well defined as those of the bulk and if there are important new phenomena to be derived therein. Although the many significant advances in surface science over the past three decades have greatly increased our understanding of surface phenomena, the incessantly decreasing size of microelectronic devices and their requisite atomic-scale surface analysis drive the development of surface science techniques. The study of surfaces is therefore interesting from both a scientific and a practical point of view.

There are three basic questions that we would like to answer about surfaces: (1) What is the elemental composition of the surface? (2) What is the structural arrangement of the surface atoms? (3) What are the electronic properties of the surface? A battery of surface analysis techniques has been developed over the past 30 years to answer these questions. These techniques are based on the interaction of particles (i.e., photons, electrons, atoms, molecules, or ions), with a surface, followed by detection of the scattered primary particle or the emitted secondary particles (i.e., electrons, photons, ions, atoms, molecules, or fragments). The experimental conditions are chosen so that the incident particles and/or the secondary particles interact primarily with the outermost atomic layers of the solid; therefore, the measured signals originate from the surface or subsurface regions.

1.3. SURVEY OF ION-SURFACE INTERACTIONS

Since this book is concerned with deriving composition and structural information about surfaces from ion scattering techniques, it is informative to survey the range of interactions accessible when using ion beams on surfaces. Ion beams provide a method of delivering unique atomic and molecular species to surfaces while controlling the interaction parameters by means of the ion kinetic energy. Consider the various phenomena that can occur when ions impinge on surfaces as illustrated[1] in Figure 1.1.

1. Electronic interactions—The incoming ions can be neutralized by electron capture from the surface. In the reverse process, electrons from incoming ions or atoms can be captured by the surface. Either process can result in excited electronic states. These processes depend on the relative energies of the filled and unfilled energy levels of the atoms and the surface.

2. Photon and electron emission—As a result of electron exchange and energy level crossings in the close collision encounters between atoms, electrons can be promoted to highly excited states that can relax by either photon or electron emission.

3. Scattering and recoiling—Ions can be scattered from the surface and atoms of the surface itself can be recoiled either into or out of the surface in positive, neutral, or negative charge states.

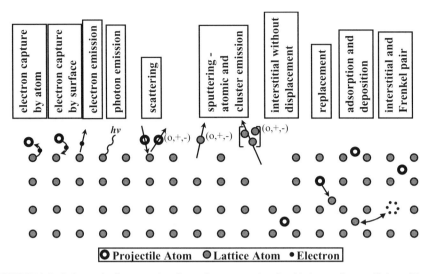

FIGURE 1.1. Schematic diagram of various phenomena involved in ion surface collisions (From Rabalais, 2001, with permission.)

4. Sputtering—The momentum imparted to surface atoms by impinging ions can cause collision cascades in the material, resulting in "sputtering" of low-energy atoms, molecules, fragments, and clusters in various charge states.

5. Adsorption, desorption, and chemical reactions—The impinging ions can be adsorbed on the surface, they can cause desorption of atoms and molecules from the surface, and chemical reactions can occur between the constituents.

6. Interstitials, displacements, and replacements—The ions can be inserted into the lattice as interstitial atoms without displacement of host atoms or they can displace host atoms, thereby creating Frenkel pairs. This can result in radiation damage.

A variety of ions with kinetic energies in the range 10^{-2}–10^7 eV can be used to probe surfaces and interfaces. Various phenomena are dominant or emphasized in different energy regions and thereby, the chemical and physical processes that are induced by the ion impacts are also controlled by this energy. The terminology that has evolved to define approximate energy ranges is as follows: <1 eV, thermal; 1–500 eV, hyperthermal; 0.5–10 keV, low energy; 10–500 keV, medium energy; and >0.5 MeV, high energy.

A schematic diagram of the various phenomena involved, along with an energy scale, is shown in Figure 1.2.[1] The equivalent translational temperature range and the common terminology for describing the various energy ranges are indicated on the lower abscissa. Particle beams in the range of 10 meV–10 eV probe the long-range atomic potentials and exhibit quantum mechanical and diffraction effects. As energy increases, the de Broglie wavelength of the particles becomes significantly

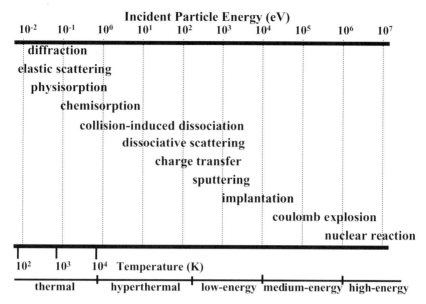

FIGURE 1.2. Selected particle–surface processes as a function of ion kinetic energy (From Rabalais, 2001, with permission).

smaller than the interatomic distances in the crystal lattice ($\sim$5 Å), and the scattering process can be treated by classical mechanics. At energies above $\sim$10 eV, the repulsive potential begins to dominate the interaction and the surface potential becomes highly corrugated. In the range of $\sim$10–100 eV, it is difficult to make ion beams of sufficiently high intensity for ion scattering due to space charge limitations. In the low-energy regime between 0.5–10 keV, the space charge limitation decreases, and the ions interact with the atomic cores of the lattice atoms. Hence, the surface appears more open and penetration can occur. Various interactions, such as sputtering, implantation, surface reactions, and atom deposition can accompany the scattering process. Ions that penetrate are efficiently neutralized. In the medium energy range of 10–500 keV, most particles penetrate into the lattice and are implanted. Energies above 0.5 MeV are used for doping of semiconductors, and energies above 1 MeV can result in nuclear fusion and fission.

1.4. HISTORICAL DEVELOPMENT OF ION SCATTERING SPECTROMETRY

The origin of scattering experiments has its roots in the development of modern atomic theory in the early 20th century. As a result of both the Rutherford experiment on the scattering of alpha particles by thin metallic foils and the Bohr theory of atomic structure, a consistent model of the atom as a small massive nucleus surrounded by

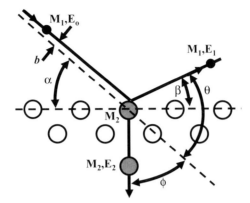

FIGURE 1.3. Atomic collisions illustrating scattering and recoiling from a surface. The angles are: incident (α), exit (β), scattering (θ), and recoiling (ϕ) angles. M_1 and M_2 are the masses of the projectile and target atoms, respectively, E_0 is the initial energy of the projectile atom, E_1 is the energy of the projectile atom after scattering, and E_2 is the energy of the recoiled target atom. b is the impact parameter.

a large swarm of light electrons was confirmed. It was then realized that the inverse process, namely analysis of the scattering patterns of ions from crystals, could provide information on composition and structure. This analysis is straightforward for atomic collisions in the keV range because the kinematics of the event are accurately described by classical mechanics. Such scattering occurs as a result of the mutual coulomb repulsion between the colliding atomic cores (i.e., the nucleus plus core electrons). The scattered primary atom loses some of its energy to the target atom which, in turn, recoils into a forward direction as shown in Figure 1.3. Here, M_1, E_0, and E_1 are the mass, initial kinetic energy, and final scattered energy of the projectile atom, M_2 and E_2 are the mass and recoiled energy of the target atom, and p is the impact parameter of the collision. The final energies of the scattered and recoiled atoms and the directions of their trajectories are determined by the masses of the pair of atoms involved and the closeness of the collision.

In the early 1960s it was shown[2,3] that a clear correlation existed between the energy loss of a scattered ion and the type of surface atoms. Low energy (1–10 keV) ion scattering spectrometry had its beginning as a modern surface analysis technique in the late 1960s with the work of Smith[4] and of researchers in the former Soviet Union. The latter work has been thoroughly reviewed in books and review articles.[5–8] In the following 30 years it has been clearly demonstrated, as noted in the extensive reference list of this book, that direct surface compositional and structural information can be obtained from ISS.

Various names and acronyms have been used for ion scattering spectrometry. The terms ion scattering spectrometry (ISS) and low-energy ion scattering spectrometry (LEIS) are general names. More specific names include TOF scattering and recoiling spectrometry (TOF-SARS),[9] scattering and recoiling imaging spectrometry (SARIS),[10] and impact collision ion scattering spectrometry (ICISS).[11]

1.5. OTHER TYPES OF ION SPECTROMETERIES

There are two types of ion spectrometeries that are related to ISS. These are secondary ion mass spectrometry (SIMS) and Rutherford backscattering spectrometry (RBS). SIMS[12] derives its information from the sputtering process. When primary ions penetrate into the solid, they undergo a sequence of collisions. During each of these collisions, target atoms can be put into motion. These recoiling atoms also generate new collisions, and collision cascades develop in the solid. The cascades develop nearly isotropically so that energy and momentum can be transferred back to surface atoms. If this transfer is sufficient, secondary particles (i.e., atoms, molecules, and fragments in neutral, positive, and negative charge states) are emitted; this is the sputtering process.[13] The energy deposited into the electronic system of the target atoms can also contribute to emission of secondary particles.[14] The energy distribution of sputtered secondary particles is broad and peaked at low energy (~ 10–30 eV) so that they are in a different energy and TOF range from the scattered and recoiled particles with keV energies.

RBS differs from ISS in that it uses ion beams in the MeV energy range. These high-energy ions penetrate deeply into materials, and the dominant energy loss is by electronic interactions rather than atomic collisions.[15] The ions lose their energy continuously along straight-line trajectories in the material. Since the kinetic energies are so high, only the large repulsive potentials encountered in near direct head-on collisions with target nuclei result in backscattering. This results in small backscattering cross sections of $\sim 10^{-5}$ nm^2 compared to ISS, where scattering cross sections are of the order $\sim 10^{-3}$ nm^2.

1.6. FEATURES OF ION SCATTERING SPECTRA

An ISS spectrum is a plot of the measured intensities of the scattered and recoiled particles versus their kinetic energies or their flight times from the sample to the detector. An energy spectrum is obtained by scanning the range of kinetic energies and plotting the number of particles exhibiting a given kinetic energy (i.e., counts per unit time versus kinetic energy). A TOF spectrum is obtained by pulsing the primary ion beam and measuring the flight times of scattered and recoiled particles from the sample to the detector. Spectra are normally displayed with the ordinate labeled as *count rate, relative intensity,* or *intensity (arbitrary units)*. The latter two designations are sufficient because the absolute count rate is a facet of many complex experimental variables, usually differing from one instrument to another, and is essentially meaningless for any given sample. However, the relative intensities are meaningful and carry information about the elemental composition and structure of the surface. A mass spectrum of the atoms in the surface can be obtained from these spectra by using the kinematic relations to convert the measured energies or TOFs of the scattered and recoiled atoms into masses.

Since the features observed in an ion scattering experiment are a function of the various angles involved in the measurement, it is necessary to specify all of the angles

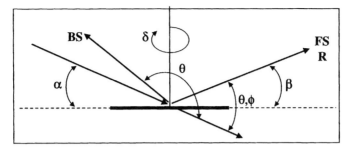

FIGURE 1.4. Definition of the various angles and terms used in ion scattering spectrometry. BS, backscattering; FS, forward scattering; R, recoiling; α, ion beam incidence angle with respect to the surface; β, exit angle; θ, scattering angle; ϕ, recoiling angle; δ, azimuthal angle of the crystal.

involved in the measurement in order to be able to identify the spectral peaks observed. The definitions of the various angles involved in a typical experiment are shown in Figure 1.4. Examples of simulated ion scattering spectra plotted with energy axes and TOF axes are shown in Figures 1.5 and 1.6. The figures exhibit spectra obtained with two different projectile atoms (i.e., He^+ and Ne^+) and two different scattering

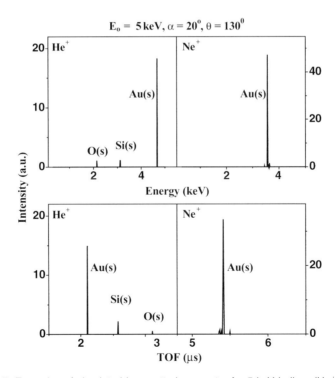

FIGURE 1.5. Examples of simulated ion scattering spectra for 5-keV helium (He^+) and neon (Ne^+) scattering from a surface composed of a random distribution of hydrogen, oxygen, silicon, and gold (H, O, Si, and Au) atoms at a backscattering angle of 130°. The scattering peaks are denoted by (s). TOF, time-of-flight.

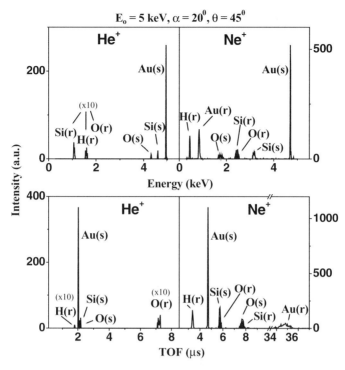

FIGURE 1.6. Examples of simulated ion scattering and recoiling spectra for 5-keV He$^+$ and Ne$^+$ scattering from a surface composed of a random distribution of hydrogen, oxygen, silicon, and gold (H, O, Si, and Au) atoms at a forward-scattering angle of 45°. The scattering and recoiling peaks are denoted by (s) and (r), respectively. TOF, time-of-flight.

angles (i.e., 130°, back scattering, and 45° forward scattering), using projectile ion energies of 5 keV. The backscattering spectra exhibit only scattering peaks, while the forward scattering spectra exhibit both scattering and recoiling peaks. The most intense features in all of the spectra are the peaks corresponding to scattering of the primary ion from Au atoms. Heavy atoms such as Au have large scattering cross sections, and light ions such as He$^+$ and Ne$^+$ retain a large portion of their kinetic energy upon scattering from such a heavy atom, resulting in energetic scattered atoms. Scattering from the light atoms O and S results in low-intensity peaks due to their low scattering cross sections.

When the yields of scattering or recoiling atoms, such as in Figures 1.5 and 1.6, are plotted as a function of the ion beam incident (α), exit (β), or crystal azimuthal (δ) angles, peaks and valleys are observed. These features are related to the shadowing and blocking features from neighboring atoms. Since these effects can be directly calculated from Newtonian mechanics, these angular anisotropies can be directly related to the relative positions of atoms with respect to each other near the surface. The shadowing and blocking effects have a range of ~1 nm. As a result, ion scattering spectrometry requires only short-range order to obtain structural information.

The number of experimental variables available in an ion scattering experiment is large. These variables are as follows: (1) incident projectile ion (i.e., mass and atomic number); (2) incident projectile ion energy; (3) incident angle (α) of the ion beam with respect to the surface; (4) crystal azimuthal angle (δ_1) that is aligned with the ion beam; (5) scattering angle (θ); (6) exit angle (β) of scattered or recoiled particles; (7) energies of the scattered and recoiled atoms; and (8) crystal azimuthal angle (δ_2) that is aligned with outgoing scattered or recoiled atoms. For most experimentalists, it is not possible to explore all of this multidimensional parameter space due to limitations imposed by the experimental apparatus. A typical experiment usually measures the scattering and/or recoiling intensity as a function of α, β, or δ while the remaining parameters are fixed. However, recent developments[16] in large-area position-sensitive detectors have made it possible to collect the spatial intensity distributions of scattered and recoiled atoms over a large range of angular space. Such data sets allow direct observation of the focusing, shadowing, and blocking patterns, providing a rich data set for structural analysis.

REFERENCES

1. J. W. Rabalais, "Low Energy Ion Beams for Surface Modification and Film Deposition," in Applications of Accelerators in Research and Industry—Sixteenth Int'l Conf., Denton, TX, Nov. 2000, J. L. Duggan and I. L. Morgan, (Eds.) pp. 911–914 (2001).

2. B. V. Panin, *Sov. Phys. JETP* **15**, 215, 1962.

3. V. Walther and H. Hintenberger, *Z. Naturforsch.* **18A**, 843, 1963.

4. D. P. Smith, *J. Appl. Phys.* **38**, 340, 1967; *ibid, Surf. Sci.* **25**, 171, 1971.

5. E. S. Mashkova and V. A. Molchanov, Medium-Energy Ion Reflection From Solids, North-Holland, New York, 1985.

6. E. S. Parilis, L. M. Kishinevsky, N. Yu. Turaev, B. E. Baklitzky, F. F. Umarov, V. Kh. Verleger, S. I. Nizhnaya, and I. S. Bitensky, *Atomic Collisions on Solid Surfaces,* North-Holland, New York, 1993.

7. H. Niehus, W. Heiland, and E. Taglauer, *Surf. Sci. Rep.* **17**, 213 (1993).

8. J. W. Rabalais, *CRC Rev. Sol. St. Mat. Sci.* **14**, 319 (1988).

9. O. Grizzi, M. Shi, H. Bu, J. W. Rabalais, and P. Hochmann, *Phys. Rev. B,* **40**, 10127 (1989).

10. C. Kim, C. Hoefner, V. Bykov, and J. W. Rabalais, *Nucl. Instrum. Meth. Phys. Res.* **B125**, 315 (1997).

11. M. Aono, Nucl. Instrum. *Meth. Phys. Res.* **B2**, 374 (1984).

12. A. Benninghoven, F. G. Rudenauer, and H. W. Werner, *in* Secondary Ion Mass Spectrometry: Basic Concepts, Instrumental Aspects, Applications, and Trends, P. J. Elving and J. D. Winefordner (Eds.), Wiley, New York, 1987.

13. P. Sigmund, in Sputtering by Particle Bombardment I, *Topics in Applied Physics,* Vol. 47, R. Behrish (Ed.) Springer, Berlin, 1981.

14. R. E. Johnson and B. U. R. Sundqvist, *Physics Today,* March 28, 1992.

15. W. K. Chu, J. W. Mayer, and M. A. Nicolet, *Backscattering Spectrometry,* Academic Press, New York, 1978.

16. V. Bykov, L. Houssiau, and J. W. Rabalais, *J. Phys. Chem. B* **104**, 6340 (2000).

2

THEORETICAL DESCRIPTION OF ATOMIC COLLISIONS

It would be extremely difficult to approach the problem of ion scattering if one had to simultaneously take into account all of the interactions of a particle with every atom in a crystal. Although many-body methods are common in solid state theory, in ion scattering one can consider only pairwise interactions of the projectile with atoms in the target to a reasonably high accuracy. For impinging particles with energies >30–50 eV (i.e., much higher than the thermal and binding energies of atoms in solids and also due to the much shorter particle–solid interaction times in comparison with phonon frequencies), the particle–solid interaction can be treated as a chain of independent single-collision events. This chapter treats only such processes. It begins with the simplest kinematical description of collisions.

2.1. KINEMATICS OF ATOMIC COLLISIONS

The kinematical description of collisions provides information about the final states of interacting particles. Their final energies are found from their scattering angles. Since the final states are reached at $t = \infty$ when particles are at infinite distances from the interaction region, the kinematical description does not give information on positions and velocities (energies) of the particles at any finite time during the interaction. For this reason, the form of the interaction potential $V(r)$ is not important. The final scattering energies can be found using only the laws of conservation of momentum and energy. For a system of two particles with the second (target) particle initially at rest in the laboratory system and the case of no energy dissipation during the collision, the conservation equations can be written as

$$\mathbf{p}_0 = \mathbf{p}_1 + \mathbf{p}_2 \tag{1a}$$

$$E_0 = E_1 + E_2, \tag{1b}$$

where $\mathbf{p}_0$, E_0 and $\mathbf{p}_1$, E_1 are the momentum and energy of the projectile before and after the interaction, and $\mathbf{p}_2$, E_2 are the momentum and energy of the target atom after the interaction.

Expressions for the energies of the two particles in their final states after the interaction as functions of their respective scattering angles are required. Moving $\mathbf{p}_1$ in Eq. 1a to the left and squaring both sides yields $p_0^2 + p_1^2 - 2p_0 p_1 \cos \theta = p_2^2$, where θ is the angle between the initial and final directions of the projectile. This becomes a quadratic equation in p_1 after substituting $p_2 = M_2(p_0^2 - p_1^2)/M_1$ from Eq. 1b, where M_1 and M_2 are masses of projectile and target particles. Correspondingly

$$p_1^2(1 + M_2/M_1) - 2p_0 p_1 \cos \theta + p_0^2(1 - M_2/M_1) = 0. \tag{2}$$

Solving Eq. 2 with respect to p_1, the momentum of the projectile after the interaction is obtained as

$$p_1 = p_0 \frac{\cos \theta \pm \sqrt{\cos^2 \theta - (1 - M_2/M_1)(1 + M_2/M_1)}}{1 + M_2/M_1}. \tag{3}$$

Since in Eq. 3 $p_1 = |\mathbf{p}_1| \geq 0$, only a "+" sign is used when $M_1 < M_2$ and both the "+" and "−" signs are used for $M_1 > M_2$. It is more common to express the final state of the scattered particle in terms of its kinetic energy rather than its momentum. Squaring Eq. 3 and dividing it by $2M_1$ produces

$$E_1 = E_0 \left(\frac{M_1}{M_1 + M_2} \right)^2 \left(\cos \theta \pm \sqrt{\left(\frac{M_2}{M_1} \right)^2 - \sin^2 \theta} \right)^2 = KE_0, \tag{4}$$

where K is called the kinematic factor. Similarly, the momentum and energy of the target (recoiled) particle is found by placing p_2 on the left side of Eq. 1a and squaring both sides:

$$p_2 = p_0 \frac{2 \cos \phi}{(1 + M_1/M_2)} \tag{5}$$

$$E_2 = E_0 \frac{4M_1 M_2}{(M_1 + M_2)^2} \cos^2 \phi = TE_0. \tag{6}$$

Here, ϕ is the recoiling angle, or the scattering angle of the target particle and T is the fraction of energy transferred to the second particle. As observed from Eqs. 4 and 6, the scattering and recoiling energies do not depend on the specific potential used, but only on the masses of the interacting particles and their scattering and recoiling angles. The kinematic factor K as a function of the scattering angle θ for several target-to-projectile mass ratios M_2/M_1 is plotted in Figure 2.1a. The dependence of the fraction of the energy transferred to the recoiling atom T on the recoiling angle ϕ is shown in Figure 2.1b. In the latter figure, only mass ratios greater than unity are shown, since T is symmetric with respect to M_1 and M_2.

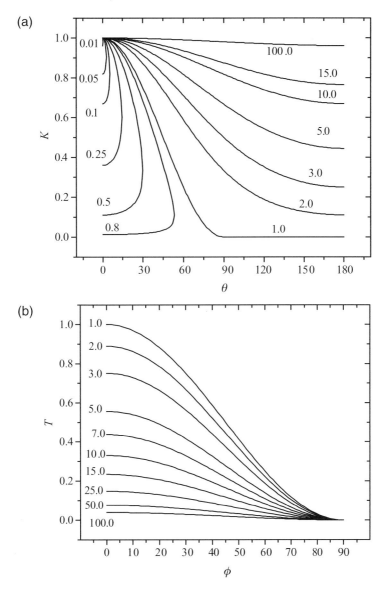

FIGURE 2.1. (a) Dependence of the kinematic factor on the scattering angle for various target-to-projectile mass ratios. (b) Dependence of the fraction of energy transferred to the target atom as a function of the recoiling angle. Values of M_2/M_1 are shown beside each curve.

It is worth noting from Eq. 4 that a particle that has experienced two collisions by angles θ_1 and θ_2 will have a greater energy than it would have after a single collision by an angle $\theta = \theta_1 + \theta_2$. This follows from the relations

$$K(\theta) < K(\theta_1)K(\theta_2) \qquad \text{and} \qquad \theta = \theta_1 + \theta_2. \tag{7}$$

This statement can be generalized for any number of collisions.

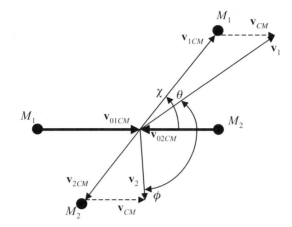

FIGURE 2.2. Vector diagram showing velocities and scattering angles of two particles in the center of mass (CM) and laboratory *l*-systems. The velocities of the first and second particle in the CM system are $\mathbf{v}_{01CM}$, $\mathbf{v}_{02CM}$ before the interaction and $\mathbf{v}_{1CM}$, $\mathbf{v}_{2CM}$ after the interaction. The CM scattering angle χ, laboratory scattering angle θ, and laboratory recoiling angle β are shown. Velocity vectors of particles in the *l*-system are obtained by adding the velocity of the center of mass to their velocities in the CM system.

Single scattering events can be illustrated by a vector diagram based on momentum conservation and the geometrical rules of vector addition. The sum of the momenta of the two particles before and after the interaction is a constant, which equals zero in a center of mass (CM) reference system. Therefore, the interaction can be graphically represented as a rotation of the initial velocity vectors of the first and second particles in the CM system $\mathbf{v}_{01CM}$ and $\mathbf{v}_{02CM}$ by an angle χ, (i.e., the CM scattering angle) (Figure 2.2). Since the second particle may not initially be at rest, a different notation will be used here, namely, a zero and a particle number subscript will be used to designate quantities before the interaction, and only a particle number subscript will be used for the quantities after the interaction. The velocities of the particles in the laboratory system (*l*-system) are obtained by adding the velocity of the CM ($\mathbf{v}_{CM}$) to their velocities in the CM system $\mathbf{v}_{1CM}$ and $\mathbf{v}_{2CM}$

$$\mathbf{v}_1 \equiv \mathbf{v}_{1l} = \mathbf{v}_{1CM} + \mathbf{v}_{CM} \tag{8a}$$

$$\mathbf{v}_2 \equiv \mathbf{v}_{2l} = \mathbf{v}_{2CM} + \mathbf{v}_{CM}. \tag{8b}$$

The angles of scattering and recoiling θ and ϕ are the angles between the velocity vectors $\mathbf{v}_1$ and $\mathbf{v}_2$ of the first and second particles after the interaction in the *l*-system and the initial direction of the first particle. They can be expressed via the CM scattering angle χ by using a momentum diagram rather than the velocity diagram. Multiplying Eqs. 8a and 8b by M_1 and M_2 correspondingly, we obtain for the momenta of the first and second particle in the l-system

$$\mathbf{p}_1 = \mathbf{p}_{1CM} + M_1 \mathbf{v}_{CM} \tag{9a}$$

$$\mathbf{p}_2 = \mathbf{p}_{2CM} + M_2 \mathbf{v}_{CM}. \tag{9b}$$

Because in the CM system momenta only change their directions but not their absolute values, the momenta can be written as

$$\mathbf{p}_{1CM} = p_{01CM}\, \mathbf{n} \tag{10a}$$

$$\mathbf{p}_{2CM} = -p_{01CM}\, \mathbf{n}, \tag{10b}$$

where $\mathbf{n}$ is a unit vector designating the new direction of the first particle in the CM system after the interaction. Noting that the velocity of the CM is $\mathbf{v}_{CM} = (M_1 \mathbf{v}_{01} + M_2 \mathbf{v}_{02})/(M_1 + M_2)$ and substituting Eq. 10 into Eq. 9, the following expressions are obtained:

$$\mathbf{p}_1 = p_{01CM}\, \mathbf{n} + M_1 (\mathbf{p}_{01} + \mathbf{p}_{02})/(M_1 + M_2) \tag{11a}$$

$$\mathbf{p}_2 = -p_{01CM}\, \mathbf{n} + M_2 (\mathbf{p}_{01} + \mathbf{p}_{02})/(M_1 + M_2). \tag{11b}$$

It is observed from Eq. 11 that the final momenta in the l-system, $\mathbf{p}_1$ and $\mathbf{p}_2$, are found by adding or subtracting $p_{01CM}\, \mathbf{n}$ from the vectors $\overrightarrow{DO} = M_1 (\mathbf{p}_{01} + \mathbf{p}_{02})/(M_1 + M_2)$ and $\overrightarrow{OB} = M_2 (\mathbf{p}_{01} + \mathbf{p}_{02})/(M_1 + M_2)$ obtained from $\mathbf{p}_{01} + \mathbf{p}_{02}$ and then taking a ratio of $M_1 : M_2$ as in Figure 2.3, so that $DO/OB = M_1/M_2$. The vector $p_{01CM}\, \mathbf{n}$ originates at the point of division O, and its end can lie anywhere on the circle with radius p_{01CM}. The exact position of the end of the vector $p_{01CM}\, \mathbf{n}$ on the circle depends on the impact parameter, initial energies, and the interaction potential between the two particles. Then $\mathbf{p}_1$ and $\mathbf{p}_2$ are represented by the lines connecting the end of $p_{01CM}\, \mathbf{n}$ with the beginning and end of $\mathbf{p}_{01} + \mathbf{p}_{02}$.

If the second particle is initially at rest in the l-system ($\mathbf{p}_{02} = 0$), the velocity of the first particle in Eq. 8a transforms simply as

$$\mathbf{v}_{01CM} = \frac{M_2}{M_1 + M_2} \mathbf{v}_{01}, \tag{12}$$

from which it follows that $OB = p_{01CM}$, and, therefore point B lies on the circle. Now vectors $\mathbf{p}_{01}$ and $\mathbf{p}_{01CM}$ are collinear, and $\mathbf{n}$ is at an angle χ with respect to $\mathbf{p}_{01}$. If the mass of the projectile is less than the mass of the target, $M_1 < M_2$, point D is inside the circle (Figure 2.4a) and if $M_1 > M_2$, point D is outside the circle (Figure 2.4b). From Figures 2.4a and 2.4b, we express the laboratory scattering and recoiling angles θ and ϕ via the CM scattering angle χ as

$$\tan \theta = \frac{\sin \chi}{M_1/M_2 + \cos \chi} \tag{12a}$$

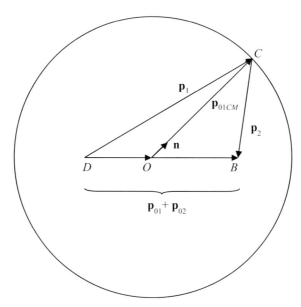

FIGURE 2.3. Momentum vector diagram for two interacting particles. Vector $\vec{DB}$ is the sum of the momenta of particle 1 and particle 2 before the interaction in the *l*-system: $\vec{DB} = \mathbf{p}_{01} + \mathbf{p}_{02}$ $\vec{DB}$. Point O divides vector $\vec{DB}$ into two vectors $\vec{DO}$ and $\vec{OB}$ so that $\vec{DO} = \frac{M_1}{M_1 + M_2} \vec{DB}$ and $\vec{OB} = \frac{M_2}{M_1 + M_2} \vec{DB}$. The momenta of the particles after the interaction in the *l*-system, $\mathbf{p}_1$ and $\mathbf{p}_2$, are found by adding or subtracting $p_{01CM} \mathbf{n}$ from $\vec{DO}$ and $\vec{OB}$, where $\mathbf{n}$ is a unit vector designating the new direction of the first particle in the CM system.

and

$$\phi = \frac{1}{2}(\pi - \chi). \tag{12b}$$

From Figure 2.4b, it follows that when the projectile atom is heavier than the target atom, there exists a maximum angle $\theta_{\max}$ at which the projectile can be scattered by the target in the *l*-system, i.e.,

$$\sin \theta_{\max} = M_2/M_1. \tag{13}$$

It should be noted that for scattering angles θ less than $\theta_{\max}$, there are always two values of χ and ϕ that correspond to the same value of θ, since line DC has two points of intersection with the circle, although only one pair of values is shown in Figure 2.4b.

(a)

(b)

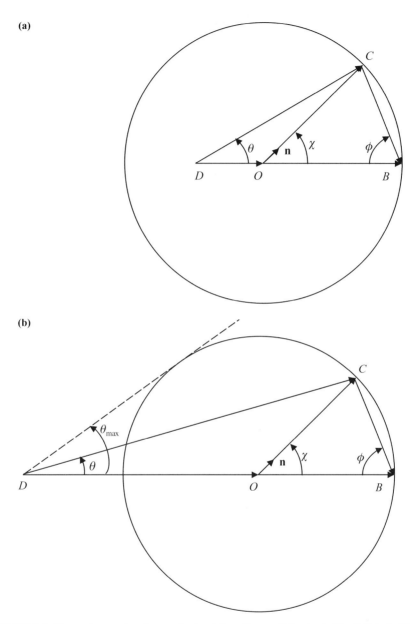

FIGURE 2.4. Momentum vector diagram for two interacting particles (as in Fig. 2.3), but with the second (target) particle initially at rest in the *l*-system. Now, point *B* lies on the circle of radius p_{01CM} and vector **n** makes angle χ with the initial projectile direction $\vec{DB}$. The *l*-system scattering and recoiling angles θ and ϕ are shown. (a) Case when the projectile atom is lighter than the target atom, $M_1 < M_2$. Point *D* lies inside the circle. (b) Case of projectile atom heavier than the target atom, $M_1 > M_2$. Point *D* is outside the circle and a maximum scattering angle exists. Note that there are two CM scattering angles corresponding to each laboratory scattering angle.

2.2. DYNAMICS OF ATOMIC COLLISIONS

2.2.1. Interaction Potential

As an energetic atom approaches another atom constituting the target, there is no interaction at large distances (i.e., of the order of dozens of atomic dimensions), and then, as the two atoms approach each other and their electronic clouds overlap, a complex interaction between the electrons and nuclei of the two atoms occurs. The detailed description of this interaction process should include all electron–electron, nucleus–electron, and nucleus–nucleus interactions. For practical purposes, one is interested in a potential function that depends only on the distance r between the nuclei, i.e., $V = V(r)$. Since both of the interacting nuclei are shielded by the electrons surrounding them, the deviation of the potential V from the coulomb interaction is accounted for by introducing the screening function $\Phi(r)$, defined as $\Phi(r) = \frac{V(r)}{C/r}$, where C/r is a purely coulombic interaction between the nuclei. The Thomas-Fermi statistical model of the atom is usually used to obtain the form of the screening function. The most common empirical screening functions used in ion scattering are based on this model.

The Thomas-Fermi statistical model assumes that the electrons in an atom form an ideal degenerate gas in which two electrons (spin up and spin down) occupy a volume of h^3 in the momentum-coordinate (phase) space. The density of electrons in the phase space is then $2/h^3$. The number of electrons per unit volume in the real space is found by integrating the momentum components out of the phase space density. In doing this, it is assumed that all of the momentum states of the electron up to $p_e = p_{e0}$ are occupied. Then, at a distance r from the nucleus, the density of the electron cloud is

$$\rho(r) = \frac{2}{h^3} \int\limits_{p_e < p_{e0}} dp_e = (2/h^3) \cdot (4/3)\pi p_{e0}^3(r). \tag{14}$$

The maximum total energy that an electron can have at a distance r from the nucleus is given by

$$E_0 = \frac{p_{e0}^2(r)}{2m} - e\varphi(r), \tag{15}$$

where m is the mass of an electron and $\varphi(r)$ is the potential created by the nucleus and all of the electrons. If $E_0 = -e\varphi_0$ is designated and p_{e0} from Eq. 15 is substituted into Eq. 14, the electron density is

$$\rho(r) = \frac{8\pi}{3h^3} (2me)^{3/2} (\varphi - \varphi_0)^{3/2}. \tag{16}$$

By substituting $\rho(r)$ into the Poisson's equation, an equation for a self-consisted field created by an atom is obtained as

$$\nabla^2(\varphi - \varphi_0) = -b_0(\varphi - \varphi_0)^{3/2}, \tag{17}$$

where $b_0 = \frac{32\pi^2 e}{3h^3}(2me)^{3/2}$.

A spherically symmetric solution of Eq. 17 in the form of a screened coulomb potential is

$$\varphi - \varphi_0 = \frac{Ze}{r}\Phi\left(\frac{r}{a}\right), \tag{18}$$

where Z is the atomic number of the atom, a is a screening parameter, and $\Phi(r/a)$ is a screening function. It is easy to see then that Eq. 17 transforms to

$$\frac{d^2\Phi}{dx^2} = x^{-1/2}\Phi^{3/2}, \tag{19}$$

known as the Thomas-Fermi equation, where $x = r/a$ and

$$a = \left(\frac{3}{32\pi^2}\right)^{2/3}\frac{h^2}{2me^2Z^{1/3}} \simeq \frac{0.88534a_0}{Z^{1/3}}. \tag{20}$$

Here $a_0 = \hbar^2/me^2$ is the radius of the first orbit of an electron in the hydrogen atom (the first Bohr radius). The boundary conditions for $\Phi(r)$ are $\Phi(0) = 1$ and $\Phi(\infty) = 0$. The first condition is for the field created by an atom to become purely coulombic at very small distances, while the second condition is for the complete screening of the coulomb field at large distances.

If the atomic number of the projectile is not negligible compared to that of the target, screening of both interacting particles should be taken into account. It can be shown that in this case Eq. 19 for the screening function still holds true, but the screening constant a has to be modified to accommodate for both the projectile and target electron screening. Firsov[1] has obtained the following expression for $a = a_F$ (known as the *Firsov screening length*):

$$a_F = \frac{0.88534a_0}{\left(Z_1^{1/2} + Z_2^{1/2}\right)^{2/3}}. \tag{21}$$

Another well-known expression for the universal screening parameter a is one used by Lindhard *et al*:[2]

$$a_L = \frac{0.88534a_0}{\left(Z_1^{2/3} + Z_2^{2/3}\right)^{1/2}}. \tag{22}$$

Although the exact analytical solution for Eq. 19 does not exist, many analytical approximations of $\Phi(r)$ have been obtained. Several approximations for $\Phi(r)$

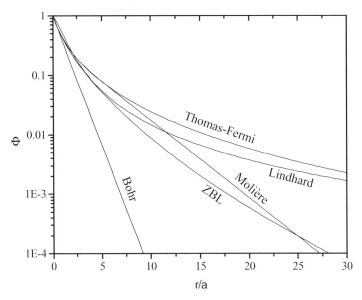

FIGURE 2.5. Approximations for the Thomas-Fermi screening function $\Phi(r)$ as a function of the reduced radius r/a. The distance on the x-axis is measured in units of a screening constant a. ZBL, Ziegler-Biersack-Littmark.

suggested by various authors are listed in Torrens *et al.*[3] and some of them, together with an exact solution, are plotted in Figure 2.5. The reduced radius r/a, having the units of a given screening constant, is used on the x-axis. The expressions for two widely used approximations, Molière and Ziegler-Biersack-Littmark (ZBL), are provided here. Molière[4] has proposed to approximate $\Phi(r)$ by a sum of three exponents

$$\Phi(r) = 0.35\exp(-0.3r/a) + 0.55\exp(-1.2r/a) + 0.1\exp(-6.0r/a), \quad (23)$$

which does not deviate from the exact solution of Eq. 19 by more than 0.2% for $0 \le r/a \le 6$. Both Eqs. 21 and 22 can be used as a screening length in Eq. 23. A modified screening length a' is sometimes used with the Molière screening function to achieve better agreement with experiment, i.e.,

$$a' = ca. \quad (24)$$

Here, c is a fitting parameter less than unity, and a is either the Firsov or Lindhard screening length. Ziegler, Biersack, and Littmark[5] have performed their own calculation of the screening function using a more complex model than that of Thomas-Fermi. Their model has explicitly included coulombic, electronic overlap, and exchange contributions in the potential along with a more realistic Hartree-Fock-Slater charge distribution around the nucleus. Averaging over potentials calculated for each of 522 randomly selected atomic pairs has yielded the universal screening function

$$\Phi(r) = 0.1818 \exp(-3.2r/a) + 0.5099 \exp(-0.9423r/a)$$
$$+ 0.2802 \exp(-0.4029r/a) + 0.02817 \exp(-0.2016r/a) \tag{25}$$

with the screening length

$$a_{ZBL} = \frac{0.88534a_0}{Z_1^{0.23} + Z_2^{0.23}}. \tag{26}$$

The potentials described provide good approximations in the keV ion scattering energy range and begin to fail, together with the binary collision approach, for energies less than 30–50 eV where more sophisticated approaches are needed. The two advantages of these simple one-variable potentials are their satisfactory agreement with the experimental data and their usefulness in simulations requiring frequent evaluations of the interaction function.

2.2.2. Scattering Integral

If the force of interaction between the two particles depends only on the distance r between them (central field), and there are no transverse forces, the system can be reduced to a problem of one particle moving in this central field. The derivation of the scattering law in this case[6] is now described. The Lagrangian $\mathcal{L}$ for a system of two particles can be written as

$$\mathcal{L} = \frac{1}{2}\left(M_1\dot{\mathbf{r}}_1^2 + M_2\dot{\mathbf{r}}_2^2\right) - V(|\mathbf{r}_1 - \mathbf{r}_2|). \tag{27}$$

It can be simplified by using the CM coordinate system instead of the l-system. In the CM system, it is convenient to choose zero at the position of the center of mass of the two particles,

$$\mathbf{R}_{CM} = M_1\mathbf{r}_1 + M_2\mathbf{r}_2 = 0, \tag{28}$$

and introduce a vector of relative displacement

$$\mathbf{r} = \mathbf{r}_1 - \mathbf{r}_2 \tag{29}$$

drawn from particle 2 to particle 1. Then the coordinates of each particle in the CM system are

$$\mathbf{r}_1 = \frac{A}{1+A}\mathbf{r}$$
$$\mathbf{r}_2 = -\frac{1}{1+A}\mathbf{r}, \tag{30}$$

where A is a target-to-projectile mass ratio $A = M_2/M_1$. In the new coordinates, $\mathcal{L}$ is

$$\mathcal{L} = \frac{1}{2}\mu\dot{\mathbf{r}}^2 - V(|\mathbf{r}|), \tag{31}$$

where $\mu = M_1 M_2/(M_1 + M_2)$ is the reduced mass. The Lagrangian in Eq. 31 is for a single particle of mass μ moving in a central field. In the future, the term *particle* is used to refer to a fictitious particle with a mass μ described by Eq. 31, and the terms *projectile*, or first atom, and *target*, or second atom, to refer to the "real" particles M_1 and M_2. It should be clear from the context what particle is referred to in each case.

The Lagrangian Eq. 31 is further simplified based on angular momentum conservation. Since the vector of the angular momentum $\mathbf{L} = \mathbf{r} \times \mathbf{p}$ is constant, $\mathbf{r}$ being perpendicular to $\mathbf{L}$, the motion of a particle is confined to one plane perpendicular to $\mathbf{L}$. Let us use polar coordinates in this plane with a zero placed in the center of the force field and label the position of the particle inside the plane with the coordinates r and φ. Eq. 31 then becomes

$$\mathcal{L} = \frac{\mu}{2}(\dot{r}^2 + r^2\dot{\varphi}^2) - V(r), \tag{32}$$

where r is a distance from the particle to the force center and φ is a polar angle. From the Lagrange's equation written for the pair $\dot{\varphi}, \varphi$,

$$\frac{d}{dt}\left(\frac{\partial\mathcal{L}}{\partial\dot{\varphi}}\right) = \frac{\partial\mathcal{L}}{\partial\varphi}, \tag{33}$$

the generalized momentum for this pair is obtained as

$$\mu r^2\dot{\varphi} = const. \tag{34}$$

It can be shown that the expression in Eq. 34 is the total angular momentum L in the CM system, i.e.,

$$L = \mu r^2\dot{\varphi}. \tag{35}$$

Instead of directly integrating the Lagrange's equation for the $\dot{r}, r$ pair, the solution of the equations of motion using the conservation of two quantities (i.e., the total angular momentum L and total energy E of the system) is obtained. To do this, $\dot{\varphi}$ from Eq. 35 is substituted into the expression for the total energy E:

$$E = \frac{\mu\dot{r}^2}{2} + \frac{L^2}{2\mu r^2} + V(r). \tag{36}$$

From Eq. 36, $\dot{r}$ is obtained as

$$\dot{r} = \frac{dr}{dt} = \sqrt{\frac{2}{\mu}(E - V(r)) - \frac{L^2}{\mu^2 r^2}}, \tag{37}$$

or integrating,

$$t = \int \frac{dr}{\sqrt{\frac{2}{\mu}(E - V(r)) - \frac{L^2}{\mu^2 r^2}}} + const. \tag{38}$$

This gives an implicit time dependence of the distance r between the particle and the center. Finally, expressing $d\varphi$ from Eq. 35 as $d\varphi = \frac{L}{\mu r^2} dt$, substituting dt from Eq. 37, and integrating, the equation for the path of the particle is obtained as

$$\varphi = \int \frac{L dr / r^2}{\sqrt{2\mu(E - V(r)) - \frac{L^2}{r^2}}} + const. \tag{39}$$

Eqs. 38 and 39 are a general solution to the problem of motion of two particles. Recalling that $L = b(2\mu E)^{1/2}$, where b is an impact parameter, Eq. 39 can be rewritten in the alternative form:

$$\varphi = \int \frac{b dr / r^2}{\sqrt{1 - \frac{V(r)}{E} - \frac{b^2}{r^2}}} + const. \tag{39'}$$

There is a property of the symmetry of the trajectory of the particle that is needed for future derivations that will now be described. It is noted that Eq. 36 has the form of the energy of a particle moving in one dimension in an effective potential

$$V_{eff}(r) = \frac{L^2}{2\mu r^2} + V(r). \tag{40}$$

When the kinetic term in Eq. 36 reaches zero, the motion comes to a turning point where the radial velocity Eq. 37 changes sign. At this point the minimum–maximum value of the distance between the particle and the center is reached and the distance starts to increase if it was decreasing previously, and vice versa. The total velocity of the particle is not zero at the turning point, however, if the angular velocity $\dot{\varphi}$ is nonzero. If the origin for the angle φ is chosen at one of the turning points, it is observed from Eq. 39 that φ will have the same values of opposite sign for the same r values on the two sides of the line $\varphi = 0$. The opposite sign of φ is a consequence of the sign change of the square root in Eq. 37 on going through the turning point.

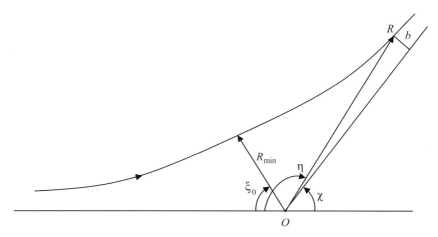

FIGURE 2.6. Scattering of a μ-particle by a center of force.

Therefore, the trajectory of the particle is symmetric with respect to the line drawn from the center to its turning point.

The scattering angle in the CM system χ can be found using Eq. 39. Since the trajectory of the particle is symmetric with respect to the turning point,

$$\chi = \pi - 2\xi_0, \tag{41}$$

where ξ_0 is an angle between the asymptote and a line connecting the center with the particle at the point of closest approach (Figure 2.6). The angle ξ_0 is found by applying Eq. 39 for any of the halves of the trajectory with corresponding integration limits R_{min} and ∞. Then

$$\chi = \pi - 2b \int_{R_{min}}^{\infty} \frac{dr/r^2}{\sqrt{1 - \frac{V(r)}{E} - \frac{b^2}{r^2}}} = \pi - bI_1. \tag{42}$$

where I_1 designates twice the integral from the second term in Eq. 42.

2.2.3. Binary Collision Approximation

In a binary collision approximation (BCA), the process of interaction of an impinging ion with a solid is divided into a series of pairwise collisions between a moving particle (called a projectile) and a particle that is usually initially at rest in the l-system (a target atom). The BCA is used in scattering simulations when the energy of the primary particle is high enough to neglect the deviations of the exact trajectories from their asymptotes at ranges comparable with the interatomic distances in the target.

Simplifications to the actual trajectories are made in the description of each collision. The projectile is considered to move along a straight line until some

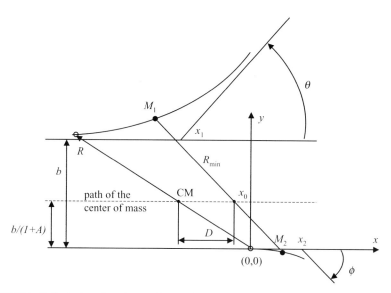

FIGURE 2.7. Trajectories of two interacting particles in the *l*-system. The second (target) particle is initially at rest and at the origin. Binary collision approximations to the outgoing trajectories are shown as straight lines starting at points x_1 (projectile) and x_2 (target). CM, center of mass.

point where an instantaneous transfer of energy to the target atom occurs and both particles start moving in new directions. In addition, the target atom is instantaneously displaced some distance in the direction of the initial projectile motion. Thus, most of the time, the particles move freely in the absence of the field between them, except for the moments when the infinite field is switched on and they deflect and exchange energy. The replacement of the continuous interaction over a finite period of time during the collision with one step-like change in the state of the particles allows one to avoid integration of the equations of motion for each collision, which significantly decreases the amount of time required for simulations. Thus, the primary purpose of the BCA is to locate the intersections of the incoming and outgoing trajectories of the particles where the changes of directions of propagation are assumed to occur. Figure 2.7 shows the actual trajectories of the particles and their BCA approximations. It is observed that the three quantities of interest in the BCA are the CM scattering angle χ and the intersections x_1 and x_2 of the outgoing projectile and target asymptotes with the lines $y = b$ and $y = 0$, correspondingly. The scattering angle χ can be found using Eq. 39 for the dependence of the angular position of the particle on the distance from the origin. In order to find the intersections of the asymptotes, we recall that the trajectory of a particle with effective mass μ is symmetric with respect to the line drawn from the center to its turning point (see section 2.2). The individual particles in the *l*-system after the collision are then located far enough from the area of the interaction so that the deviations of the trajectories from the asymptotes are negligible.

   The coordinates of the projectile and target atoms in the CM system at a certain moment of time when the separation between the particles has increased after the interaction to some large value $|\mathbf{r}| = R$ must be found. Because $\mathbf{r}$ is a relative position of the particles, rather than their individual coordinates, it coincides with the radius-vector of the particle with mass μ (described in section 2.2). Using the symmetry of the trajectory of the particle with respect to the line drawn from the particle to the center at the closest approach (Figure 2.6), the angle η between R and the initial projectile direction can be expressed as

$$\eta = \chi + \arcsin(b/R) \tag{43}$$

and the coordinates of the particle as

$$x_{CM} = R\cos\eta = \sqrt{R^2 - b^2}\cos\chi - b\sin\chi$$
$$y_{CM} = R\sin\eta = \sqrt{R^2 - b^2}\sin\chi + b\cos\chi. \tag{44}$$

The positions of the projectile and target atoms in the CM system are obtained from the coordinates of the particles by substituting Eq. 44 into Eq. 30. Finally, their coordinates in the l-system can be found by recalling that in the l-system the center of mass is moving rectilinearly along the line $y = b/(1+A)$, where A is the target-to-projectile mass ratio, $A = M_2/M_1$. At the moment of interest, the coordinates in the l-system are connected with the CM by

$$x_l^R = x_{CM} + D + x_0$$
$$y_l^R = y_{CM} + \frac{b}{1+A}, \tag{45}$$

where x_l^R, y_l^R are coordinates of the projectile in the l-system when the particles are at separation R, D is the distance that the center of mass travels when the separation $r = |\mathbf{r}|$ increases from the minimum value R_{min} at the turning point to R, and x_0 is the coordinate of the center of mass at the moment of the closest approach. D can be found from the approaching part of the symmetrical trajectory as the distance the center of mass travels when the separation between the particles decreases from R to R_{min}:

$$D = V_{CM}t_D, \tag{46}$$

where

$$V_{CM} = \sqrt{\frac{2E}{\mu}\frac{1}{1+A}} \tag{47}$$

is the constant velocity of the center of mass and t_D is the time it takes the center of mass to travel the distance D. Using Eq. 38, t_D can be expressed as

$$t_D = \int_{R_{min}}^{R} \frac{dr}{\sqrt{\frac{2}{\mu}(E - V(r)) - \frac{L^2}{\mu^2 r^2}}} = \sqrt{\frac{\mu}{2E}} \int_{R_{min}}^{R} \frac{dr}{g(r)}, \tag{48}$$

where

$$g(r) = \sqrt{1 - V(r)/E - b^2/r^2}. \tag{49}$$

When the particles are at a distance R from each other, it is observed that the x-coordinate of the center of the mass is $-\sqrt{R^2 - b^2}/(1 + A)$. During the time t_D, as the particles approach each other and the separation between them diminishes from R to R_{min}, the center of mass travels the distance D along the x-axis, its new x-coordinate being x_0, so that

$$x_0 \approx D - \frac{\sqrt{R^2 - b^2}}{1 + A}. \tag{50}$$

Eq. 50 becomes exact when $R \to \infty$. Substituting Eq. 46 into Eq. 50 with V_{CM} from Eq. 47 and t_D from Eq. 48 and letting $R \to \infty$, x_0 is

$$x_0 = \frac{1}{1 + A} \lim_{R \to \infty} \left(\int_{R_{min}}^{R} \frac{dr}{g(r)} - \sqrt{R^2 - b^2} \right) = -\frac{\tau}{1 + A}, \tag{51}$$

where

$$\tau = \sqrt{R_{min}^2 - b^2} - \int_{R_{min}}^{\infty} dr \left(\frac{1}{g(r)} - \frac{1}{\sqrt{1 - b^2/r^2}} \right) \tag{52}$$

is a so-called time integral. Its value can be found only numerically for the case of screened coulomb potentials used in ion scattering. Using simple geometrical considerations, the asymptote intersections can be expressed via the time integral and the CM scattering angle χ by locating the particles after the collision far enough from each other so that their subsequent interaction can be neglected. At separation R, their l-system coordinates are x_l^R, y_l^R, and connected to their CM coordinates by Eq. 45. Once the particles in the laboratory system have been located, it is easy to find their asymptotes.

For instance, the asymptote of the projectile in an l-system is a line with an equation

$$y - y_l^R = \tan\theta \left(x - x_l^R \right), \tag{53}$$

where θ is the l-system scattering angle. In order to find the intersection of the projectile asymptote x_1, set $y = b$ in Eq. 53. After substituting $\tan\theta$ from Eq. 12a, transforming x_l^R, y_l^R according to Eq. 45, and substituting D, x_0 from Eqs. 50 and 51,

Eq. 53 transforms to

$$\frac{1 + A \cos \chi}{A \sin \chi} \left[b - \frac{1}{1 + A} \left(A\sqrt{R^2 - b^2} \sin \chi + Ab \cos \chi + b \right) \right]$$

$$= x_1 - \frac{1}{1 + A} \left(A\sqrt{R^2 - b^2} \cos \chi - Ab \sin \chi + \sqrt{R^2 - b^2} - 2\tau \right). \tag{54}$$

From Eq. 54 the intersection x_1 is found to be

$$x_1 = \frac{b(1 - A)\tan(\chi/2) - 2\tau}{(1 + A)}. \tag{55}$$

Similarly, for the intersection of the target asymptote,

$$x_2 = b\tan(\chi/2) - x_1. \tag{56}$$

2.2.4. Small Angle Approximation

Another approximation used in finding a projectile scattering angle is the momentum or small-angle approximation. Unlike the BCA, it allows derivation of a complete analytical solution for the scattering angle in the l-system without resorting to the CM system. The momentum approximation assumes that the laboratory and CM projectile scattering angles θ and χ are small enough (<1) to neglect the projectile deflection and calculate the sideways momentum transferred to the projectile during the interaction as if it were moving along a straight line.

Suppose that an impinging particle has impact parameter b with a target particle. If there is no displacement of the target particle during the interaction, and the projectile is considered to move along a straight line, then the sideways force acting on the projectile can be written as

$$F_x = -\frac{dV(r)}{dr}\frac{b}{r} = -\frac{\partial V(\sqrt{z^2 + b^2})}{\partial b}, \tag{57}$$

where $r = \sqrt{z^2 + b^2}$ is the distance between the projectile and target, and z is a coordinate on the line along which the projectile moves. The transferred sideways momentum is obtained by integrating F_x with respect to time, and the l-system scattering angle is found as the ratio of the transferred and initial momenta as

$$\theta \approx \frac{1}{M_1 v_{01}} \int F_x dt = -\frac{1}{M_1 v_{01}^2} \int_{-\infty}^{\infty} dz \frac{\partial}{\partial b} V(\sqrt{z^2 + b^2}). \tag{58}$$

where v_{01} is the initial velocity of the projectile. The integral in Eq. 58 can be evaluated analytically using a modified Bessel function of the second kind if $V(r)$ is a screened coulomb potential with a screening function represented by the sum of exponents.[7]

In the small-angle approximation, the trajectory of the scattered particle is then given by

$$x = b + z\theta, \tag{59}$$

where z is a coordinate along the initial projectile direction and x is a transverse coordinate.

2.2.5. Differential Scattering Cross Section

In experiments one usually deals with large numbers of particles scattered in short periods of time, thus one needs a suitable description that takes into account this multiple nature of the scattering process. The collision dynamics alone are insufficient in this case, since it describes only single-collision events. In order to account for large numbers of particles and to consider the scattering process in general, rather than individual collisions, a quantity called the differential scattering cross section is used.

Consider a target bombarded by a constant flux of identical particles of the same velocity. The differential scattering cross section is defined as the ratio of the number of particles $dN_{d\Omega}$ scattered by the target into the solid angle $d\Omega$ to the number N of the particles per unit area impinging on the target:

$$d\sigma = \frac{dN_{d\Omega}}{N} = \sigma_{diff}\, d\Omega. \tag{60}$$

The incoming flux N is assumed to be constant for all impact parameters with the target. It is observed from Eq. 60 that σ_{diff} has the dimensions of area per solid angle and represents the relative angular intensity of the scattered particles. It is called a differential scattering cross section and is an important characteristic of the scattering process.

It is instructive to obtain an expression for $d\sigma$ in terms of the function describing the dependence of the scattering angle on the impact parameter with a target. The CM system will be used, because it is easier to operate on the CM quantities due to their connection with the central force problem solved earlier. For simplicity, consider a target consisting of only one atom. As a consequence, the system will have axial symmetry. The analysis can then be generalized for a target with an arbitrary geometry or composition (see Appendix), although the simplest expression for an axially symmetric system is sufficient for most cases encountered in practice.

First assume that there exists only one value of the impact parameter that corresponds to each scattering angle, i.e., the function $b(\chi)$, where χ is a CM scattering angle and b is a single-valued function. In this case, only the particles with the impact parameters between b and $b + db$ will be scattered at angles between χ and $\chi + d\chi$. From geometrical considerations as shown in Figure 2.8, it follows that the total number of the particles having their impact parameters in the interval $(b, b + db)$ equals the number N of particles impinging on the unit area times the

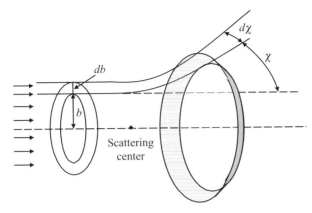

FIGURE 2.8. Scattering of a parallel beam of incident particles by a center of force.

area of the strip of length $2\pi b$ and width db containing all impact parameters in the range between b and $b + db$:

$$dN_{b, b+db} = dN_{\chi, \chi+d\chi} = N \cdot 2\pi b\, db. \tag{61}$$

Then, by definition, the scattering cross section is

$$d\sigma = \frac{dN_{\chi, \chi+d\chi}}{N} = 2\pi b\, db = 2\pi b \left| \frac{db}{d\chi} \right| d\chi = b \left| \frac{db}{d\chi} \right| \frac{d\Omega}{\sin \chi}. \tag{62}$$

In the last equation, the expression $d\Omega = 2\pi \sin \chi\, d\chi$ was used for the solid angle formed by the two cones with vertical angles χ and $\chi + d\chi$. If there are several values of impact parameter corresponding to the same scattering angle, all contributions should be taken into account by summing expressions like Eq. 62 for each b. The modulus of the derivative $db/d\chi$ has been taken since the scattering cross section is always positive, whereas the derivative may be, and usually is, negative. From Eq. 62 it is observed that in order to find the scattering cross section, one has to calculate the derivative of the deflection function $d\chi/db$. Differentiating Eq. 42 with respect to b yields[8]:

$$\frac{d\chi}{db} = I_1(W^{-1} - 1) + (WE)^{-1} \int_{R_{\min}}^{\infty} \frac{dr}{r^4 g(r)^3} \left[V'(R_{\min}) R_{\min}^3 - V'(r) r^3 \right], \tag{63}$$

where

$$W = 1 - \frac{R_{\min}^3 V'(R_{\min})}{2b^2 E}, \tag{64}$$

$$g(r) = \sqrt{1 - \frac{V(r)}{E} - \frac{b^2}{r^2}}. \tag{65}$$

Using Eq. 62 together with Eq. 63 allows exact numerical calculation of the scattering cross section for any potential.

The expression for the scattering cross section in the l-system for the projectile or target atom is obtained by expressing χ in terms of the projectile scattering angle θ or target recoiling angle ϕ from Eqs. 12a and 12b and substituting it into Eq. 62. The scattering cross section of the projectile is then

$$d\sigma_{scat} = 2\pi b \left| \frac{db}{d\chi} \frac{d\chi}{d\theta} \right| d\theta = b \left| \frac{db}{d\chi} \frac{d\chi}{d\theta} \right| \frac{d\Omega}{\sin\theta}, \tag{66}$$

and the scattering cross section of the target particle (recoiling cross section) is then

$$d\sigma_r = b \left| \frac{db}{d\chi} \frac{d\chi}{d\phi} \right| \frac{d\Omega}{\sin\phi} = 2b \left| \frac{db}{d\chi} \right| \frac{d\Omega}{\sin\phi}. \tag{67}$$

Figure 2.9 shows examples of cross sections calculated using Eqs. 66 and 67 for various projectile–target combinations using the ZBL potential. The projectile scattering cross section decreases with increasing scattering angle, while the trend for the recoiling cross section is the opposite.

2.3. MULTIPLE COLLISIONS

2.3.1. Directional Effects in Ordered Solids

The bombarding atoms enter the half-space enclosing the target at random points at the target surface. As they enter, their motion begins to be influenced by the presence of the electrons and nuclei in the vicinity of their paths. The former do not affect the direction of propagation of the projectile and cause only inelastic energy losses, whereas the latter are responsible for both the elastic energy losses and momentum transfer. The range of energies considered in this book (1 keV to 10s of keV) is well below the limits for which electronic energy losses play a significant role as an inelastic energy loss mechanism. Therefore, the electronic energy losses are small and can be safely neglected both as a source of energy loss and direction change of the incoming particle. In a first approximation, it is necessary only to consider nuclear stopping. If there is no order in the target, the motion of the energetic particle is governed by random collisions with the nuclei of the atoms forming the target and, in a chain of collisions, each atom is independent of the others. On the contrary, in the case of scattering of energetic particles by well-ordered solids such as crystals, the collisions experienced by a particle moving from one atom to another are correlated due to the order existing in the crystal. Depending on the initial direction of propagation, these correlations can significantly affect the resulting trajectory of a particle as compared to the trajectory in the random medium. The elements affecting the trajectory are atomic strings and planes.

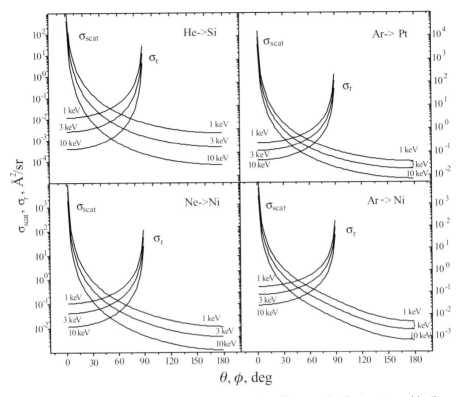

FIGURE 2.9. Scattering and recoiling cross sections for different projectile–target combinations using a Zigler-Biersack-Littmark (ZBL) potential as functions of the respective scattering and recoiling angles. He, helium; Si, silicon; Ar, argon; Pt, platinum; Ne, neon; Ni, nickel.

Scattering by an Atomic String If the interaction of a particle with an atomic string is considered without paying attention to the structural details of the string, the potential created by the points (atoms) on the string can be replaced with one created by a continuous line. Suppose that atoms forming an infinite string belong to the same species and are at the same distance d from each other. In order to remove structural irregularities associated with the points of location of the atoms, each atom is replaced by a line of length d adjacent to the lines formed by the two neighboring atoms so that these lines form an infinite string.[9] If the potential of a single atom is designated by $V(r)$, the potential created by one line of length d is

$$U_d(r) = \frac{1}{d} \int_{-d/2}^{d/2} V(\sqrt{r^2 + z^2})dz, \tag{68}$$

where z is a coordinate along the line. Therefore the potential of an infinite string is

$$U(r) = \frac{1}{d} \int_{-\infty}^{\infty} V\left(\sqrt{r^2 + z^2}\right) dz. \tag{69}$$

If $V(r)$ is a Thomas–Fermi-type potential, it is reasonable to assume that after the integration, the dependence of $U(r)$ on r in Eq. 69 will have one power of $1/r$ less than $V(r)$ so that

$$U(r) = \frac{Z_1 Z_2 e^2}{d} \xi(r/a). \tag{70}$$

Here, Z_1 and Z_2 are the atomic numbers of the projectile and chain atom, e is the charge of an electron, and $\xi(r/a)$ is a new screening function that is approximated as

$$\xi\left(\frac{r}{a}\right) \approx \log\left[\left(\frac{C_1 a}{r}\right)^2 + 1\right], \tag{71}$$

where C_1 is a fitting constant $\sim \sqrt{3}$.

Consider the conditions when the string approximation for a chain of atoms is valid and the potential of Eq. 69 can be used. It has been shown[9] that at high primary energies, the string approximation holds true for angles ψ between the direction of a particle and a chain such that

$$\psi < \psi_{c1} = \sqrt{\frac{2s}{d}}, \tag{72}$$

where ψ_{c1} is the critical Lindhard angle for high energies and s is a collision diameter in the l-system

$$s = \frac{Z_1 Z_2 e^2}{E}. \tag{73}$$

For the particle to be considered highly energetic, ψ_{c1} should satisfy the following condition

$$\psi_{c1} \leq \frac{a}{d}. \tag{74}$$

If this inequality is not satisfied (i.e., in case of low-projectile energy, where $\psi_{c1} > \frac{a}{d}$), the critical angle for channeling is given by

$$\psi_{c2} = \left(\frac{C_1 a}{d\sqrt{2}} \psi_{c1}\right)^{1/2}. \tag{75}$$

These expressions provide the conditions to test whether the motion of a particle will be highly correlated or, on the contrary, random. Usually, the critical angles for channeling are smaller than critical angles for the continuum approximation.

2.3.2. Shadowing and Blocking Cones

Shadowing and blocking are specific phenomena characteristic of scattering by repulsive potentials. They are manifested as redistributions of the scattered flux and appearance of regions in space close to the target atoms in which the scattering atoms cannot penetrate due to the mutual repulsion between the projectiles and target particles. If a target consists of only one atom, such a region behind this atom that is "banned" for penetration by the incoming particles is called a shadow cone. For a target containing two atoms, the scattering particles directed from the first atom towards the second atom and repelled by its potential, create a trajectory pattern known as a blocking cone. Other regions become filled with the trajectories deflected from the shadowing and blocking areas, giving rise to a local increase in the number of trajectories (i.e., a phenomenon known as *focusing*).

Shadow Cone   If a single atom is placed in a parallel flux of incoming particles that are lighter than that atom, the particles within a certain area of small-impact parameters will have their scattering angles large enough to prevent them from penetrating the region immediately behind the atom. The atom will cast a "shadow" behind itself (i.e., an area free of incoming particles will be formed). This area is called a shadow cone. The trajectories repelled by an atom will concentrate at the edges of the cone. Figure 2.10 shows a shadow cone simulation for a 10-keV monoenergetic parallel beam of helium (He) ions impinging on a single platinum (Pt) atom using a ZBL interaction potential. The simulation gives an exact shape of a shadow cone for any specific potential used, and therefore in such a simulation any divergence from an actual shape comes only from the inexactness of the potential. Approximations are made to obtain the expressions for the shape of the shadow cone in an analytical form. Using a small-angle approximation and an unscreened coulomb potential, the following formula for the radius of the shadow cone R_{sh} is obtained[9]:

$$R_{sh} = 2\sqrt{\frac{Z_1 Z_2 e^2 z}{E}}, \tag{76}$$

where z is the distance behind the target atom, E is the primary energy, e is the charge of an electron, and Z_1 and Z_2 are atomic numbers of the projectile and target atoms, respectively. Using the parameter s from Eq. 73 that determines the strength of the interaction, Eq. 76 can be rewritten as

$$R_{sh} = 2\sqrt{sz}. \tag{76'}$$

As observed from Eq. 76, the dependence of the radius of the shadow cone on the distance behind the atom for the coulomb potential is parabolic. It has been shown[10]

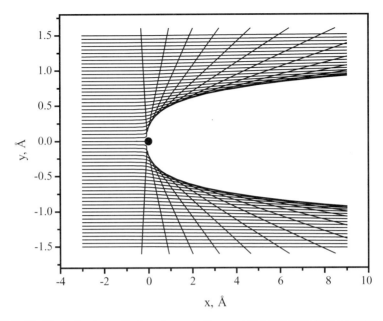

FIGURE 2.10. Simulation for scattering of a parallel beam of 10-keV helium from a single platinum atom using a ZBL potential illustrating a shadow cone.

that in the small-angle approximation, the ratio of the shadow cone radius for the screened Molière potential to that for the coulomb potential is a universal function of a single dimensionless parameter κ

$$\kappa = 2\sqrt{sz}/a = R_{sh}/a. \tag{77}$$

The following empirical expression[11] can be used to calculate this ratio with an accuracy of better than 1%:

$$\frac{R_{sh}^{Moli\grave{e}re}}{R_{sh}} = \begin{cases} 1.0 - 0.12\kappa + 0.01\kappa^2 & 0 \le \kappa \le 4.5 \\ 0.924 - 0.182\ln\kappa + 0.0008\kappa & 4.5 \le \kappa \le 100 \end{cases}. \tag{78}$$

Here, $R_{sh}^{Moli\grave{e}re}$ is the radius of the Molière shadow cone in the small-angle approximation.

The knowledge of the shape of the shadow cone is applied to determine interatomic distances between the surface atoms in a crystal. In a technique known as impact-collision ion scattering spectroscopy (ICISS),[12] an observation angle close to 180° is used so that the detected scattered particles experience collisions with a near-zero impact parameter. By varying the incidence angle of the impinging beam, the edge of the shadow cone created by one atom is directed towards another atom so that most of the focused particles impact the second atom within a small interval of impact parameters near $b = 0$ (Figure 2.11). At this critical incidence angle α_{cr}, a sharp

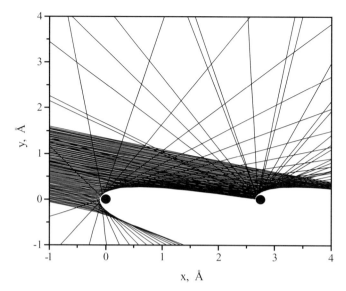

FIGURE 2.11. Simulation for scattering of a parallel beam of 10-keV helium from two platinum atoms at a distance 2.75 Å from each other. The critical incident angle α_{cr}, i.e. the angle between the direction of the beam and the line connecting the two target atoms is $\alpha_{cr} = 14.1°$. The α_{cr} is chosen so as to position the edge of the shadow cone created by the first atom on the center of the second atom. Upon exiting the shadow area, the second atom begins to backscatter the particles focused onto it by the first atom, producing a sharp rise in the number of scattered particles collected by the detector.

increase in the number of particles striking the detector is observed. The distance between the atoms is then easily determined from α_{cr} and the geometry of the shadow cone. For scattering angles less than $180°$, the fact that the impact parameter with the second atom is not zero should be taken into account.

Blocking Cone Blocking is another example of the appearance of a region free of scattered particles behind an atom. The presence of at least two atoms is required for blocking. When a beam of ions with parallel trajectories is directed at an atom, the angular intensity distribution of scattered particles is nearly isotropic for scattering angles not very close to zero, as shown by the dashed line of Figure 2.12. If there is a second atom in the vicinity of the first atom, it will block part of the particles scattered by the first atom so that there will be no scattered trajectories behind the second atom. This results in a hyperboloid-like pattern called a blocking cone, with apex on atom 1 and centered approximately on the line connecting the two atoms as shown in Figure 2.13. The trajectories repelled by the second atom will concentrate at the edges of the blocking cone, yielding the maximum scattering intensity. The solid line in Figure 2.12 shows the scattering intensity distribution in the plane containing the two atoms and the primary beam, while the two maxima correspond to the edges of the blocking cone.

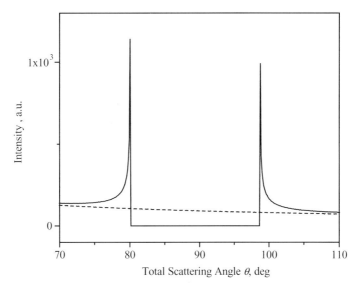

FIGURE 2.12. Scattering intensity for a 10-keV helium (He) beam impinging on a single platinum (Pt) atom in the plane containing the Pt atom and incident He beam in the vicinity of a 90° scattering angle (dashed line). Scattering intensity after a second Pt atom is added in the plane at a distance of 5 Å from the first atom along the 90° scattering direction (solid line). The two maxima are at the edges of the blocking cone.

The two important characteristics of the blocking cone are the direction of interatomic axis Q and the size ψ defined as the angle between Q and the edge of the blocking cone. In a crystalline target, blocking cones are created by atomic pairs lying along low-index directions. The experimentally measured quantity is the projection of the edge on the detector plane (YZ plane in Figure 2.13). Because a blocking cone is aligned with the interatomic axis, the direction of the axis can be found from the position of the blocking cone projection. This can provide the information about relaxation or reconstruction of the surface. Measuring the size of the blocking cone gives information about the interatomic distance between the pair of atoms. The size of the blocking cone in the low-energy range has been estimated by Lindhard[9] to be

$$\psi = \left(\frac{2.828\psi_{c1} C_1 a}{d} \right)^{1/2} = 2\psi_{c2}, \tag{79}$$

where ψ_{c1} and ψ_{c2} are the Lindhard critical angles (Eqs. 72 and 75), d is the interatomic distance, a is Lindhard's screening length (Eq. 22), and C_1 is a fitting constant $\sim \sqrt{3}$. Molecular dynamics simulations have been performed to obtain the blocking cone size for various projectile–target combinations.[13] They revealed azimuthal asymmetry in the size of the blocking cone with respect to the interatomic axis, as shown in Figure 2.14. The minimum value of ψ is at the top of the blocking cone (on the side facing the primary beam), whereas the maximum ψ is reached at the bottom. By fitting

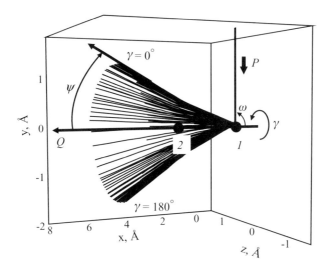

FIGURE 2.13. A blocking cone created by scattering a parallel beam of 10-keV helium from two platinum atoms at a distance $d = 3$ Å from each other. The direction of the primary beam is designated by a straight arrow and letter P, and the angle between the primary beam and interatomic axis is ω. The angle of the blocking cone is ψ, and the zero for the azimuthal angle γ is chosen at the top of the cone.

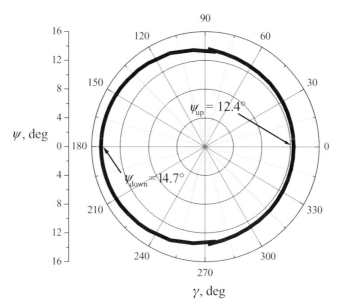

FIGURE 2.14. Projection of the blocking cone from Figure 2.13 onto the YZ plane showing the asymmetry of the blocking cone. The minimum value $\psi_{min} = 12.4°$ reached at $\phi = 0°$, and the maximum value $\psi_{max} = 14.7°$ is at $\phi = 180°$.

calculated points in the space of the parameters of the interacting atomic species, a universal formula for calculating the blocking cone size for arbitrary energies and interacting species has been derived:

$$\psi_{up} = \exp(F_0 + F_1 x + F_2 x^2 + F_3 xy + F_4 y + F_5 y^2), \tag{80}$$

where

$$F_i = f_i + h_i \frac{M_1}{M_2} + g_i \ln(Z_1 Z_2) + e_i \ln(\omega), \tag{81a}$$

for $i = 0, 3–5$, and

$$F_i = f_i + h_i \frac{M_1}{M_2} + g_i \left(Z_1^{0.23} + Z_2^{0.23}\right) + e_i \ln(\omega) \tag{81b}$$

for $i = 1, 2$. Here, ω is the angle between the primary beam direction and interatomic axis, M_1, Z_1 are the mass and atomic number of the projectile, and M_2, Z_2 are masses and atomic numbers of the two target atoms. The numerical values of the coefficients $f_i - e_i$ are given in Table 2.1. The lower part of the blocking cone can be estimated from $\psi_{down} = k\psi_{up}$, where k is in the range 1.0–1.3 and is dependent on the same variables as ψ_{up}. Table 2.2 presents the fitting coefficients $F_0' - F_5'$ of the expression for k

$$\ln k = F_0' + F_1' \ln d + F_2' (\ln d)^2 + F_3' \ln d \ln E + F_4' \ln E + F_5' (\ln E)^2 \tag{82}$$

for the case of four projectile–target combinations. The blocking cone asymmetry is appreciable for energies and interatomic distances encountered in low-energy ion scattering. For example, in the case of 10-keV He scattered from two Pt atoms at a distance $d = 3$Å from each other, the asymmetry is $\sim 15\%$. It is more pronounced the lower the primary energy and the shorter the interatomic distance. This asymmetry

TABLE 2.1. Numerical values of parameters in Eqs. 81(a,b) found from fitting molecular dynamics calculations of the blocking cone size (ψ_{up})

	0	1	2	3	4	5
f_i	1.34 ± 0.04	0.13 ± 0.03	-0.074 ± 0.003	-0.060 ± 0.004	0.0096 ± 0.0009	-0.0404 ± 0.0009
h_i	0.90 ± 0.01	-0.24 ± 0.01	0.0255 ± 0.0006	0.0084 ± 0.0008	0.047 ± 0.002	-0.0010 ± 0.0001
g_i	0.174 ± 0.001	-0.133 ± 0.003	0.0093 ± 0.0002	0.0043 ± 0.0001	0.0167 ± 0.0004	0.00253 ± 0.00003
e_i	0.36 ± 0.01	-0.078 ± 0.007	0.0053 ± 0.0005	0.0140 ± 0.0009	-0.069 ± 0.003	0.0032 ± 0.0002

TABLE 2.2. Numerical values of parameters in Eq. 82 for finding the coefficient k used in determining the size of the lower part of the blocking cone for four common systems. In all calculations, $\omega = 90$[a]

	F'_0	F'_1	F'_2	F'_3	F'_4	F'_5
He–Pt	0.4578	−0.1075	0.00945	0.01535	−0.1085	0.00510
	± 0.0002	± 0.0001	± 0.00003	± 0.00003	± 0.0002	± 0.00003
Ne–Pt	0.1749	−0.1941	0.0212	0.0253	−0.1653	0.0077
	± 0.0001	± 0.0001	± 0.0002	± 0.0002	± 0.0009	± 0.0002
Ne–Ni	0.2341	0.0042	−0.00536	0.01017	−0.0980	0.00481
	± 0.0002	± 0.0001	± 0.00002	± 0.00003	± 0.0001	± 0.00002
He–Si	0.1494	−0.0150	−0.00051	0.00667	−0.0747	0.00800
	± 0.0001	± 0.0001	± 0.00001	± 0.00001	± 0.0001	± 0.00001

[a] He, helium; Pt, platinum; Ne, neon; Ni, nickel; Si, silicon.

should be taken into account when drawing conclusions about the direction of the interatomic axis based on the experimentally measured shape of the blocking cone.

APPENDIX 2.1. GENERALIZED DEFLECTION FUNCTION

If the target is not axially symmetric, the two-dimensional Jacobian describing the connection of the coordinates b and φ in the plane perpendicular to the direction of the incoming particles with the angular scattering coordinates χ and δ should be used. The number of particles scattered into the differential element of the solid angle $d\Omega = \sin \chi \, d\chi \, d\delta$ will be

$$dN_{\chi, \chi + d\chi; \delta, \delta + d\delta} = Nb \, db \, d\varphi = NbJ \, d\chi \, d\delta = NbJ \frac{d\Omega}{\sin \chi} \tag{A1}$$

with

$$J^{-1} = \left| \frac{\partial(\chi, \delta)}{\partial(b, \varphi)} \right| = \left| \begin{array}{cc} \partial\chi/\partial b & \partial\delta/\partial b \\ \partial\chi/\partial\varphi & \partial\delta/\partial\varphi \end{array} \right| = \left| \frac{\partial\chi}{\partial b} \frac{\partial\delta}{\partial\varphi} - \frac{\partial\chi}{\partial\varphi} \frac{\partial\delta}{\partial b} \right|. \tag{A2}$$

For a system possessing axial symmetry, it is convenient to chose $\delta = \varphi$, so that $\partial\delta/\partial\varphi = 1$ and $\partial\delta/\partial b = 0$. Then J^{-1} in (A2) becomes a modulus of the derivative $d\chi(b)/db$ of the deflection function $\chi(b)$, and (A1) divided by N becomes Eq. 62.

REFERENCES

1. O. B. Firsov, *Sov. Phys. – JETP* **5**, 1192 (1957).

2. J. Lindhard, V. Nielsen, and M. Scharff, *Fys. Medd. Dan. Vid. Selsk.* **36** n.10, 1 (1968).

3. I. M. Torrens, *Interatomic Potentials,* Academic Press, NY, 1972.

4. G. Molière, *Z. Naturforsch* **2a,** 133 (1947).

5. J. F. Ziegler, J. P. Biersack, and U. Littmark, *The Stopping and Range of Ions in Solids,* J. F. Ziegler (Ed.), Pergamon Press, NY, 1985.

6. L. D. Landau and E. M. Lifshitz, *Mechanics,* Pergamon Press, NY, 1966.

7. B. Hird, *Can. J. Phys.* **69,** 70 (1991).

8. G. E. Ioup and B. S. Thomas, *J. Chem. Phys.* **50,** 5009 (1969).

9. J. Lindhard, *Mat. Fys. Medd. Dan. Vid. Selsk.* **34** n.14, 1 (1965).

10. O. S. Oen, in Proc. 7th Intern. Conf. on Atomic Collisions in Solids, Vol. 2, Yu. Bulgakov and A. F. Tulinov (Eds.), Moscow State University Publishing House, Moscow, 1981, p. 124.

11. O. S. Oen, *Surf. Sci.* **131,** L407 (1983).

12. M. Aono, Y. Hou, C. Oshima, and Y. Ishizawa, *Phys. Rev. Let.* **49,** 567 (1982).

13. A. Kutana, I. L. Bolotin, J. W. Rabalais, *Surf. Sci.* **495,** 77 (2001).

3

EXPERIMENTAL METHODS

This chapter addresses the basic design elements of ion scattering spectrometers, experimental procedures, and factors affecting experimental spectra. Many new developments in instrumentation have appeared in recent years and have resulted in novel spectrometer designs. It is not possible to discuss all of these new developments here. Hence the objective of this chapter is to focus attention on some of the most commonly used instrumentation and procedures rather than to be encyclopedic.

3.1. GENERAL DESCRIPTION OF AN ION SCATTERING SPECTROMETER SYSTEM

3.1.1. Basic Components

The basic components of an ion scattering spectrometer are shown in Figure 3.1. Ion scattering spectrometer systems are usually constructed from both custom-made and commercially available components. They consist of an ultrahigh vacuum (UHV) scattering chamber with some or all of the following components: (1) primary ion beam that is either pulsed (for time-of-flight [TOF] analysis) or unpulsed (for electrostatic energy analysis [ESA]); (2) detector; (3) precision manipulator on which to mount the sample; (4) ~1-m TOF drift region (for TOF analysis); (5) electrostatic analyzer (for energy analysis); (6) sputter ion gun for sample cleaning; (7) mechanism for heating and annealing the sample; (8) pumping system to achieve UHV conditions; (9) pulse generating, timing, detection, and control electronics; and (10) computer with data acquisition and data handling capabilities. The UHV chamber often houses other surface analysis techniques, such as low-energy electron diffraction (LEED), x-ray photoelectron spectroscopy (XPS), and Auger electron spectroscopy (AES). The combination of these techniques provides a powerful surface-analysis system.

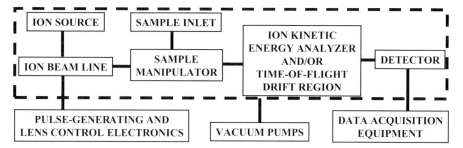

FIGURE 3.1. Block diagram of the basic components of an ion scattering spectrometer.

There are many variations and configurations of vacuum chambers and associated components that have been built and successfully used to provide valuable information about surfaces. Conventional ion scattering spectrometry uses planar configurations (i.e., the ion source, sample, and detector are in the same plane) and detectors that collect small solid angles ($<10^{-4}$ sr) of the scattered flux. Structural information is obtained from the anisotropic scattered ion flux distribution by rotating either the sample or the detector with respect to the ion beam incident direction. For example, a typical azimuthal angle scan covering $90°$ of azimuthal angle space in $2°$ increments would require ~ 135 min and a total ion dose of $\sim 10^{14}$ ions/cm^2. Another method of collecting scattering data is to use a large, gated, position-sensitive microchannel plate (MCP) detector and TOF methods to capture images of scattered and recoiled atoms in both planar and nonplanar directions simultaneously. This method greatly decreases the data collection time.

The basic components of an ion scattering system are described in general in this section. This is followed by detailed descriptions of specific instrumentation in the later sections. Section 2 details the use of a small-area detector with both TOF and kinetic energy analysis and a continuously variable scattering angle for TOF analysis, a technique known as TOF scattering and recoiling spectrometry (TOF-SARS).[1] Section 3 describes a coaxial scattering spectrometer.[2,3,4] A large area position-sensitive detector with TOF detection, known as scattering and recoiling imaging spectrometry (SARIS), is described in section 4.[5] The use of sequential deflection pulses (SDP) for mass and charge selection of pulsed ion beams is described in section 5.[6] Section 6 provides a description of a technique for ion scattering from liquid surfaces.[7]

3.1.2. Vacuum Chamber and Pumping System

The vacuum chambers for ion scattering are normally fabricated entirely of stainless steel with permanent joints made by inert gas welding and demountable joints made with copper gaskets. These chambers are usually cylindrically shaped with many ports for introducing instrumentation. For TOF measurements, long drift tubes radiate out from the chamber walls.

Rough pumping from atmospheric pressure is accomplished by sorption pumps or oil-free mechanical pumps. UHV conditions can be achieved by using any number of different pumps such as turbomolecular, ion, cryogenic, and titanium sublimation pumps. The gas load in such chambers can be high when inert gas ion beams are used. In such cases, turbomolecular pumps are desirable to handle the large influx of inert gases. The entire system can be baked by means of rubber strip heaters that are glued to the chamber walls in order to evenly disperse the power over the system and/or by quartz lamp heaters positioned inside the chamber. Base pressures of 1×10^{-10} Torr are obtainable after baking. A gas manifold is attached to the chamber through a bakeable leak valve for introduction of gases for chemisorption or surface treatments.

3.1.3. Sample Manipulation and Cleaning

Samples are mounted on a precision manipulator that allows $0°$–$360°$ polar beam incident angle rotations, $0°$–$180°$ sample azimuthal angle rotations, translations along the three orthogonal axes, and adjustment of the sample tilt angle with respect to the incident ion beam. Both the polar incident and azimuthal rotations are controlled by stepper motors that are interfaced to a minicomputer. The beam incident angle and scattering angles can be aligned by means of a laser beam to an accuracy of $1°$. The azimuthal angle can be determined crudely from the LEED pattern and then precisely by monitoring the surface semichanneling[8] effects of keV ions.

Due to the extreme surface sensitivity of ISS, it is important to carefully prepare the sample surfaces. Single crystals can be polished to achieve smooth surfaces. Powders can be pressed into pellets to reduce the macroscopic roughness. Samples can be cleaned in a vacuum by means of sputtering by inert gas ions, annealing by electron bombardment, and chemical reactions. The sputtering beam incident angle can be varied from grazing to $45°$ incidence. Sputtering can affect the atomic structure and composition of the surface and surface layers through preferential sputtering effects. Annealing temperatures above $2500°C$ are obtainable by direct electron bombardment from a tungsten ribbon filament mounted in back of the sample. Temperature can be measured by means of a thermocouple attached to the edge of the sample or through the viewport by means of a portable infrared thermometer. Exposing the sample to reactive gases while at high temperature can induce chemical reactions that lead to desorption of impurities. For example, at elevated temperatures, surface hydrocarbons and hydrogen can generally be removed by exposure to O_2 and surface O can generally be removed by exposure to H_2.

Ion scattering can be performed on samples that are conducting and semiconducting without sample charging problems. For insulating samples, the ion beam induces a positive charge on the sample surface. For a good insulator, this charging can continue until the surface potential reaches the kinetic energy of the beam and no further ions can reach the target. In this case, primary ions are deflected by the surface electric field and may be registered with the energy of the unscattered primary beam. For a surface potential less than the primary ion kinetic energy, the energy of scattered particles will be determined by their charge state after the interaction with the surface. If the ion is neutralized in the scattering collision, the scattering peak will be shifted towards lower energy due to deceleration on the incoming part of the trajectory by the field of the sample. If neutralization is avoided, an ion that was decelerated by the sample field upon approaching will be accelerated by the same field on the way out. The deceleration results in a smaller energy loss in the binary encounter. Thus, in the case of a partially charged surface, the final energy of the detected ion can be even higher than in the case of a surface with no charge. Insulating samples can also be studied by using a low-energy (e.g., <10 eV) electron flood to maintain surface neutrality during scattering. Surface charging can also accelerate the sputtered ions. Since these sputtered ions from a conducting surface have low kinetic energies that peak below 10 eV, the kinetic energies of sputtered ions provide a good measure of the degree of sample charging. For rough surfaces, the surface charging can be inhomogeneous, resulting in a very broad energy distribution of the scattered ions.

3.1.4. Ion Source and Beam Line

There are three critical requirements for the ion source: (1) a small energy spread, (2) no fast neutrals in the beam, and (3) ions must be differentially pumped to reduce the pressure in the scattering region. The most common types of ion sources used in scattering are those that produce 1–20-keV noble gas or alkali metal ions with narrow energy distributions. Due to the natural line width of the scattering peaks, an energy resolution of $\Delta E/E$ of about 1% is adequate for most applications. For TOF analysis, the mass separated ion beam is pulsed by means of a square-wave voltage applied to two pairs of beam pulsing plates. Ion pulses with time widths of 5–50 ns, measured as the full-width-at-half-maximum (FWHM) intensity, have been reported.[9,10,11]

Noble gas ions are usually created by electron impact ionization in the source, which has an off-axis filament to reduce the amount of fast neutrals directed towards the sample. Alkali ions are usually generated by heating alkali-impregnated aluminosilicate wafers.[12] There are three major differences in the scattering of noble gas and alkali metal ions. First, noble gas ions are neutralized with high probability during scattering collisions, whereas alkali metal ions have higher survival probabilities. Therefore, for ESA analysis, alkali ions provide a higher sensitivity than rare gas ions, although rare gas ions provide the highest first-layer sensitivity due to their enhanced neutralization rate from subsurface layers. Although alkali ions have higher survival probabilities, significant variations still remain in their scattered ion yields that can affect quantitative analysis. For TOF analysis with detection of both neutrals and ions, such variations in the neutralization probabilities are not important. Second, due to the high neutralization rates for scattering of noble gas ions, ESA analysis requires rather large primary ion fluences ($\sim 10^{13}$–10^{14} ions/cm^2) to obtain reasonable counting statistics. When using TOF analysis with detection of both neutrals and ions, this fluence is reduced to $\sim 10^{11}$ ions/cm^2. Finally, alkali ions are deposited on the surface during scattering, an effect that can change the surface work function and reconstruction among other things.

3.1.5. Other *in situ* Analysis Techniques

Other *in situ* surface analytical techniques are desirable to have in the chamber for additional surface characterization. These include an electron gun, x-ray source, and ultraviolet source for use with an electrostatic analyzer for AES, XPS, and UPS measurements. A LEED system is invaluable for determining the surface symmetry when doing structural analysis.

3.1.6. Analyzers

Both TOF and ESA are used in ion scattering. TOF provides velocity analysis of both neutrals and ions with moderate resolution. An ESA provides energy analysis of only the ions with high resolution. An ESA can be used to verify the masses of the ejected particles by measuring the TOF of particles traveling through the analyzer while the analyzer is set to pass a specific kinetic energy. Both the velocity and energy of the transmitted particles are known, thereby providing a unique mass determination.

Time-of-flight Analysis   A TOF analyzer is simply a long drift region. If a deflector plate, an acceleration tube lens, and a retarding grid analyzer are included in the TOF region, four different types of spectra can be obtained: (1) neutrals plus ions—obtained with all elements at ground potential; (2) neutrals only—obtained with voltage on the deflector; (3) neutrals and accelerated ions—obtained with accelerating voltage on the tube lens; (4) neutrals plus ions with specific charges—obtained with the retarding grid at a preselected voltage. This grid consists of a biasable mesh sandwiched between two grounded meshes that can be used to transmit the scattered and recoiled neutrals plus ions with specific charges as follows. Doubly charged ions A^{2+} at kinetic energy E can be separated from singly charged ions A^+ at kinetic energy E by placing a potential V on the grid so that $E/2e < V < E/e$. The length of the flight path is usually chosen so that it provides the required energy resolution at sufficient transmission. When the mass composition of the particles being analyzed is known, the energy of the particles E is connected with the flight time by the relation

$$E = ML^2/2t^2, \tag{1}$$

where M is the particle mass and L is the flight length (i.e., the distance between target and detector). Differentiating this expression gives the relation between the experimentally measured flight time distribution of particles, dN/dt, and their energy distribution, dN/dE. The resulting expression is

$$dN/dE = -(dN/dt)t^3/ML^2. \tag{2}$$

The ratio of the ion pulse width to the flight time determines the energy resolution for a given length of flight path. For example, consider 3-keV Ne^+, a pulse width of 50 ns, and a flight path of 1 m, $\Delta t/t = 8.5 \times 10^{-3}$ corresponding to an energy resolution of $\Delta E/E = 2\%$. Increasing the path length and decreasing the ion pulse width can improve this resolution considerably.

For the typical limited energy regions observed in ion scattering, it is usually not necessary to correct the shape of the time spectrum from a TOF measurement. In order to transform a TOF spectrum $F(t)$ into an energy spectrum $N(E)$, it is necessary to convert constant time increments into constant energy increments by using Eq. 1 and

$$N(E) = F(t)t^3/ML^2, \tag{3}$$

where t is the measured time for a particle of mass M to travel through the potential-free drift tube.

Electrostatic Analysis Energy separation of ions in an electrostatic analyzer is made by spatial dispersion of the charged-particle trajectories in a known electrical field. The field is produced by an analyzer whose shape and dimensions are carefully designed to provide good optical properties. Different analyzers have been used for ISS: cylindrical-mirror analyzer (CMA),[13] hemispherical analyzer (HSA)[14] with or without preretardation lens system, 127° cylindrical analyzer,[15] and toroidal systems.[16] Schematic drawings of a CMA and HSA are shown in Figure 3.2. These analyzers are either fixed in the UHV chamber or mobile so that the scattering and

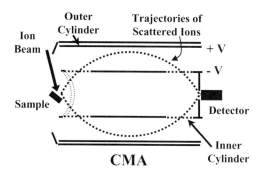

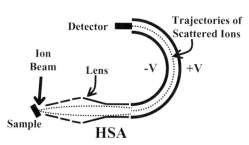

FIGURE 3.2. Simplified schematic drawings of a cylindrical mirror analyzer (CMA) and a hemispherical analyzer (HSA) (From Hellings *et al.*, 1985 and Bergmans *et al.*, 1993, with permission).

recoiling angle can be continuously varied from $0°$ to $\sim180°$. Except for very precise energy-loss measurements, energy resolution of the order of $\sim1\%$ is enough for ISS surface composition analyses. Figure 3.3 shows a cross section of a toroidal electrostatic analyzer that is rotationally symmetric.[16,17] A two-dimensional detector allows the simultaneous detection of 100 energy channels, thus enhancing the sensitivity by three orders of magnitude over conventional ESAs.

For these ESAs, the filtered energy E is proportional to the potential difference V applied across the electrodes,

$$E = eFV, \tag{4}$$

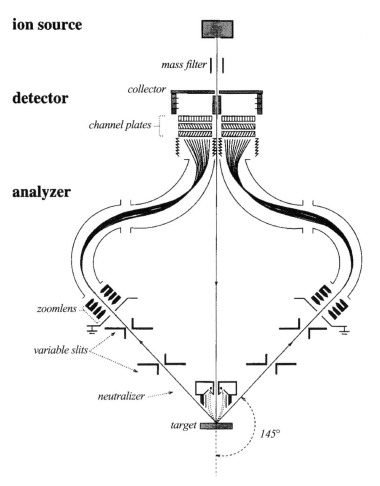

FIGURE 3.3. Cross section of a toroidal electrostatic analyzer that is rotationally symmetric (From Hellings *et al.*, 1985 and Bergmans *et al.*, 1993, with permission).

where F is a form factor depending on the analyzer shape and size. As an example, for a CMA analyzer (Figure 3.2), which has been used extensively in ISS due to its large acceptance angle (a full annular solid angle is accepted),

$$F = 0.763 \ln(r_2/r_1), \tag{5}$$

where r_2 and r_1 are the radii of the outer and inner cylinder electrodes, respectively. With the CMA, a second-order focusing of the trajectories making an angle $\psi = 42.3°$ with respect to the spectrometer axis is obtained. For ISS, if the ion beam is aligned inside the CMA along the analyzer axis, the surface can be bombarded at normal incidence, and the scattering angle is fixed to $180° - \psi = 137.7°$ as in Figure 3.2. The width of the input and output slits situated in the inner cylinder fixes the solid angle and the accepted angular range $\Delta\psi(\Delta\theta)$ so the energy resolution of the CMA is[18]

$$R = \Delta E/E = 0.18s/r_1 + 1.39(\Delta\psi)^3. \tag{6}$$

The solid angle $\Delta\Omega$ is given by

$$\Delta\Omega = 2\pi[\cos(\psi + \Delta\psi/2) - \cos(\psi - \Delta\psi/2)]. \tag{7}$$

The CMA collects scattered ions in an annular ring about the analyzer axis. A rotating aperture can be used to limit the angular range of collection. CMAs are most commonly used for elemental analysis rather than structure determination due to the limited range of scattering angles available.

3.1.7. Spectrometer Transmission Function

For quantitative ISS analysis, it is important to understand the influence of the experimental parameters on the intensity of the measured signal. For that purpose the role of each component of the spectrometer has to be detailed. A spectrometer consists of four basic parts: analyzer, detector, detection electronics, and computer/multichannel analyzer (MCA). When bombarding the solid with primary ions, only the current (I_θ) reemitted into the solid angle $\Delta\Omega$ making an angle θ with the ion beam line direction is accepted for analysis. The role of the analyzer is to determine the energy distribution $n_\theta(E)$ of the input current I_θ, where

$$I_\theta = \int_{\Delta\Omega} \int_0^\infty n_\theta(E)dEd\Omega. \tag{8}$$

The output current I_A of the analyzer tuned at a filtered energy E_i is given by

$$I_A(E_i) = \int_{\Delta\Omega} \int_0^\infty n_\theta(E)T(E_i - E)dEd\Omega, \tag{9}$$

where $T(E_i - E)$ is the transmission function of the analyzer itself.

The analyzer output current $I_A(E_i)$ is the convolution of the initial distribution and the transmission function. This current is converted by the detector into a series of pulses with an efficiency γ that is a function of the ion energy E. These pulses, if above a fixed threshold, are amplified and shaped by the amplification/discriminator electronics. They are then counted during a specified dwell time Δt fixed by the timer and finally stored in the ith memory channel of the MCA working in the multiscaling mode. After this cycle, the analyzer energy is incremented by ΔE_{MCA} and the MCA switches to the $(i + 1)$th memory channel so that

$$E_{i+1} = E_i + \Delta E_{MCA}. \tag{10}$$

The previous procedure is repeated so as to produce a histogram as schematized in Figure 3.4. The number of counts in channel i equals

$$N_i = \Delta t \int_{\Delta\Omega} \int_0^\infty n_\theta(E) T(E_i - E) \gamma(E) dE d\Omega. \tag{11}$$

If the energy distribution to be measured (scattering or recoiling peaks) is narrow with respect to the transmission function and is centered at an energy E_i characteristic of the phenomenon, the following approximation can be applied:

$$n_\theta(E) \cong I_\theta \delta(E - E_i), \tag{12}$$

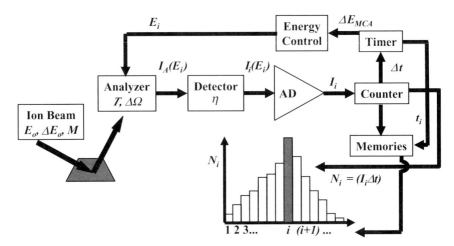

FIGURE 3.4. Principle of numerical acquisition of a scattered and/or recoiled ion signal. MCA, multichannel analyzer; AD, analog-to-digital.

where δ is the Dirac function. If the accepted solid angle is small, Eq. (11) can be simplified as

$$N_i = \Delta t I_\theta T(E_i - E_1)\gamma(E_i)\Delta\Omega. \tag{13}$$

When the area A of the peak is measured in the histogram, we have

$$A = \sum_i N_i \cong I_\theta \Delta t / \Delta E_{MCA} \Delta\Omega \int T(E_i - E_1)\gamma(E_i)dE_i. \tag{14}$$

The summation can be replaced by an integration if ΔE_{MCA}, the energy increment of the MCA, is chosen small enough. The integration is made in the energy region of interest. If, in that region, the efficiency of the detector is constant and assuming a Gaussian transmission function,

$$T(E_i - E) = t \exp\{-[(E_i - E)/\sigma]^2\}, \tag{15}$$

where t is supposed to be close to one and the variance σ is related to the analyzer resolution R by the relation

$$\sigma = k\Delta E_i = kE_i/R \tag{16}$$

with

$$k = \frac{1}{2}(\pi/\ln 2)^{1/2} \tag{17}$$

and

$$\begin{aligned}\int T(E_i - E_1)\gamma(E_i)dE_i &\cong \gamma(E_1)\int_0^\infty \exp\{-[(E_i - E)R/(kE_i)]^2\}dE_i \\ &\cong k\gamma(E_1)E_1/R,\end{aligned} \tag{18}$$

Eq. 14 becomes

$$A \cong I_\theta(\Delta t/\Delta E_{MCA})\Delta\Omega k\gamma(E_1)E_1/R. \tag{19}$$

It is observed that $En(E)$ is in fact measured. For ion energies lower than a few keV, the detector efficiency is not constant but decreases as the ion energy is lowered.[19]

3.1.8. Detectors: Channel Electron Multipliers and Microchannel Plates

Channel electron multipliers (CEM) and microchannel plate (MCP) detectors are commonly used for detection of the small fluxes of particles, such as ions, atoms, or electrons, resulting from ESA or TOF analysis. A CEM is a thin tube that is

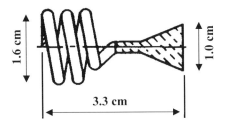

FIGURE 3.5. Example of a channel electron multiplier (CEM) (From Burle Industries, Inc., with permission).

~2 mm in diameter and ~1–3 cm long with a ~5-mm cone on the front end, as shown in Figure 3.5.[20] The internal surfaces of the tube and cone are coated with a high-resistance, electron-emissive material. A field of ~2 kV is applied by means of electrodes connected at each end. A charged particle or a sufficiently energetic neutral particle striking the cone causes ejection of secondary electrons. These secondary electrons are accelerated through the tube where they make collisions with the walls, as shown in Figure 3.6.[20] Each collision results in emission of more secondary electrons, thereby creating an avalanche of electrons. These electrons are detected as a "pulse" by a collector at the exit end of the tube. Gains as high as 10^9 can be obtained with these CEMs. In order to minimize acceleration of the ions, and hence greater detection efficiency over neutrals, the CEM can be biased with the cone at ground potential and the anode at high positive voltage. The counting equipment is standard and consists of a preamplifier, amplifier, and multichannel analyzer.

The MCP detectors are composed of arrays of CEMs that comprise a two-dimensional image intensifier, converting particles to amplified electron pulses, as shown in Figure 3.6. For detection in the single-pulse counting mode, two MCPs in series (called a chevron configuration) are often used to increase amplification. In order to obtain an electrical signal output, an anode is installed on the output side

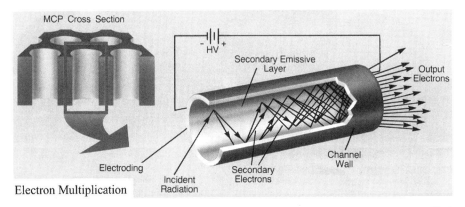

FIGURE 3.6. Schematic drawing of the conversion of individual particle collisions into amplified electron pulses by a CEM detector and their arrangement in a microchannel plate (MCP) array (From Burle Industries, Inc., with permission).

(a)   (b)

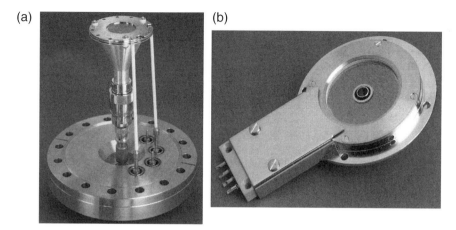

FIGURE 3.7. (a) Example of a multichannel plate (MCP) mounted on a vacuum flange. (b) Example of a MCP with an aperture for coaxial ion scattering (From Burle Industries, Inc., with permission).

of the MCP. This anode collects electron pulses coming through the channels of the MCP and encodes their positions by distribution of the charge signal among a number of output electrodes. In this manner, these MCPs provide position-sensitive detection, (i.e., the spatial distribution of the impinging particles). A typical MCP set with 10-μm pores on 12-μm centers provides a spatial resolution of ~18 line pairs/mm. Figure 3.7(a)[20] shows an example of an MCP mounted on a vacuum flange.

The CEMs and MCPs can be baked within the vacuum chambers to keep them free of adsorbed impurities. They are quickly destroyed if the background pressure rises to the level for arching to occur within the device. They are sensitive to both ions and fast neutrals. For energies above ~1 keV, the sensitivity to ions and neutrals is about equal and is a slowly increasing function of kinetic energy. For energies below ~1 keV, the sensitivity decreases dramatically and scales with the particle velocity. Since hydrogen atoms have relatively high velocities at low energies, they can be detected at energies even below 100 eV. Experimental efficiency versus velocity curves have been analyzed for a CEM. The resulting expression for the detector efficiency, ε, is a function of the ion atomic number and velocity, v, and can be written as $\varepsilon(v,Z) = a(v - b/Z^c)Z^n$, with a, b, c, and n being constants.[21] The detection efficiencies of CEMs and MCPs for ion-scattering applications have been studied.[22,23]

3.2. TIME-OF-FLIGHT SCATTERING AND RECOILING SPECTROMETER

3.2.1. Vacuum Chamber

A schematic drawing of the spectrometer chamber for TOF-SARS measurements with continuous variation of the scattering and recoiling angles is shown in Figure 3.8.[1]

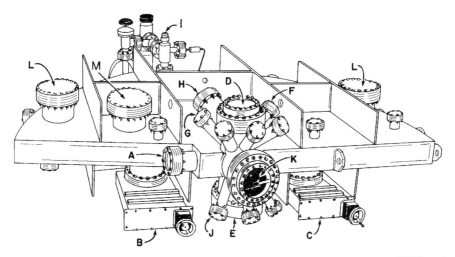

FIGURE 3.8. Spectrometer system designed for TOF analysis of neutrals and ions and ESA analysis of ions that are scattered and recoiled along with conventional Auger electron spectroscopy (AES), x-ray photoelectron spectroscopy (XPS), low-energy electron diffraction (LEED), ultraviolet photoelectron spectroscopy (UPS) analysis. A, pulsed ion beam; B, turbomolecular pump; C, ion pump; D, sample manipulator; E, detector precision rotary motion feed-through; F, x-ray source; G, electron gun; H, 180° electrostatic hemispherical analyzer; I, sorption pumps; J, sputter ion gun; K, viewport or reverse view LEED optics; L, titanium sublimation pump; M, cryopump (From Grizzi et al., 1990, with permission).

The vacuum chamber consists of two semicircular plates with a radius of 40 in. that are separated by a 3.0-in. spacing. The plates and spacing bars are 0.5 in. thick. A 6.0-in. "tee" with 8.0-in. o.d. flanges is welded into the plate structure at the center of the semicircle. The top flange of the tee is used for the sample manipulator. The sample is at the center of the large semicircle, as shown in Figure 3.9.[1] The bottom flange of the tee is used for a rotary motion manipulator for rotating the detector through a 165° angular range. The middle flange of the tee is used for a viewport or reverse view LEED optics. A slice of 13° is cut out from the semicircle in order to accommodate the ion beam line. Eight small flanges, which are directed to the center, protrude from the tee above and below the plates and at 45° angles to the plates. These ports are used for an electron gun, x-ray source, UV source, ion sputtering gun, small viewpoints, gas inlets, and so on. A flange protruding from the intersection of the tee and top plate at a 30° angle to this plate houses an ESA. This ESA can be used for XPS, UPS, and AES. Several ports with different size flanges protrude from the top and bottom plates; these are used for vacuum pumps, ion gauges, and access to the rotatable detector. In order to prevent bending of the large plates due to the pressure differential, reinforcing bars (0.37 × 4.0 in.^{2}) are spaced across the outside of the top and bottom plates. Small ports project radially around the arc of the semicircle to allow for longer flight paths at specific angles.

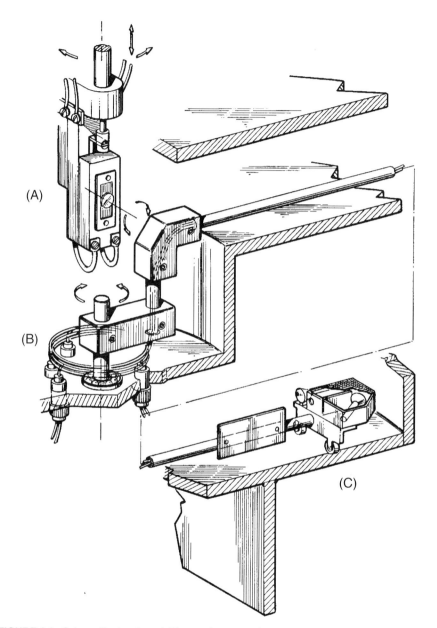

FIGURE 3.9. Schematic drawing of (A) sample mounted on the manipulator, (B) rotary motion feed through for moving the detectors, and (C) detector mounting. Note that there are two channel electron multipliers (CEM), one with direct line-of-sight and one indirect (From Grizzi *et al.*, 1990, with permission).

3.2.2. Ion Beam Line

A schematic drawing of an ion beam line, detector, and associated electronics is shown in Figure 3.10[1]. The ions are accelerated to the desired voltage, mass analyzed by a Wien filter, and collimated by a tubular Einzel lens through two 0.8-mm diameter circular apertures. The Wien filter consists of a magnet and a pair of electrostatic deflection plates mounted such that the electric field E is perpendicular to the magnetic field B. When a beam of charged particles with velocity v and charge e pass through the filter, they are deflected in one direction by an electrostatic force $F_e = Ee$ and in an opposite direction by a magnetic force $F_m = Bev$. When these two opposing forces are equal, the charged particles with velocity v pass undeflected through the filter. In practice, it is convenient to use a small fixed magnet and vary the electric field between the deflection plates. For applications where a continuous beam is required, such as measurements of ion-induced Auger electron emission and energy-analyzed ion scattering spectrometry, continuous currents up to $0.2\ \mu A$ with a spot size <0.8-mm diameter and angular dispersion <1° are obtainable with 4-keV Ar^+ and Ne^+ ions. For TOF measurements that require a pulsed ion beam, a pulse generator is connected to a deflector plate located 4 in. from the first circular aperture. Ion pulse widths <30 ns with average currents of 0.5 nA and pulse rates of 10–40 kHz can

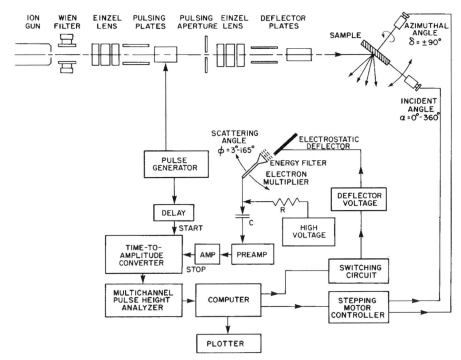

FIGURE 3.10. Schematic drawing of a pulsed ion beam line, sample, detector, and associated electronics (From Grizzi *et al.*, 1990, with permission).

be obtained in the pulsing mode. The beam current can be monitored by means of a Faraday cup with a 0.5-mm aperture mounted on the sample holder.

3.2.3. Pulsing System

The pulsing system consists of a pair of electrostatic plates and an aperture (Figure 3.10). The aperture facilitates collimation and differential pumping of the beam; if placed off-axis, it eliminates fast neutrals from the beam. A pulsed beam is obtained by applying a voltage pulse to one of the plates in order to rapidly sweep the beam across the aperture. A second ion pulse that would normally appear on the falling edge of the voltage pulse can be removed by applying a delayed voltage pulse to the second set of orthogonal deflection plates. In practice, the duration of the ion pulse through the slit has been varied from 5–100 ns with ion energies of a few keV. Generation of a pulsed beam consists of producing a continuous beam of the order of 1–50 nA and then adjusting the deflector so that 10^3–10^4 ions/pulse go through the slits and reach the target. Using a repetition frequency of 5–50 kHz, an average ion current of 0.1–0.5 nA/cm^2 can be obtained at the target. This current density is small enough that beam damage during the measurement period is negligible. TOF spectra are typically collected with a dose of $<10^{-3}$ primary ions/surface atom. For a sample area of 0.1 cm^2 with 2×10^{15} atoms/cm^2 in a monolayer, at an average ion current of 0.3 nA/cm^2, it would take ~30 h of continuous bombardment to sputter a monolayer, assuming a sputtering yield of unity.

3.2.4. Channel Electron Multiplier Detector

The CEM is supported on a long 0.5-in. o.d. tube that is attached to a precision rotary motion feed-through the bottom of the tee, allowing continuous scattering-angle variation over the angular range 3–165°. A typical time resolution, measured as FWHM, of the ion bombardment induced photon peak is <30 ns. In order to measure neutral particles alone, the ions are deflected by an electrostatic deflector plate located near the CEM housing. In order to minimize the effect of beam fluctuations when measuring ion fractions [ions/(neutrals + ions)], the TOF spectra of (neutrals + ions) and (neutrals only) are collected in alternating time intervals by switching the deflector voltage every 20 s. The voltage and signal cables to the detectors are channeled through the supporting tube and are accessed by electrical feed throughs on the bottom flange surrounding the rotary motion feed-through (Figure 3.9). The timing electronics and pulsing sequence are shown in Figure 3.10. The trigger output of the pulse generator, delayed by the time necessary for the pulsed beam to travel from the pulse plate to the sample, starts a time-to-amplitude converter (TAC). The TAC is stopped from the signal output of a particle reaching the detector. The output of the TAC yields a histogram of the distribution of particle flight times. The data are collected into a multichannel pulse height analyzer and stored in a minicomputer.

Electrons and ions can be energy analyzed by means of a 180° hemispherical electrostatic analyzer with an energy resolution of $<1\%$. This analyzer can be used for electrons or ions ejected within a narrow cone at 30° elevation angle from the large plates and $75°$ from the ion beam line. This arrangement allows simultaneous measurements of scattered and recoiled ion or ejected electron energies with the electrostatic analyzer while velocity distributions (or TOF) are measured with the rotatable CEM.

3.3. COAXIAL SCATTERING SPECTROMETER

Spectrometers have been designed[2,3,4] for collecting scattered atoms that have made head-on collisions; the impact parameter of the collision is $p \sim 0$ and the scattering angle $\theta \sim 180°$ as shown in Figure 3.11.[4] A schematic drawing of such a coaxial scattering instrument is shown in Figure 3.12.[4] The primary ions are directed through

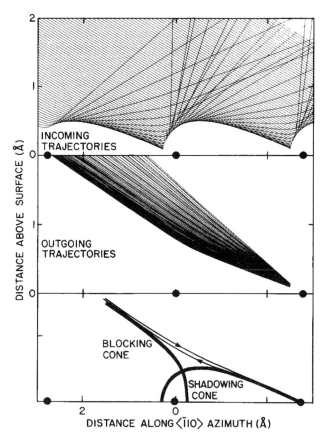

FIGURE 3.11. Trajectory simulations demonstrating coaxial scattering of 5-keV Ne$^+$ along the $\langle \bar{1}10 \rangle$ azimuth of Pt(110). The incoming and outgoing neon trajectories are shown separately. The incoming trajectories are impinging from the left and the outgoing trajectories are exiting to the left also. The bottom figure shows a schematic of the shadowing and blocking cones and the incoming and outgoing trajectories (From Wang *et al.*, 1992, with permission).

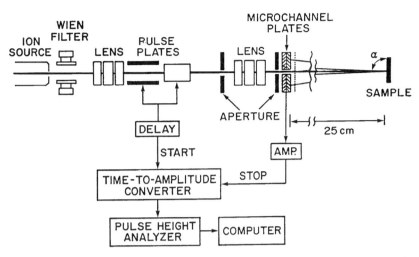

FIGURE 3.12. Schematic diagram of a coaxial scattering apparatus and timing electronics (From Wang *et al.,* 1992, with permission).

a series of focusing and lens elements, through an aperture in an MCP detector (Figure 3.7b), and onto a sample surface. Ions and fast neutrals that are backscattered near 180° are detected directly by the MCP. These particles are velocity-analyzed by measuring their flight times from the sample to the MCP. The advantage of coaxial scattering is that due to the nearly identical incoming and outgoing trajectories, it is capable of probing deeper layers than noncoaxial scattering. Using ~5-keV incident ion energies, it is possible to probe down to at least the sixth atomic layer.[24] It is also possible to detect coaxial scattering from an atom directly below a first-layer atom, thus facilitating measurement of interlayer spacings and the relative orientation of first-second-layer atoms. However, for surfaces with open channels such as the (1×2) and (1×3) reconstructed phases of Pt(110), such measurements are complicated by the exposure of two or three atomic layers to the surface. From a practical point of view, coaxial scattering is simple because both the primary ion beam and detector are attached to the chamber by means of a single port. This makes it useful as a probe for surface deposition or modification processes. The large area of the MCP detector results in high counting rates and short acquisition times, allowing real-time monitoring of surface processes. The advantages and disadvantages of the coaxial scattering geometry have been compared to those of noncoaxial scattering.[4]

3.4. SCATTERING AND RECOILING IMAGING SPECTROMETER

3.4.1. General Description

SARIS[5] overcomes the limitations of small-area detectors by using a large, gated, position-sensitive MCP detector and TOF methods to capture images of both ions and fast neutrals that are scattered and recoiled from a surface. Due to the large solid angle subtended by the MCP, atoms that are scattered and recoiled in both planar and nonplanar directions are detected simultaneously. For example, an MCP that has a 75×95-mm active area that is situated at a distance of 16 cm from the sample spans a solid angle of ~ 0.3 sr corresponding to an azimuthal range of $26°$. Using a beam current of ~ 0.1 nA/cm^2, the four images required to make up a $90°$ azimuthal angle range can be collected in ~ 2 min with a total ion dose of $\sim 10^{11}$ ions/cm^2. The gating of the MCP provides resolution of the scattered and recoiled atoms into time frames as short as 10 ns, thereby providing element-specific spatial-distribution images. These SARIS images contain features that are sharply focused into well-defined patterns as a function of both space and time by the crystal structure of the target sample. These patterns are highly specific to the types of trajectories involved (e.g., single- and multiple-scattering, recoiling, surface or subsurface scattering, etc.). The SARIS technique provides spatial- and time-resolved, element-specific images from surfaces that directly expose the three-dimensional anisotropy of keV-scattered and recoiled atom trajectories.

Two related large-solid-angle detection instruments were developed prior to SARIS.[25,26] These experiments were successful in that they demonstrated that valuable structural information could be obtained. However, due to the lack of mass and energy selection and detector gating capabilities, it was not possible to resolve scattering events from recoiling events nor to obtain element-specific images from compound surfaces. In order to overcome this limitation, SARIS uses TOF analysis and selective gating of the MCP. This method collects both ejected ions and fast neutrals and disperses scattered and recoiled particles according to their velocities as a function of the projectile–target atom masses and deflection angle. The scattering and recoiling patterns are spatially resolved by the MCP in 256 time-resolved frames. Good statistics are obtained in a total acquisition time of seconds. SARIS provides structural information without the need for stepping through incident and azimuthal angles as in TOF-SARS, thereby reducing the data acquisition time by a factor $> 10^3$. These images combine atomic scale microscopy and spatial averaging simultaneously since they are created from a macroscopic surface area (~ 1 mm^2), but they are directly related to the atomic arrangement of the surface. Both SARIS and TOF-SARS are sensitive to short-range order (i.e., individual interatomic spacings up to distances of ~ 10Å).

3.4.2. Vacuum Chamber and Pumping

The stainless steel UHV chamber is in the shape of a large half-cylinder (radius 61 cm and height 74 cm) with 2.5-cm thick walls and a wedge cut out of one side for the ion beam (Figure 3.13.)[5] Various ports are placed around the chamber for mounting ionization gages, pumps, and heating lamps and for access to the detector and goniometer. Reinforcing bars are welded at various positions on the outside of the chamber walls to prevent bending. A small (15.2-cm radius and 28-cm height) sample treatment chamber is mounted on the top center flange and separated from the main chamber by means of a gate valve. This chamber houses reverse view LEED optics, an ion-sputtering source, and leak valves for gas dosing. Sorption pumps are used

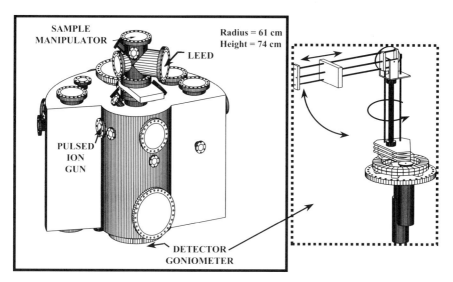

FIGURE 3.13. Schematic drawing of a scattering and recoiling imaging spectrometer (SARIS) chamber and detector goniometer. LEED, low-energy electron diffraction. (From Kim *et al.,* 1998, and Yao, *et al.,* 1998, with permission).

for roughing from atmospheric pressure, and the main chamber and small sample chamber are pumped by means of turbomolecular pumps. The instrument is baked by means of quartz lamp heaters positioned in both chambers. The pulsed ion beam line is baked by means of strip heaters that are glued to the walls.

3.4.3. Sample Manipulator and Detector Goniometer

A long-stroke precision sample manipulator is mounted on top of the treatment chamber. The manipulator allows positioning of the sample in either the upper or lower chamber, rotation about the vertical axis, rotation about the sample surface normal (azimuthal rotation), translation along three orthogonal axes, and adjustment of the sample tilt angle with respect to the vertical. These motions are controlled by stepping motors. The sample azimuthal angle is determined qualitatively from the LEED pattern; quantitative determination is obtained from the SARIS images and from surface semichanneling effects.

A triple-axis goniometer (Figure 2.13)[27] is used for positioning the MCP detector within the UHV chamber. This allows the MCP to be independently rotated in horizontal (ϕ) and vertical (θ) planes as well as translated with respect to a fixed sample position. The angle ϕ can be varied from 0–160° in order to change the scattering angle for SARIS. The angle θ can be varied from 0–80°, allowing the MCP to be moved out of the scattering plane. Translation allows the MCP to be moved relative to the sample position over a range of 10–60 cm. All of the movements are driven by computer-controlled stepping motors.

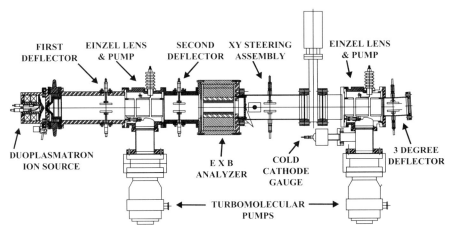

FIGURE 3.14. Schematic drawing of a pulsed, mass-selected, differentially pumped ion beam with a duoplasmatron ion source and energy range of 3–25 keV. E × B, (crossed electrical and magnetic fields). (From Chmara, with permission).

3.4.4. Pulsed Ion Beam

The pulsed, mass-selected ion beam is shown in Figure 3.14.[28] It has a duoplasmotron ion source that produces beam spot sizes down to 1 mm^2 with energies variable over the range 3–25 keV and a final energy spread of <50 eV. A two-step pulsing system produces pulsed beam widths of <10 ns and an average beam current of 10–100 pA (0.1–1.0 μA dc current before pulsing). Countdown circuitry permits pulse repetition rates over a range of 5–20 kHz.

3.4.5. Microchannel Plate Detector

A schematic drawing of the detector and associated electronics is shown in Figure 3.15.[5] A rectangular (75 × 95 mm) chevron-type MCP is used with a resistive anode encoder (RAE) plate mounted behind the MCP. Each particle that strikes the MCP generates an electron charge that dissipates in the RAE. The four signals A, B, C, and D generated at each corner of the RAE are picked up by a set of capacitors and are transmitted inside the chamber by a set of 50-ohm coaxial cables. These signals are amplified by a set of preamplifiers mounted on the vacuum chamber. The amplitude of these signals depends on the position (X,Y) at which the electron charge from the MCP strikes the RAE (i.e., the position of the particle hitting the MCP).

Determination of the position (X,Y) and time-of-flight (TOF) information of each particle striking the MCP represents the basis of the SARIS instrument. The X and Y data are calculated by an analog computer using the A to D signals and the following formulae: $X = x/a = (B + C)/(A + B + C + D)$ and $Y = y/b = (A + B)/(A + B + C + D)$ as illustrated in Figure 3.16. The X and Y analog signals are converted into digital format by flash type analog-to-digital converters. There are six bits for each

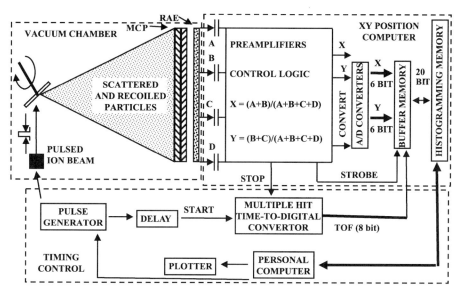

FIGURE 3.15. Schematic drawing of a scattering and recoiling imaging spectrometer (SARIS) detection system. MCP, microchannel plate detector; RAE, resistive anode encoder (From Kim *et al.,* 1998, with permission).

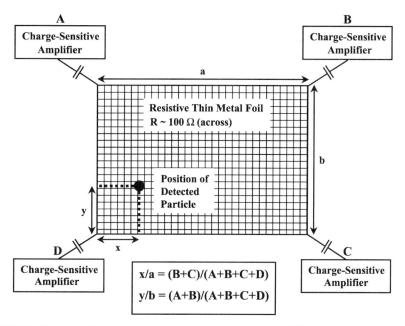

FIGURE 3.16. Schematic drawing of a resistive anode encoder (RAE) for determination of the X,Y position of a particle striking the microchannel plate (MCP) (From Kim *et al.,* 1998, with permission).

X and Y coordinate, giving a spatial resolution of 1.2 and 1.5 mm for the X and Y axis, respectively. A set of logic circuits, mounted in the preamplifier box, controls the validity of events detected. If the amplitude of the summing signal $(A + B + C + D)$ falls outside of a set range or if the time difference between successive events is too small, the events are neglected. The preamplifier and control logic circuits have a dead time of ~300 ns, giving a maximum data rate of 3 Mcts/sec. Data can be collected in 256 time frames, each with 64×64 pixels.

3.4.6. Multiple Stop Time to Digital Converter

A multiple-hit time-to-digital converter (TDC)[29] is used to measure the TOF data. The timing diagram is shown in Figure 3.17.[5] The START signal is produced by the pulse generator that is used for pulsing the primary ion beam. This START is delayed by the time required for the ions to travel from the pulse plates to the sample (typically a few μs). When a valid event is detected, a STOP signal, generated by the logic circuits, is sent to the TDC. The STOP signal is the same signal used for converting X,Y data to digital format (indicated by the CONVERT signal in Figure 3.15). The TDC consists of two synchronous four-bit binary counters connected in series to give eight-bit time information; the TDC runs off a TTL clock oscillator. The STOP signal transfers the time information from the counter to a temporary register; it does not stop the counter

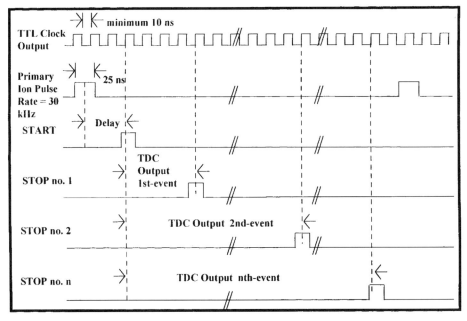

FIGURE 3.17. Timing diagram for scattering and recoiling imaging spectrometer (SARIS). TDC, multiple-hit time-to-digital converter (From Kim *et al.*, 1998, with permission).

(i.e., it leaves the counter ready to determine the TOF of the following events). The system has a maximum time resolution of 10 ns.

A certain time after generating the stop signal, a strobe signal is generated to transfer the X,Y and TOF data to a first-in-first-out (FIFO) buffer memory. The buffer size is 16 words and each word is 20 bits wide, divided as follows: first 6 bits, X data; second 6 bits, Y data; and last 8 bits, TOF data. These data are continuously transferred to a histogramming buffer memory of 1 M (2^{20} = 1 M memory address) words and each word is 20 bits wide, allowing a maximum accumulation of 64 kcts for each specific X, Y, and TOF combination. The data in the histogramming memory can be transferred to a host computer at the end of or during data accumulation. Although the latter mode allows real-time monitoring of the data, it increases the time of data acquisition. A typical acquisition time for one image using the first mode is ~10 s, which is at least a factor of 10^3 faster than TOF-SARS systems.

3.4.7. Ion Fraction Images

Images of only the scattered and recoiled neutrals can be obtained by means of the deflection system[30] shown in Figure 3.18. The system consists of a set of three high-transmission deflection grids between the sample and the MCP. The two external grids are grounded and the central grid is biased to +5 kV. The difference in the number of particles measured with a grounded central grid (neutrals + ions) and with a retarding deflection voltage on the grid (only neutrals) gives the absolute number of ions.

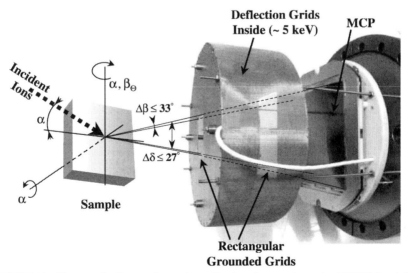

FIGURE 3.18. Photograph of scattering and recoiling imaging spectrometer (SARIS) ion deflection grids and microchannel plate (MCP) with a schematic drawing of the sample showing the angular range subtended by the MCP (From Vaquila *et al.*, 2002, with permission).

3.5. MASS AND CHARGE SELECTION OF PULSED ION BEAMS USING SEQUENTIAL DEFLECTION PULSES

Mass and charge selection of ion beams for ion scattering is usually accomplished by means of a magnetic sector, quadrupole filter, or Wien filter. An alternative method of preparing ion beams of known mass and charge using sequential deflection pulses (SDP) is described in this section.[6] In the standard version, ion pulses are formed by supplying a voltage deflection pulse from a pulse generator (PG1) to one of two plates labeled D in Figure 3.19.[6] In the SDP mode, a second pulse generator (PG2) is added in order to pulse the deflector plates labeled F at a later point along the ion trajectories. PG2 is triggered by PG1 following a variable delay time. Mass and charge selection is accomplished by separation of the species according to the different flight times of the individual pulses traveling from the entrance of plates D to the entrance of plates F. The flight time of the species of interest sets the delay time for PG2. These times, t_1 and t_2, are related to the masses m_i and kinetic energy E by

$$\Delta m = m_2 - m_1 = 2E\left(t_2^2 - t_1^2\right)/x^2 \tag{20}$$

and

$$m = \frac{1}{2}(m_1 + m_2) = E\left(t_2^2 + t_1^2\right)/x^2, \tag{21}$$

where E is the kinetic energy and x is the distance traveled. The mass resolution is related to the flight times by

$$m/\Delta m = \frac{1}{2}\left[\left(t_1^2 + t_2^2\right)/\left(t_2^2 - t_1^2\right)\right]. \tag{22}$$

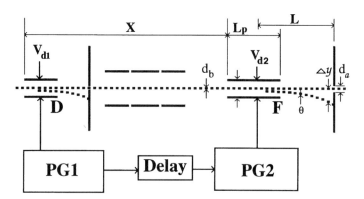

FIGURE 3.19. Detailed schematic of the pulsing region of the time-of-flight and scattering and recoiling spectrometer (TOF-SARS) ion beam line used in the sequential deflection pulse (SDP) mode. PG, pulse generator. (From Sung *et al.*, 1994, with permission).

If the difference in flight times $\Delta t = t_2 - t_1$ is small, then $(t_1 \cdot t_2) \sim t^2$, where $t = \frac{1}{2}(t_1 + t_2)$, and Eq. 22 reduces to

$$m/\Delta m = t/2\Delta t. \tag{23}$$

In order to separate pulses of two different species going through the plates F, their time difference Δt must be equal to or longer than the time, termed Δt_d, needed to deflect one of the species to a minimum lateral position (Δy) necessary to prevent the ions from passing through the aperture. The schematic diagram of Figure 3.19[6] illustrates the details. The minimum deflection Δy is expressed as

$$\Delta y = \frac{1}{2}(d_a + d_b), \tag{24}$$

where d_a and d_b are the diameters of the aperture and the ion beam, respectively. The deflection angle θ can be expressed as[5]

$$\tan \theta = \Delta y/L_d = V_{d2}L_{p,m}/2dV, \tag{25}$$

where L_d is the deflection length defined by the intersection of the extrapolated final-beam trajectory with the beam axis and aperture, V_{d2} is the deflection voltage, $L_{p,m}$ is the minimum deflection plate length, d is the spacing between the deflection plates, and $V = E/e$. The minimum time necessary for deflection of species i is $\Delta t_{d,i} = L_{p,m}/v_i$, where v_i is the velocity of species i. Assuming that there are only two species to be separated and using

$$m/\Delta m = t/2\Delta t_{d,i} \qquad \text{where} \qquad t = \frac{1}{2}(1/v_i + 1/v_j)x, \tag{26}$$

the mass resolution can be expressed as

$$m/\Delta m = [(v_i/v_j) + 1]xL_dV_{d2}/8\Delta ydV. \tag{27}$$

Since $v_j \sim v_i$, Eq. 27 can be simplified to

$$m/\Delta m = xL_dV_{d2}/4\Delta ydV. \tag{28}$$

From Eq. 28, we note that the mass resolution can be improved by increasing the distance x between the two sets of pulse plates D and F, the distance L from the second set of pulse plates F to the aperture d_a, and the deflection voltage V_{d2} and by decreasing the required deflection distance Δy, the spacing d between the pulse plates F, and the kinetic energy of the ions E.

Consider two cases, depending on whether the faster or slower of the two particles is deflected.

Case 1   Transmission of Faster Species and Deflection of Slower Species   The deflection length L_d can be expressed as

$$L_d = [L + (L - L_p/2)]/2 = L - \frac{1}{4}L_p, \tag{29}$$

where L is the distance from the center of the deflector of length L_p to the aperture. Using the values x = 210 mm, L_d = 32.5 mm, V_d = 70 V, $\Delta y = (1 + 1)/2 = 1$ mm, d = 10 mm, and V = 4 kV, the calculated resolution is $m/\Delta m \sim 3$.

Case 2   Transmission of Slower Species and Deflection of Faster Species   The deflection length L_d can be expressed as

$$L_d = [L + (L + L_p/2)]/2 = L + \frac{1}{4}L_p. \tag{30}$$

Using the same parameters as above and L_d = 57.5 mm, the calculated resolution is $m/\Delta m \sim 5$. For the parameters used, the resolution $m/\Delta m$ varies between 3 and 5, depending on which species is transmitted through the aperture. If the faster particle is transmitted through the aperture, then the slower particle will be deflected in the region of the plates closer to the aperture, leading to a shorter L_d. On the other hand, if the slower particle is transmitted through the aperture, then the faster particle has already been deflected in the region of the deflector plates farthest from the aperture, leading to a longer L_d. Accordingly, if C^+ and O^+ ions are in a beam, it is possible to obtain a pure O^+ but not a pure C^+ beam.

3.5.1. Example of Sequential Deflection Pulses

An example of the use of SDP with a 4-keV Ar^+ beam is shown in Figure 3.20.[6] Since doubly or triply charged ions have kinetic energies of two or three times the acceleration voltage, the times during which the singly and multiply charged ions travel through the F plates are sufficiently different to allow discrimination by means of selective placement of the PG2 pulse. The spectra of Figure 3.20 were obtained by using 70 V on one of the F plates. Ions are transmitted through the F plates only when a pulse of 70 V is supplied by PG2; pulse widths of 0.5 μs were used. The TOF distributions were obtained for the unscattered primary Ar^+ beam itself by directing the beam directly into the channel electron multiplier ($0°$ scattering angle). Using a delay of 0.9 μs, both the 4-keV Ar^+ and 8-keV Ar^{2+} are transmitted through the aperture. With a delay of 0.3 μs, only the faster Ar^{2+} species is within the F plates for the full time of the pulse, resulting in deflection of Ar^+. Using a longer delay of 1.4 μs, only the slower Ar^+ is within the F plates for the full time of the pulse and the Ar^{2+} is deflected.

Example TOF spectra of a 4-keV Ar ion beam scattering from a clean indium phosphide (InP) surface using a beam incidence angle of $15°$ and a forward-scattering angle of $30°$ are shown in Figure 3.21.[6] Using a 0.9-μs delay to transmit both 4-keV

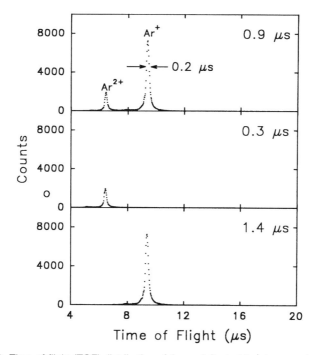

FIGURE 3.20. Time-of-flight (TOF) distribution of the undeflected Ar^+ beam using an acceleration voltage of 4 kV and a flight path of 1 m. The numbers alongside each spectrum represent
the delay time between pulse generator 1 (PG1) and PG2. Note that the TOF is measured from
the entrance of plate D to the detector (From Sung *et al.*, 1994, with permission).

Ar^+ and 8-keV Ar^{2+}, peaks due to Ar^+ and Ar^{2+} scattering from In and P atoms and
Ar^+ and Ar^{2+} recoiling In and P atoms are observed. These peaks are identified by
applying the classical binary collision approximation to calculate the flight times of
the scattered and recoiled particles. The flight times of Ar^{2+} scattering from In and
those of P atoms recoiling from Ar^{2+} collisions are separated by only 0.1 μs, resulting
in a single peak for these two processes.

The advantage of SDP is that it eliminates the need for the conventional mass and
charge-selection devices, resulting in a shorter beam line. SDP is particularly suitable
for the rare gases, where it is only necessary to separate ions of the same mass and
different charges. For example, since doubly charged ions A^{2+} are accelerated to
twice the kinetic energy of singly charged ions A^+, they will be selected at a mass
equal to one-half that of the singly charged ions. However, it can also be used in
cases where the number of different masses and charges in the beam is low. Examples
are as follows: (1) alkali ion beams where the main impurities are other alkali ions;
(2) beams of reactive atomic and diatomic ions formed from the diatomic parent, such
as $N^+ + N_2^+$ and $O^+ + O_2^+$; (3) beams of atomic and polyatomic ions formed from
parent molecules whose fragmentation patterns are relatively simple, such as CO^+
from CO or CO_2, CH^+ from C_2H_2.

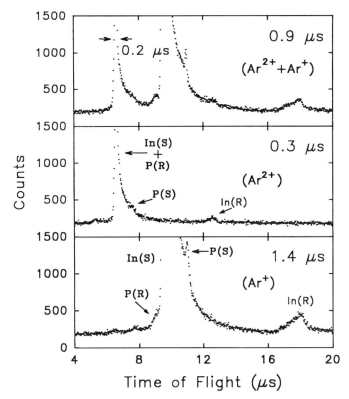

FIGURE 3.21. Time-of-flight (TOF) distribution of Ar^+, accelerated at 4 kV, scattering from an indium phosphide (InP) surface. Upper frame, transmission of both Ar^{2+} and Ar^+. Center frame, transmission of only Ar^{2+}. Lower frame, transmission of only Ar^+. The peaks, identified from the binary collision approximation (BCA), correspond to: In(S) and P(S), Ar scattering from In and P atoms, respectively; In (R) and P (R), In and P atoms, respectively, recoiled by the primary Ar atoms. The numbers alongside each spectrum represent the delay times between pulse generator 1 (PG1) and PG2 (From Sung *et al.*, 1994, with permission).

3.6. ION SCATTERING AND RECOILING FROM LIQUID SURFACES

Ion beams can be used to probe the surface composition and molecular orientation of liquid surfaces.[7] TOF analysis of the kinetic energy of scattered inert gas ions and recoiled atoms ejected from the surface reveals the identity of atoms in the topmost atomic surface layer of a liquid. Analysis of the peak intensities as a function of the experimental parameters can be used to infer average molecular orientations in the surface. The apparatus developed for scattering from liquid surfaces uses a pulsed ion beam and detector that are similar to those used for TOF-SARS of solid surfaces. The beam is focused through a reentrant differentially pumped small-diameter aperture onto the liquid surface in order to minimize the flight path of the ions through the sample vapor. The scattered or recoiled species similarly exit via differentially pumped small-diameter orifices. Calculations and experiments indicate that for liquids with vapor pressures less than 1 m Torr, the scattering from the vapor is negligible. Fresh liquid surfaces are produced by a slowly rotating (typically 0.5 Hz) wheel enclosed in a box that only exposes the liquid to the vacuum on a small area of the wheel. The wheel drags a layer of liquid past a knife edge that leaves a fresh liquid film on the wheel approximately 0.2 mm thick. The rotating wheel helps to minimize contamination by adsorption and sample charging by continuously creating a fresh surface. A recirculating water pump system provides cooling for both the liquid reservoir in the base of the cell and the (in-vacuum) motor used to rotate the wheel. The system is pumped by large diffusion pumps that have no difficulty in maintaining the pressure in the scattering chamber at $\sim 10^{-6}$ Torr with glycerol in the liquid cell. This technique has been applied to studies of the surfaces of several liquids.[31,32]

REFERENCES

1. O. Grizzi, M. Shi, H. Bu, and J. W. Rabalais, *Rev. Sci. Instrum.* **61,** 740 (1990).

2. M. Aono, Nucl. Instrum. *Meth. Phys. Res.* **B2,** 374 (1984).

3. Y. Wang, M. Shi, and J. W. Rabalais, *Nucl. Instrum. Meth. Phys. Res.* **B62,** 505 (1992).

4. M. M. Sung, J. W. Rabalais, *Nucl. Instrum. Meth. Phys. Res. B.* **108,** 389 (1996).

5. C. Kim, C. Hoefner, A. Al-Bayati, and J. W. Rabalais, *Rev. Sci. Instrum.* **69,** 1676 (1998).

6. M. M. Sung, A. H. Al-Bayati, C. Kim, and J. W. Rabalais, *Rev. Sci. Instrum.* **65,** 2953 (1994).

7. M. Tassotto, T. J. Gannon, and P. R. Watson, *J. Chem. Phys.,* **107,** 8899 (1997).

8. C. Kim and J. W. Rabalais, *Surf. Sci.* **395,** 239 (1998).

9. T. M. Buck, Y. S. Chen, G. H. Wheatley, and W. F. van der Weg, *Surf. Sci.* **47,** 244 (1975).

10. S. B. Luitjens, A. J. Algra, E. P. Th. M. Suurmeijer, and A. L. Boers, *Surf. Sci.* **99,** 652 (1980).

11. H. Niehus and G. Comsa, *Nucl. Instrum. Meth.* **B15,** 122 (1986).

12. E. Bauer and T. von dem Hagen, in "Chemistry and Physics of Solid Surfaces" VI. R. Vanselow and R. Howe, (Eds.), Springer, Berlin, 1986, p. 547.

13. M. Z. Sar-El, *J. Appl. Phys.* **38,** 340 (1967).

14. P. Bertrand, F. Delannay, and J. M. Streydio, *J. Phys. E: Sci. Instr.* **10,** 403 (1977).

15. D. P. Smith, *J. Appl. Phys.* **38,** 340, 1967; *ibid, Surf. Sci.* **25,** 171, 1971.

16. G. J. A. Hellings, H. Ottenvanger, S. W. Boelens, C. L. C. M. Knibbeler, and H. H. Brongersma, *Surf. Sci.* **162,** 913 (1985).

17. R. H. Bergmans, A. C. Kruseman, C. A. Severijns, and H. H. Brongersma, *Appl. Surf. Sci.* **70/71,** 283 (1993).

18. J. C. Riviere, in "Practical Surface Analysis, Auger and X-Ray Photoelectron Spectroscopy," D. Briggs and M. P. Shea (Eds.), Wiley, Chichester, 1990, p. 75.

19. C. N. Burrows, A. J. Lieber, and V. T. Zaviantseff, *Rev. Sci. Instrum.* **38,** 1477 (1967).

20. Burle Industries, Inc., 1000 New Holland Ave., Lancaster, PA 17601–5688.

21. M. Tassotto and P. R. Watson, *Rev. Sci. Instrum.,* **71,** 2704 (2000a).

22. J. N. Chen, M. Shi, S. Tachi, and J. W. Rabalais, *Nucl. Instrum. Meth. Phys. Res.* **B16,** 91 (1986).

23. H. Verbeek, W. Eckstein, and F. E. P. Matschke, *J. Phys. E* **10,** 944 (1977).

24. H. Bu, M. Shi, and J. W. Rabalais, *Nucl. Inst. Meth.* **B61,** 337 (1991).

25. J. A. Yarmoff and R. S. Williams, *Rev. Sci. Instrum.* **57,** 433 (1986).

26. A. Niehof and W. Heiland, *Nucl Instrum. Meth. Phys. Res.* **B48,** 306 (1990).

27. J. Yao, C. Kim, and J. W. Rabalais, *Rev. Sci. Instrum.* **69,** 306 (1998).

28. F. Chmara, Peabody Scientific, P. O. Box 2008, Peabody, MA 01960.

29. C. Kim, A. Al-Bayati, and J. W. Rabalais, *Rev. Sci. Instrum.* **69,** 1289 (1998).

30. I. Vaquila, I. L. Bolotin, T. Ito, B. N. Makarenko, and J. W. Rabalais, *Surf. Sci.* **469,** 187 (2002).

31. T. J. Gannon, G. Law, P. R. Watson, A. J. Carmichael, and K. R. Seddon, Langmuir **15,** 8429, (1999); T. J., Gannon, M. Tassotto, and P. R. Watson, *Chem. Phys. Lett.* **300,** 163 (1999).

32. M. Tassotto and P. R. Watson, *Surf. Sci.* **464,** 251 (2000b).

4

GENERAL FEATURES OF ION SCATTERING AND RECOILING SPECTRA

As described in Chapter 2, the kinetic energy loss by an ion due to scattering from a surface is primarily due to momentum transfer to surface atoms. Inelastic energy loss processes may result in additional energy losses; however, the magnitude of these losses is typically less than 1% of the beam energy and can be neglected for most practical purposes. This chapter presents ion scattering spectra (ISS) from a variety of different samples in order to familiarize the reader with the appearance of typically observed spectra. Energy spectra are presented first since these are the simplest spectra to interpret. This is followed by time-of-flight (TOF) spectra from a variety of samples. More examples of TOF spectra are presented here than energy spectra because it is felt that TOF analysis is generally superior to energy analysis for ISS spectra due to its higher sensitivity and better quantification as a result of detection of both ions and neutrals. This is followed by a comparison of ISS with other surface analysis techniques.

4.1. ENERGY SPECTRA

Scattering of rare gas ions from a metal surface typically produces a single intense peak with shoulders on both the high- and low-energy sides of this main peak. The intense peak is due to quasi-single scattering (SS), and the shoulders are due to impinging ions that have experienced multiple scattering (MS) collisions. The word *quasi* is used for SS because impinging ions can suffer very minor deflections along the incoming and outgoing trajectories that result in very small energy losses that are indistinguishable from true single collisions. An example is shown in Figure 4.1 for scattering of 3-keV ^{4}He$^+$ from an aluminum surface.[1] The energy spectrum of the scattered ions at a particular scattering angle is recorded for a fixed energy of the incident ions.

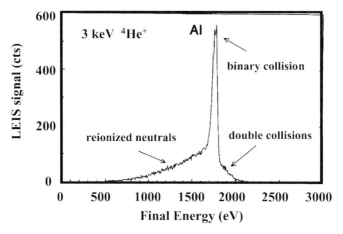

FIGURE 4.1. Ion scattering spectrum of 3-keV He$^+$ scattering from an aluminum (Al) surface (From Jacobs, 1995, with permission). LEIS, low-energy ion scattering.

The energies of the scattered ions are then characteristic of the masses of the target atoms. The intense peak at about 1700 eV corresponds to SS of helium (He) from aluminum (Al). Inelastic energy loss processes typically shift the peak maximum to slightly lower energy values than that predicted by the binary collision approximation (BCA) (i.e., Eqs. 2 and 6, Chapter 2). The high-energy onset of the peak is generally chosen as a good measure for the elastic energy value. This is the value used to calculate the mass of the surface atoms. The shoulder observed on the high-energy side of the SS peak results from MS ions that have experienced two or more collisions and, as a result, have higher kinetic energies than SS ions. From Eq. 7, Chapter 2, it is observed that an atom that is scattered through a single collision into an angle θ has lower kinetic energy than an atom that is scattered through two angles θ_1 and θ_2 for which $\theta_1 + \theta_2 = \theta$. Therefore, MS shoulders typically appear on the high-energy side of the SS peak. A broad background is also observed on the low-energy side of the SS peak. This background is due to atoms that have penetrated below the surface layer, scattered from a subsurface atom, and reemerged from the surface. Such atoms have lost energy in a series of several collisions or deflections. MS ions that penetrate below the surface are generally neutralized; however, they can be reionized in collisions as they depart the surface. Reionization occurs as a result of electron exchange when inner shell electronic levels of the incident atom and target atom overlap (see Chapter 8). The energy threshold for this process is dependent on the distance of closest approach between the two atoms. The threshold observed in Figure 4.1 is near 500 eV.

The spectrum of Figure 4.1 represents a clean Al surface. Surface contamination by adsorbed gases can be observed as recoils in the forward-scattering direction. Adsorption of typical residual gases such as hydrogen (H_2), water (H_2O), oxygen (O_2), carbon monoxide (CO), and hydrocarbons results in a decrease in the SS peak intensity and an increase in the H, C, and O recoil intensity. Surface hydrogen cannot be observed by scattering because its mass is less than that of the lightest rare gas

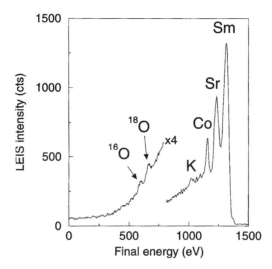

FIGURE 4.2. Ion scattering spectrum of 1.5-keV ^{4}He$^+$ scattering from ^{16}O/^{18}O exchanged $Sm_{0.8}Sr_{0.2}CoO_3$ (From Jacobs, 1995, with permission). LEIS, low-energy ion scattering.

atom He; however, it can be easily observed as a recoiled atom. All other atoms in the periodic table can be detected by scattering and/or recoiling.

The second example, presented in Figure 4.2, is a spectrum of He$^+$ scattering from an ^{18}O-exchanged Sr-Sm-Co-oxide.[1] This is a complicated sample that requires high resolution for separation of the peaks of the individual elements. Peaks due to scattering from strontium (Sr), samarium (Sm), cobalt (Co), potassium (K), ^{18}O, and ^{16}O are observable. The K is present as an impurity. Typical TOF spectra do not have sufficient resolution to separate all of these peaks.

The third example, presented in Figure 4.3, shows typical ISS spectra for some metal oxides and the corresponding alloys.[1] The spectra of clean alloys and metals exhibit significantly sharper peaks than spectra from the oxides of those metals. One reason for this is that the surfaces of the oxides are usually rougher than those of the metals and alloys. This roughness contributes significantly to MS collisions that result in broadening of the peaks. A linear background subtraction can be used to quantify the surface peak areas for the metals and alloys due to the nearly linear background. The backgrounds from the oxides are usually larger and irregular. For such cases, careful fitting and deconvolution of the peaks is necessary.[1]

4.2. TIME-OF-FLIGHT SPECTRA

4.2.1. Spectral Interpretation

Example TOF spectra[2] for 3-keV Ar$^+$ scattering from a Si sample that has some residual oxygen and hydrogen adsorbed on the surface are shown in Figure 4.4 for

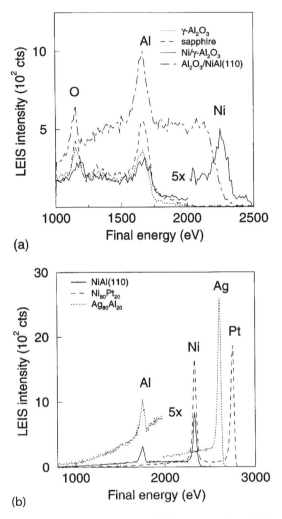

FIGURE 4.3. Typical ion scattering spectroscopy (ISS) spectra of 3-keV ^{4}He$^+$ scattering from (a) oxide and (b) alloy surfaces. (From Jacobs, 1995, with permission).

several values of the polar beam incident angle α and scattering angle θ. These angles are defined in the inset of Figure 4.4; for demonstration purposes they have been adjusted to specular conditions in the range $\theta = 4.5$–$85.0°$. Only atomic species are observed with high, discrete energies in the scattering and recoiling events due to the violent nature of the collisions. These species are unambiguously identified through the following steps: (1) A TOF spectrum of the scattered and recoiled neutrals plus ions is measured, providing the velocities (v) of the species. (2) An electrostatic analyzer (ESA) is used for an energy (E) analysis of the charged species. (3) A TOF spectrum of ions focused through an ESA that is set at the specific E of one of the

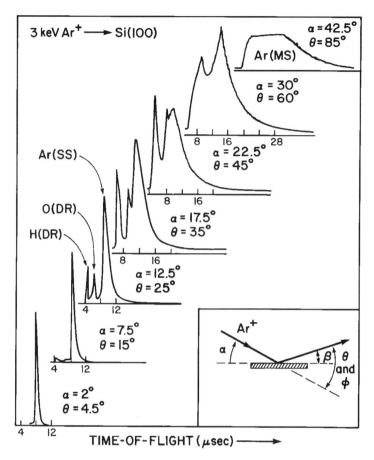

FIGURE 4.4. TOF spectra for 3-keV Ar$^+$ scattering from a Si(100) surface with hydrogen and oxygen adsorbates at several α and θ values. SS, single scattering; MS, multiple scattering; DR, direct recoiling (From Grizzi *et al.*, 1990, with permission).

peaks can be obtained. Measurements (1) and (2) together provide an unambiguous mass determination. Measurement (3) provides both ν and E for specific peaks, and thereby unambiguous mass determination of all species within a specific energy window. Since the ions and neutrals have very similar TOFs (energy losses due to neutralization processes are typically <20 eV), energy analysis of only the ions is sufficient for identification of the peaks. After some experience, in practice the TOF peaks are identified by calculating the E's and corresponding TOFs for specific atoms using the equations of Chapter 2, as will be shown below. The small inelastic losses, such as excitation and bond breaking during collision, can be neglected for this purpose.

The sharp, intense peak observed at low values of α and θ in Figure 4.4 (near grazing incidence and exit) corresponds to Ar projectiles scattered with minor energy loss from combinations of single and multiple collisions with target atoms. In this angular region and for atomically flat surfaces, projectiles moving along azimuths of high symmetry collide with several target atoms before leaving the surface with exit angles similar to the incident angle. These surface-channeling effects are useful for aligning the azimuthal angle of the sample (as will be shown in Chapter 5). Since $M_1 > M_2$ in this case, there is a critical angle $\theta_c = \sin^{-1}$ (M_2/M_1) above which only multiple scattering can occur. Since $\theta_c = 44.8°$ for Ar $\rightarrow$ Si collisions, a broad and featureless structure results at large θ, as observed in Figure 4.4.

Both H and O recoil peaks are present in the spectra of Figure 4.4. For low θ they appear at the short TOF side of the scattering peak, whereas for $\theta > 35°$ the O recoil and the scattering peak overlap. The angular range where this overlap occurs is determined by Eqs. 4 and 6 of Chapter 2 and by the width of the peaks, which has a nonlinear relation with the recoiled energy E_r (it is proportional to $E_r^{-3/2}$). A Si recoil peak is not observed because it is overlapped by the far more intense Ar scattering peak; Si recoils can be observed by using Ne projectiles where there is a good separation between the recoiling and scattering peaks. In general, the best resolution and sensitivity for different projectile energies and types of recoils are obtained experimentally by varying the observation angles, making continuous variation of the scattering angle θ a very useful feature in analysis of recoiling particles.

4.2.2. Spectral Intensities

The dependence of the scattering peak intensities I(S) on θ is illustrated[2] in Figure 4.5 for 4-keV Ar$^+$ scattering from a clean and well-annealed (to 2200 K) tungsten W(211) single crystal. The incident angle was fixed at $\alpha = 20°$ (i.e., high enough to avoid shadowing effects), and θ was varied in the range 20–150°, which corresponds to an exit angle range of $\beta = 0$–130°. The experimental intensities I(S), obtained as the integrated area of the scattering peak, are shown in Figure 4.5 together with the scattering cross sections calculated with the BCA (Chapter 2) and normalized to the experiment at $\theta = 125°$. At high θ values, both curves show the same angular dependence, indicating that the major contribution to the scattering peak originates from single collisions. In the range $\theta = 28$– 45°, I(S) is higher than the calculated BCA cross section because (1) multiple collisions, not taken into account in the BCA, have cross sections that approach those of SS collisions, and (2) focusing of the ion trajectories at the edge of the blocking cones cast by neighboring target atoms can considerably enhance the scattered intensities. For $\theta < 30°$ ($\beta < 10°$), there is a sharp decrease of I(S) because the scattered projectiles that have mass M_1 smaller than that of the target atom masses M_2 cannot penetrate the target atom blocking cones. This blocks the exit trajectories of the projectile ions at such grazing angles. This sharp I(S) decrease is a measure of the size of the blocking cones; it can be used to obtain the interatomic distance along the projectile direction (as will be shown in Chapter 5).

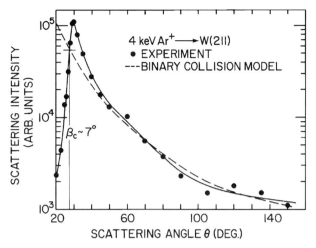

FIGURE 4.5. Scattering intensities I(S) as a function of θ for 4-keV Ar$^+$ incident on a clean W(211) surface. The dashed line represents the scattering cross sections calculated from the binary collision approximation (BCA) and normalized to the experimental values at $\theta = 125°$. The sharp decrease in I(S) for $\theta < 30°$ is due to blocking, and β_c is the critical blocking angle (From Grizzi *et al.,* 1990, with permission).

4.2.3. Spectral Sensitivity

TOF-SARS has extremely high sensitivity to the outermost first and second atomic layers. This can be observed[2] from Figure 4.6, in which the oxygen recoil intensity I(R) and the oxygen KLL intensity from Auger electron spectroscopy (AES) on a W(211) surface, calibrated in terms of oxygen coverage, are plotted as a function

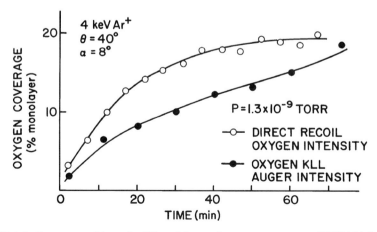

FIGURE 4.6. Oxygen recoil intensity I(R) and Auger electron spectroscopy (AES) KLL intensity, calibrated in terms of oxygen coverage, as a function of exposure time to residual gases at 1.3×10^{-9} Torr (From Grizzi *et al.,* 1990, with permission).

of exposure time to the residual gases of the vacuum chamber at a pressure of 1.3×10^{-9} Torr. I(R) rises sharply and reaches a plateau at an oxygen coverage of <20% of a monolayer, whereas the AES intensity has a more gradual slope and has not reached a plateau even after an exposure time of 75 min. Cleanliness of a surface with respect to TOF-SARS is estimated to be <0.01 monolayer of surface contamination; the "clean" surface condition is taken to be the case of complete absence of recoiled hydrogen, carbon, and oxygen signals in the TOF spectra.

4.2.4. Examples of TOF Spectra

Examples of typical TOF spectra obtained from 4-keV Ar[+] impinging on a Si(100) surface are shown[3] in Figure 4.7. Peaks due to scattered argon and recoiled hydrogen, carbon, oxygen, and silicon are observed on the uncleaned surface. These TOF peaks are identified by application of the BCA model. Only the scattered Ar and recoiled Si peaks are observed after cleaning the surface in ultrahigh vacuum (UHV). The positions of these peaks correspond to those of quasi-SS. They are not sensitive to the incident α and azimuthal δ orientations of the sample, indicating that the major contributions to these peaks are from single collision events. The broadening of the scattered Ar peak on the long TOF side is due to MS. Note the improved sharpness of the peaks from the clean surface compared to the unclean surface.

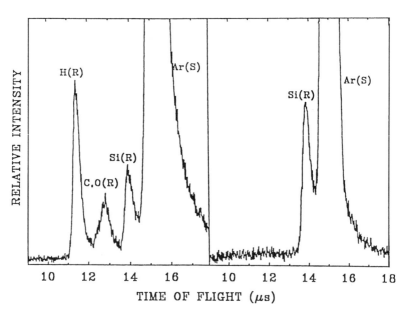

FIGURE 4.7. Time-of-flight scattering and recoiling spectroscopy (TOF-SARS) spectra of an unclean (left) and clean (right) Si(100) surface showing the scattered argon (Ar) and recoiled hydrogen (H), carbon (C), oxygen (O), and silicon (Si) features. Conditions: 4-keV Ar[+], $\theta = 28°$, $\alpha = 8°$, and $\delta = 15°$ (From Wang *et al.*, 1993, with permission).

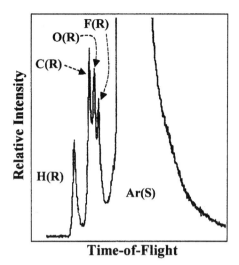

FIGURE 4.8. Forward-scattering time-of-flight scattering and recoiling spectroscopy (TOF-SARS) spectrum of hexafluorobenzene (C_6F_6) chemisorbed on a polycrystalline platinum (Pt) surface using 7.5-keV Ar^+ and a scattering angle of $\theta = 22°$.

A forward-scattering TOF-SARS spectrum of hexafluorobenzene (C_6F_6) chemisorbed on a polycrystalline Pt surface using 7.5-keV Ar^+ and a scattering angle of $\theta = 22°$ is shown in Figure 4.8. The intense Ar scattering peak is preceded by recoil peaks of hydrogen, carbon, oxygen, and fluorine. The hydrogen and oxygen peaks are due to H_2O chemisorption from the background gases. This example illustrates that it is possible to separate light atom recoils using TOF techniques.

Examples of TOF-SARS spectra for 7-keV Ne^+ scattering from a LiCu alloy surface that was (A) annealed to 500°C and (B) sputtered with 5×10^{16} ions/cm^2 of 3-keV Ar^+ are shown[4] in Figure 4.9. Both spectra exhibit an intense peak due to Ne scattering. The lower intensity peaks labeled (DR) originate from collisions of the primary Ne^+ with surface lithium (Li) and copper (Cu) atoms, resulting in direct recoil of these surface atoms into the forward-scattering angle. These spectra show that the sample can be cleaned to a condition where it was free of impurity carbon and oxygen recoils (<0.01 monolayer on the surface) by either sputtering or annealing. A small amount of hydrogen impurity is observed as a recoil on the short TOF side of the Li(DR) peak. Note that although both Li(DR) and Cu(DR) are observed on the sputtered surface, there is no evidence of Cu(DR) on the annealed surface. This indicates that the Cu concentration in the outermost atomic layer of the annealed surface is <0.01 monolayers (i.e., it is almost a pure Li surface). Annealing induces diffusion of Li to the alloy surface.

Examples of experimental and simulated TOF-SARS spectra from a clean Ni(111) surface are shown[5] in Figure 4.10. The simulations are from the scattering and recoiling imaging code (SARIC)[6] (see Chapters 6 and 7 for examples). Peaks due to both scattered Ar and recoiled Ni are observed. These peaks are identified from the ion trajectories in the SARIC simulations. There is good agreement between the

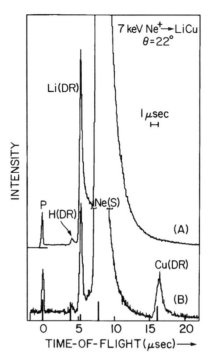

FIGURE 4.9. Time-of-flight scattering and recoiling spectroscopy (TOF-SARS) spectra of 7-keV Ne$^+$ scattering from a lithium copper (LiCu) alloy surface that was (A) annealed to 500°C and (B) sputtered with 5×10^{16} ions/cm^2 of 3-keV Ar$^+$ with $\theta = 22°$; P, photon pulse; DR, direct recoil, Ne, neon. Note the lack of a Cu(DR) peak in the annealed sample (From Shi *et al.*, 1990, with permission).

experimental and simulated spectra. The three Ar scattering peaks observed, SS, MS, and double scattering (DS), are typical of cases where the mass of the projectile ion is close to the mass of the target atoms. For spectra such as those of Figure 4.10, it is imperative to have trajectory simulations to obtain a complete assignment of the observed peaks.

Multiple recoil peaks from a single element can sometimes be observed when light atoms are recoiled at low recoiling angles. An example is shown[7] in Figure 4.11 for recoiling of H atoms from a W(211)-H surface using 4-keV Ne$^+$ with $\theta = 45°$. For $\alpha = 5°$, a single H recoil peak is observed. At $\alpha = 6°$ the recoil structure splits into two peaks, with one peak drifting towards lower TOF as α increases. At higher α values, this low TOF peak drifts back toward the stationary high TOF peak, coalescing into a single peak again near $\alpha \sim 42°$. This drifting peak is a result of a surface recoil (SR) sequence. The light H atom can scatter from W with negligible energy loss (maximum energy loss is 2%). On the other hand, the energy of a direct recoil (DR) hydrogen atom is strongly dependent on the recoil angle ϕ or impact parameter p. For low α, specific SR sequences such as low-angle H recoil (ϕ near 0°) followed by H scattering from W (θ near 45°) can produce H that is significantly

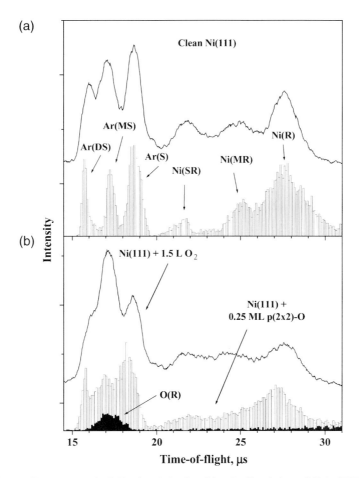

FIGURE 4.10. Experimental (solid line) and simulated (vertical bars) time-of-flight (TOF) spectra for 3-keV Ar$^+$ impinging on a clean (a) and O$_2$ dosed (b) Ni(111) surface. The peaks labeled Ar(S), Ar(DS), and Ar(MS) correspond to scattered Ar atoms: (S), quasi-single, (DS), quasi-double in-plane, (MS), multiple out-of-plane; peaks labeled Ni(R), Ni(SR), and Ni(MR) correspond to recoiled Ni atoms: R, quasi-direct recoil; SR, surface-recoiled; (MR), recoiled after multiple out-of-plane scattering; and the peak labeled O(R) corresponds to recoiled O atoms. The scattering conditions are scattering angle $\theta = 60°$; beam incident angle from surface $\alpha = 30°$; and crystal azimuthal angle from the [1$\bar{2}$1] direction $\delta = 0°$. The vibrational amplitude used in the simulations was 0.06 Å (From Bolotin *et al.*, 2001, with permission).

faster than that produced from a simple DR event with $\phi = 45°$. The SR and DR peaks can be positively identified by simple classical calculations. As an example,[7] consider the data from Figure 4.12. The TOFs of the two recoil peaks are plotted as a function of ϕ for $\delta = 0°$ and 90° and $\alpha = 10°$ and 16°; calculated TOF curves for DR of H by Ar in a single binary collision, according to Eq. 6, Chapter 2, are also shown. The TOF of the true DR peak is always within 0.25 μsec of the calculated TOF for recoiling by single collisions. The TOF of the SR peak is considerably less

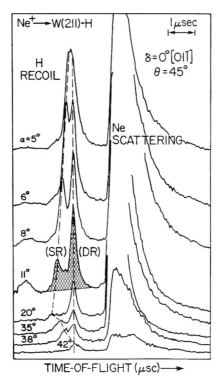

FIGURE 4.11. Series of forward-scattering time-of-flight (TOF) spectra for 4-keV Ne$^+$ scattering from a W(211)-H surface for different incident angles α. The beam direction was along the [011] ($\delta = 0°$) azimuth, and the scattering θ and recoiling ϕ angle was 45°. The area measured as representative of the H recoil intensity I(R) [both direct (DR) and surface (SR) recoil] is shown hatched for the $\alpha = 11°$ spectrum (From Shi *et al.*, 1989, with permission).

sensitive to changes in ϕ and appears at shorter TOF than that of the single-collision recoils.

The surface of LiTaO$_3$ (000$\bar{1}$) has been studied by TOF-SARS as a function of temperature.[8] The variations in surface voltage induced by the pyroelectric changes of spontaneous polarization were estimated by measuring spectral peak shifts and by computing the effects of surface voltage. The observed spectral signals in Figure 4.13 include: (a) Recoil hydrogen, H(r), at 2.57 μs, indicating the presence of surface hydrogen, either as adsorbates or from proton exchange with Li. (b) Recoil lithium, Li(r), at 3.29 μs. A minor peak in the tailing region of H(r) indicates the presence of a small amount of surface Li. The depletion of Li is attributed to proton exchange. (c) Recoil carbon, C(r), at 3.91 μs; this signal level is low, indicating that C contamination is not significant. (d) Scattered neon, Ne(s), from Ta, O, and Li is responsible for the broad peak at 4.5–6.5 μs. Recoil oxygen, O(r), at 4.39 μs is not resolved from the main peak. The surface composition was changed by prolonged 3 keV Ar$^+$ sputtering. This resulted in removal of H, and the scattering signals associated with Li and O, relative to those associated with Ta, were greatly reduced. Hence, the data

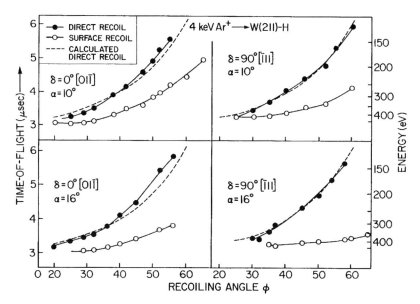

FIGURE 4.12. Plots of time-of-flight (TOF) versus recoiling angle ϕ for the hydrogen direct (DR) and surface recoil (SR) peaks observed in 4-keV Ar^+ scattering from a W(211)-H surface for different scattering and recoiling angles. Conditions: $\alpha = 16°$; $\theta = 20–50°$. The calculated TOF for (DR) of H by 4-keV Ar^+ in a single binary collision is shown as a dashed line (From Shi *et al.,* 1989, with permission).

also reveal the effects of preferential sputtering, leading to a Ta-rich surface. Such results provide a description of surface voltage and charge compensation mechanisms in response to pyroelectric changes during sputtering of $LiTaO_3$.

TOF-SARS spectra can be obtained from many insulating surfaces, provided that there is a sufficient amount of conductivity to maintain the surface potential near neutrality. Examples are shown[9] in Figure 4.14 and 4.15 for Ar^+ scattering from a polycrystalline cesium bromine (CsBr) sample. Ar scattering from both the Cs and Br target atoms is observed. The CsBr sample in Figure 4.14 contains adsorbed H_2O on the surface. The TOF spectra of such a contaminated surface at $42°$ and $105°$ clearly demonstrate the DR and SR processes. The H(DR) and O(DR) peaks present in the forward-scattering spectrum are of course absent in the backscattering spectrum. Note, however, that H(SR) and O(SR) are observed in the backscattering spectrum. The TOF spectra for 3-keV Ar^+ and 6-keV Ar^{2+} bombardment of clean polycrystalline CsBr at $47°$ in Figure 4.15 exhibit Ar scattering peaks from both Cs and Br as well as recoiling peaks of both Cs and Br at longer TOF. The Cs is recoiled primarily as positive ion, whereas the Br is recoiled primarily as a neutral atom. The Cs^+/Cs and Br^+/Br ion fractions are ~ 0.90 and ~ 0.08, respectively. The Ar^+ ion fractions (both Ar^+ and Ar^{2+}) surviving the Cs and Br collisions are ~ 0.20 and ~ 0.09, respectively, indicating preferential neutralization in collisions with Br atoms.

Another example of spectra from an insulating sample is shown[10] in Figure 4.16 for 5-keV Ar^+ scattering from a polycrystalline LiF surface with $\theta = 42°$. Spectra are

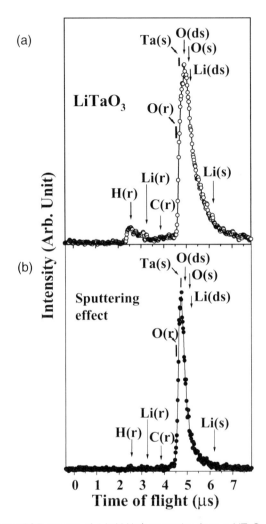

FIGURE 4.13. Typical TOF spectra of 4 keV Ne$^+$ scattering from a LiTaO$_3$ (000$\bar{1}$) single crystal surface with scattering angle $\theta = 20°$ and flight path L = 90 cm from (a) a cleaned and fully charge-compensated surface and (b) a surface after prolonged sputtering with 3 keV Ar$^+$. The calculated peak positions for the relevant scattering and recoiling signals are marked in the figure as a reference. "r" = direct-recoil, "s" = quasi-single-scattering, and "ds" = quasi-double-scattering. It was found that the pyroelectric voltage could be altered by ion irradiation induced electron and ion emission, which in turn depended on the surface charge condition and stoichiometry. (From Fang, *et al.*, 2002, with permission.)

shown for the total neutrals plus ions (N + I), neutrals only (N), and ions only (I). The broad argon (Ar) scattering peak is a result of scattering from both lithium (Li) and fluorine (F) atoms, although the cross section for F collisions is much larger than that for Li collisions. Recoil peaks of both Li and F, as well as impurity H, are observed. In the spectrum of ions only, note that the intensity of scattered Ar$^+$ is very low (i.e., most

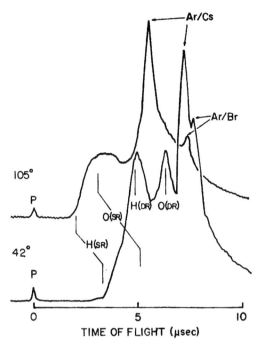

FIGURE 4.14. Time-of-flight (TOF) spectra of scattered and sputtered neutrals and ions for 3-keV Ar^+ bombardment of cesium bromine (CsBr) at a forward (42°, 170-cm path) and backscattering (105°, 45-cm path) angle. P represents a bombardment induced photon pulse, Ar/Cs and Ar/Br are scattering peaks for binary elastic collisions of Ar with Cs and Br. H(SR) and O(SR) are surface recoiled hydrogen and oxygen, and H(DR) and O(DR) are direct recoiled H and O (From Rabalais *et al.,* 1983, with permission).

of the scattered Ar is in a neutral charge state). This example clearly shows the advantage of using TOF-detection techniques. The sensitivity to scattered Ar with an electrostatic analyzer would be extremely low. The peak labeled P corresponds to ion-induced photon emission and serves as an excellent calibration for the zero of TOF.

4.3. RECOILING SPECTRA WITHOUT SCATTERING SPECTRA

The most intense peak in a typical TOF spectrum corresponds to the incident particles scattered (S) from the surface; it is normally one or two orders of magnitude more intense than the recoil (R) peaks. This (S) peak is useful in many applications; however, it can obscure the observation of a (R) peak if the (S) and (R) particles have similar velocities. For such cases, it is advantageous to suppress the intense (S) peak. Typical primary ions used in scattering are rare gas or alkali ions. The use of alkali ions has the advantage in that their neutralization probabilities are low; therefore, most of them survive the scattering process as ions. The combination of the high ion fraction of the (S) alkali particles with the low ion fraction of the (R) atoms allows a tremendous enhancement of the (R)/(S) ratio by deflecting the ions before they reach the detector,

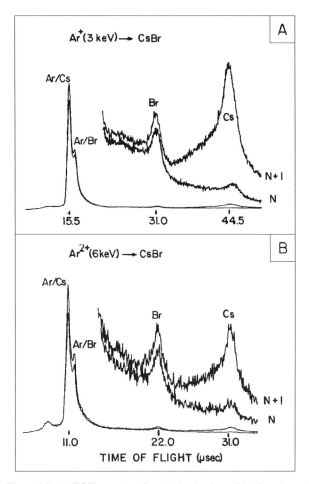

FIGURE 4.15. Time-of-flight (TOF) spectra of neutrals plus ions (N + I) and neutrals only (N) at a 47° scattering angle (170-cm path) for 3-keV Ar^+ and 6-keV Ar^{2+} on cesium bromine (CsBr). The (N) spectra are obtained by using an electrostatic deflector to remove the ions in the flight tube (From Rabalais *et al.*, 1983, with permission).

thus acquiring the neutrals only (N) spectra. Under certain scattering conditions, it is even possible to totally eliminate the scattering signal.

Figure 4.17 (left) shows the neutral-only (N) spectra[11] of hydrogen, carbon, and sulfur-contaminated Pt(111) induced by a 4-keV Rb^+ beam at increasing incidence angles α. The detection angle is fixed at $\phi = 21°$. The (DR)/(S) ratio for recoiling of (H + C + S) stays constant between $\alpha = 7°$ and $16°$ in the (I + N) spectrum, whereas it increases by a factor of two for the (N) spectrum. The scattering peak could not be totally eliminated in the Rb^+/Pt system at $\phi = 21°$. This may be due to the very large scattering cross section of Rb^+ on Pt at $\phi = 21°$. A detection angle of $\phi = 33°$ was necessary in order to obtain a scatter-free direct recoil spectrum.

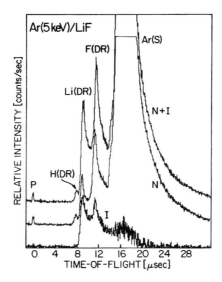

FIGURE 4.16. Time-of-flight (TOF) spectra of neutrals plus ions (N + I), N, and I for 5-keV Ar$^+$ scattering from lithium fluorine (LiF) at a 22° angle. Ne(S) and Li(DR) represent scattered Ne and directly recoiled Li, respectively. P is a photon pulse resulting from the collision (From Chen *et al.,* 1987, with permission).

In order to better understand the influence of the scattering conditions on the intensity of the neutral (N) scattering peak, the relative recoiling and scattering intensities are presented[11] in Figure 4.17 (right) for K$^+$ incident on a Si surface that contains a small amount of contamination. Spectra for three different incident angles α using a fixed scattering/recoiling angle θ are presented. The Si(R)/K(S) intensity ratio increases by a factor of 15 from $\alpha = 7°$ to 16°, where a near scatter-free recoil spectrum is obtained. This dramatic increase in the (R)/(S) ratio occurs only in the (N) spectrum. In the (I + N) spectrum, this ratio increases only by a factor of 1.4 between $\alpha = 7°$ and 16°.

4.4. SAMPLING DEPTH

Due to the detection of both ions and fast neutrals, TOF methods are sensitive to scattering from both the first- and subsurface layers. The sensitivity to various layers can be controlled through the scattering parameters and the ion type and energy. For example, small incident and detection angles emphasize first-layer scattering, whereas subsurface layers can be probed by using large incident and detection angles. Also, first-layer scattering is emphasized with low-energy, heavy ions and subsurface-layer scattering dominates for high-energy, light ions. The ability to probe the first- to subsurface layers is necessary for studies of surface relaxation and reconstruction and interfaces of thin films with substrates. On the other hand, scattering contributions from subsurface layers can be the major source of uncertainty in analysis of the

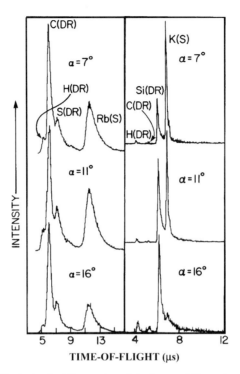

FIGURE 4.17. (left) Neutral-only (N) spectra of hydrogen (H), carbon (C), and sulfur-(S)-contaminated platinum Pt(111) induced by a 4-keV rubidium (Rb$^+$) beam at increasing incidence angles α. The detection angle is fixed at $\phi = 21°$. (right) Neutral-only (N) spectra of clean Si(100) induced by a 4-keV potassium (K$^+$) beam at increasing incidence angles α. The detection angle is fixed at $\phi = 21°$ (From Masson *et al.*, 1988, with permission). DR, direct recoil.

outermost surface layer structure. It is therefore important to choose the conditions appropriately to either maximize or minimize subsurface-layer scattering depending on the application.

The sampling depth is determined by the sizes of the shadowing and blocking cones, which are controlled by the type and energy of the primary ions as well as the target atom mass. The shadow cones for 4-keV Ar$^+$, Ne$^+$, and He$^+$ and 2-keV Ne$^+$ scattering from a single Ir atom, as calculated from the Zigler-Biersack-Littmark (ZBL) potential, are shown[12] in Figure 4.18. The radii of the 4-keV Ar$^+$ and 2-keV Ne$^+$ cones are within $\sim$0.05 Å of each other, while the radii of the 4-keV Ne$^+$ and He$^+$ cones are $\sim$0.15 Å and $\sim$0.45 Å, respectively, at a distance of 5 Å behind the target atom. The sizes of the blocking cones vary in the same manner as the shadow cones.

The sampling depth dependence on the type and energy of the primary ion can be demonstrated by scattering from the Ir(110) surface. This clean surface is reconstructed at room temperature with dominant (1×3) domains and minor (1×1) patches; a schematic drawing of this surface is shown[12] in Figure 4.19. The results of Figure 4.20 show[12] that the number, intensity, and position of the observed peaks

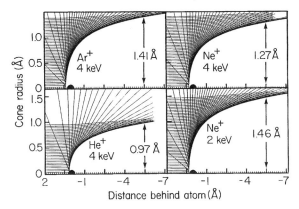

FIGURE 4.18. Shadow cones for 4-keV argon (Ar^+), neon (Ne^+), and helium (He^+) and 2-keV neon (Ne^+) computed from the envelope of classical trajectories using the Zigler-Biersack-Littmark (ZBL) potential function (From Bu *et al.,* 1991, with permission).

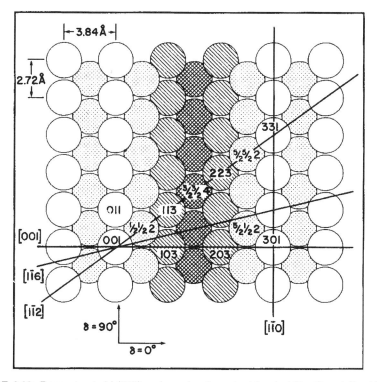

FIGURE 4.19. Reconstructed Ir(110) surface showing mixed faceted (1 × 3) and (1 × 1) structures. Open circles, first layer; dotted circles, second layer; dashed circles, third layer. The nomenclature for the atomic positions is indicated (From Bu *et al.,* 1991, with permission).

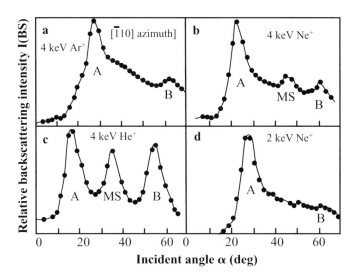

FIGURE 4.20. Backscattering intensity at $\theta = 163°$ versus incident angle α scans along the [1$\bar{1}$0] azimuth for the Ir(110) surface in the (1×3) phase for different primary ions and energies. The structures are labeled according to the identities of the two atoms involved in the shadowing and scattering as follows: A = (001–011) and B = (001–013). MS, multiple scattering sequence (From Bu *et al.*, 1991, with permission).

in the incident angle α scans along the [1$\bar{1}$0] azimuth are strongly dependent on the energy and type of primary ion. For 4-keV Ar$^+$ and 2-keV Ne$^+$, the structures are very similar. The intense peak at $\alpha \sim 28°$ results from focusing of trajectories by first-layer atoms onto their first-layer neighbors, such as the (001)–(011) interaction. This notation identifies the shadowing and scattering atoms involved in contributing to the intensity. The low-intensity peak at $\sim 60°$ results from first- to third-layer interactions, such as (001)–(013). For 4-keV He$^+$, the intensities of the (001)–(013) peak and the subsurface MS scattering peak are comparable to that of the (001)–(011) peak. For the series Ar$^+$, Ne$^+$, and He$^+$, the α positions of the peaks shift consistently towards lower values due to the smaller shadow cone radii. The α scans along the [001] azimuth shown in Figure 4.21[12] exhibit a variety of peaks that have been assigned according to shadow cone simulations and the known Ir(110) structure. The most intense structures for 4-keV He$^+$ and Ne$^+$ appear at high α values due to the dominant subsurface scattering. These results show that the precise layers from which scattering occurs is dependent on a number of variables, including the type and energy of the primary ion, the incident α and azimuthal δ angles, and the structural features of the surface.

4.5. ATTRIBUTES OF THE ION SCATTERING TECHNIQUE

There are no individual surface analysis techniques that can provide all of the information that is desirable about a surface. For this reason, surface science is usually

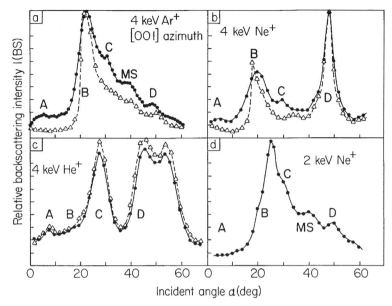

FIGURE 4.21. Backscattering intensity versus incident angle α scans along the [001] azimuth for the Ir(110) surface in the (1 × 3) phase (solid curves) and the (1 × 1) phase (dashed curves) for different primary ions and energies. The structures are labeled according to the identities of the two atoms involved in the shadowing and scattering as follows: A = (001–301); B = (001–303), (301–401), and ($\frac{5}{2}$ $\frac{1}{2}$ 2–$\frac{7}{2}$ $\frac{1}{2}$ 2); C = (001–203) and ($\frac{1}{2}$ $\frac{1}{2}$ 2 − $\frac{5}{2}$ $\frac{1}{2}$ 4); D = (001–103) and ($\frac{1}{2}$ $\frac{1}{2}$ 2 − $\frac{3}{2}$ $\frac{1}{2}$ 4) (From Bu *et al.,* 1991, with permission).

approached by using a variety of different techniques, preferably all within the same vacuum chamber. These different techniques all provide different types of information about a surface, and together they can sustain a more complete and accurate description of a surface than any of the individual techniques alone. The advantages and disadvantages of ISS compared to other well-known surface analysis techniques will be described herein.

The classical mechanics that underlie low-energy ion scattering are sufficiently well understood (Chapter 2) to provide a quantitative description of the atomic collision processes occurring. As a result, the "real-space" interpretations of ISS data are relatively straightforward compared to some of the other surface techniques. ISS can be the method of choice for specific surface elemental analyses and element-specific surface structural determinations. The major advantages of ISS are as follows:

1. Outermost monolayer or subsurface sensitivity can be obtained by judicious choice of the primary ion and angles involved. When using an electrostatic analyzer with a noble gas ion beam, first-layer sensitivity is extremely high due to the high neutralization probability (>98%) of noble gas ions that penetrate and scatter from subsurface layers. With alkali ions that have a lower

neutralization probability, or with TOF detection, first-layer or subsurface layer analysis can be achieved by judicious choice of the ion beam incidence angle and the detection exit angle.

2. ISS is element-specific and sensitive to all elements, including hydrogen, in the outermost surface layers with a sensitivity of $<10^{-2}$ of a monolayer. The limited resolution ($\Delta M/M \sim 0.03$) of the TOF technique precludes separation of heavy atoms with similar masses; however, such separations are accessible by electrostatic analysis.

3. Mass-selective real-space surface structure analyses are possible through the use of simple triangulation procedures in some cases and classical ion trajectory simulations in more complicated cases. Measurements of interatomic spacings to an accuracy of $<0.01\,\text{Å}$ can be obtained by comparison to classical ion trajectory simulations.

4. ISS is applicable to most materials, including metals, semiconductors, and insulators. The positive charge developed on insulator surfaces during positive ion irradiation can be neutralized by electrons from an incandescent filament near the sample surface.

5. Measurements of scattered and recoiled ion fractions can provide information on the various ion-surface charge exchange processes and their relative probabilities (see Chapter 8).

6. The observed anisotropic scattering features result from the ordered surface regions, whereas defects and steps produce an isotropic background. Thus the peak-to-background intensity ratios can provide information on the degree of order in a surface.

4.6. COMPARISON TO OTHER SURFACE ELEMENTAL ANALYSIS TECHNIQUES

For general purpose elemental analysis of a variety of samples, techniques such as X-ray photoelectron spectroscopy (XPS), AES, and secondary ion mass spectrometry (SIMS) are superior to ion scattering techniques. Since XPS and AES sample some 15–30 Å of the surface layers, they provide an averaged composition over this region, and it is difficult to obtain first-layer sensitivity. SIMS performed in the static mode can be monolayer sensitive. All three techniques yield information on the chemical environment of elements in the surface, and in many cases it is possible to identify molecules or compounds. The advantages of ISS over these techniques are its high first-layer sensitivity, its ability to quantitatively detect hydrogen, and its high sensitivity to surface structure.

4.6.1. Diffraction Methods

Diffraction methods such as low-energy electron diffraction (LEED) and reflection high-energy electron diffraction (RHEED) sample the outermost atomic layers of a surface and provide information based on reciprocal-space measurements. Long-range order ($>100\,\text{Å}$) is required and the data are not element-specific. The LEED and RHEED techniques provide simple, quick visual inspection of the surface symmetry. Complex superstructures can be delineated, which is not possible for real-space techniques such as ISS. Detailed structural information can be obtained from current-voltage (or I-V) profiles coupled with an intense theoretical input that is based on complicated quantum and solid-state theories. The results of these calculations are not always unambiguous.

The technique of surface X-ray diffraction (XRD) has been greatly enhanced by the development of high-intensity synchrotron sources. The data can be directly interpreted through the kinematic scattering model and is ideal for studying systems where long-range order or finite size effects are important (e.g., surface phase transitions and defects). The large penetration depths of X rays allows studies of buried interfaces. The low-scattering cross sections from elements of low atomic number make XRD less sensitive to surface contaminants and allow studies of surfaces in high-pressure environments. A disadvantage of these attributes is the difficulty in resolving scattering features from first-, second-, and third-layer scattering and the low sensitivity to light elements.

4.6.2. High-Energy Ion Scattering

Medium energy ion scattering (MEIS) and Rutherford backscattering (RBS) can have large sampling depths due to the small size of the shadow-cone radii at high energies. The locations of impurities and lattice defects in crystals can be obtained as well as relaxations in the outermost layers. Evaluation of the spectra is by the same classical ion trajectory simulation methods described herein. The major differences are the decreased surface sensitivity of MEIS and RBS and the high cost of the instrumentation.

4.6.3. Helium Atom Scattering

Helium atom scattering (HAS) depends on surface charge density and has the advantage of being sensitive to low atomic number elements, including hydrogen. Since the primary beam is composed of neutral atoms, it can be used to study insulating surfaces. The main disadvantage of HAS is the difficulty of inverting the measured corrugation amplitudes back to the ion core positions.

4.6.4. Scanning Microscopy

Scanning tunneling microscopy (STM) and atomic force microscopy (AFM) provide unique, real-space, local atomic position information with atomic-scale resolution. STM is sensitive to the local properties of the electron density of states modulated by the electron tunneling probabilities, rather than the positions of nuclear cores. Time-resolved measurements at a level of <1 s are possible by STM, whereas SARIS images (Chapter 6) provide time-averaged properties. Element-specificity by STM is by indirect methods involving changes in the charge distributions. AFM provides excellent information on surface morphology and defects. STM and AFM clearly reveal surface defects, steps, and abnormalities, whereas SARIS images are more sensitive to the ordered surface regions. Scanning electron microscopy (SEM) obtains atomic resolution in only special cases and requires elaborate sample preparation. Field ion microscopy (FIM) has atomic scale resolution; however, it is limited to only specific classes of materials.

4.6.5. Electron Scattering

Electron scattering techniques such as electron energy loss spectroscopy (EELS) and high-resolution EELS provide unique information on the vibrational properties of adsorbates on surfaces and thereby their bonding sites.

4.6.6. Ionization and Bond-Breaking Techniques

These techniques include photon- and electron-stimulated desorption (PSD and ESD), thermal desorption spectrometry (TDS), XPS, AES, SIMS, and surface-extended X-ray absorption fine structure (SEXAFS). These methods provide chemical and structural information since the ionization or bond-breaking processes depend on the chemical environment of the elements.

REFERENCES

1. J.-P. Jacobs, "Catalytically-Active Oxides Studied by Low-Energy Ion Scattering," Thesis Technische Universiteit Eindhiven, The Netherlands, 1995.
2. O. Grizzi, M. Shi, H. Bu, and J. W. Rabalais, *Rev. Sci. Instrum.* **61,** 740 (1990).
3. Y. Wang, M. Shi, and J. W. Rabalais, *Phys. Rev.* **B48,** 1678 (1993).
4. M. Shi, O. Grizzi, and J. W. Rabalais, *Surf. Sci.* **235,** 67 (1990).
5. I. L. Bolotin, A. Kutana, B. Makarenko, and J. W. Rabalais, *Surf. Sci.* **472,** 205 (2001).
6. Bykov V., C. Kim, M. M. Sung, K. J. Boyd, S. S. Todorov, and J. W. Rabalais, *Nuc. Instrum. Meth. Phys. Res.* **114,** 371 (1996); V. Bykov, L. Houssiau, and J. W. Rabalais, *J. Phys. Chem.* B **104,** 6340 (2000).
7. M. Shi, O. Grizzi, H. Bu, J. W. Rabalais, R. R. Rye, and P. Nordlander, *Phys. Rev.* **B40,** 10163 (1989).
8. Z. L. Fang, K. M. Lui, W. M. Lau, B. Makarenko, and J. W. Rabalais, *J. Vac. Sci. Technol.,* in press (2002).
9. J. W. Rabalais, J. A. Schultz, R. Kumar and P. T. Murray, *J. Chem. Phys.* **78,** 5250 (1983).
10. J. N. Chen, M. Shi, and J. W. Rabalais, *J. Chem. Phys.* **86,** 2403 (1987).
11. F. Masson, S. Aduru, C. S. Sass, and J. W. Rabalais, *Chem. Phys. Lett.* **152,** 325 (1988).
12. H. Bu, M. Shi, and J. W. Rabalais, *Nucl. Instrum. Meth. Phys. Res.* **B61,** 337 (1991).

5

STRUCTURAL ANALYSIS FROM TIME-OF-FLIGHT SCATTERING AND RECOILING SPECTROMETRY

Three basic types of data can be obtained from time-of-flight scattering and recoiling spectrometry (TOF-SARS) experiments: (1) TOF spectra, (2) incident angle α scans, and (3) azimuthal angle δ scans. Figure 5.1 illustrates these three types of data for 3-keV Kr^+ impinging on a cadmium sulfide CdS(0001) surface with chemisorbed H_2O. First, TOF spectra of the type displayed in Chapter 4 and Figure 5.1(a) are obtained by measuring the velocity distributions of scattered and recoiled particles using a channel electron multiplier detector with a small acceptance solid angle. The pulsed ion beam direction and the detector are in the same plane, and the angle between the beam and detector is fixed. Elemental analyses are achieved by converting the velocity distributions into energy distributions and relating those to the masses of the target atoms through the kinematic relationships that describe classical scattering and recoiling. Second, incident angle α scans as in Figure 5.1(b) are obtained by rotating the sample about an axis that goes through its plane while monitoring the intensity of a specific peak in the TOF-SARS spectrum. Third, azimuthal angle δ scans as in Figure 5.1(c) are obtained by rotating the sample about an axis that goes through the surface normal, while monitoring the intensity of a specific spectral peak. Since the sample rotation changes both the incident beam direction and the detector direction, the spectra are affected by both shadow cones and blocking cones if a forward-scattering angle is used. In order to reduce blocking cone effects for simpler interpretations, high-angle backscattering can be used. The anisotropic features in the α- and δ-scans are interpreted by means of shadow cone and blocking cone analyses.

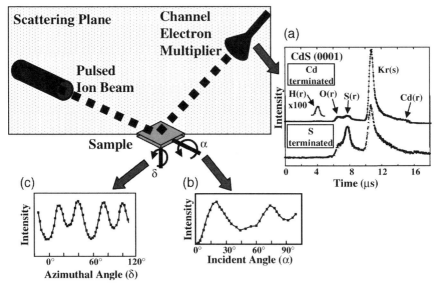

FIGURE 5.1. Schematic diagram of time-of-flight scattering and recoiling spectrometry (TOF-SARS) illustrating the plane of scattering formed by the ion beam, sample, and detector and the three different modes of data collection: (a) TOF spectra, (b) incident angle α scans, and (c) azimuthal angle δ scans. The data are for 3-KeV Kr^+ impinging on a cadium sulfide CdS(0001) surface that has a small amount of H_2O chemisorbed on the surface.

5.1. ATOMIC COLLISIONS IN THE keV RANGE

The dynamics of keV atomic collisions are well described as classical binary collisions between the incident ion and surface atoms (as described in Chapter 2). When an energetic ion makes a direct collision with a surface atom, the surface atom is recoiled into a forward direction as shown in Figure 5.2. Both the scattered and recoiled atoms have high, discrete kinetic energy distributions as described by Eqs. 4 and 6 of Chapter 2. As a result of the energetic nature of the collisions, only atomic species are observed as direct recoils, and their energies are independent of the chemical bonding environment.

Although scattering in the keV range is dominated by repulsive potentials, it is not simply a hard sphere or billard ball collision where there is a clean "hit" or "miss." The partial penetration of the ion into the target atom's electron cloud results in bent trajectories even when there is not a "head-on" collision. As detailed in Chapter 2, this type of interaction can be described by a screened coulomb potential function that allows one to determine the relationship between the scattering angle θ and the impact parameter b. Considering a large number of ions with parallel trajectories impinging on a target atom, the ion trajectories are bent by the repulsive potential such that there is an excluded volume, called a shadow cone, in the shape of a paraboloid formed behind the target atom as shown in Figure 5.2. Ion trajectories do not penetrate into the shadow cone, but instead are concentrated at its edges much like rain pours off an umbrella. Atoms located inside the cone behind the target atom are shielded

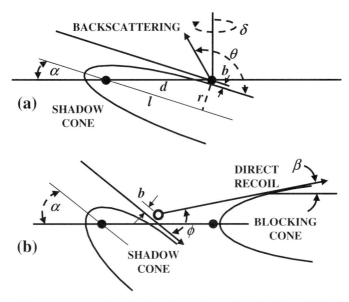

FIGURE 5.2. Schematic illustration of (a) backscattering and shadowing and (b) direct recoiling with shadowing and blocking.

from the impinging ions as shown[1] in Figure 5.3. Similarly, if the scattered ion or recoiling atom trajectory is directed towards a neighboring atom, that trajectory will be blocked. For a large number of scattering or recoiling trajectories, a blocking cone will be formed behind the neighboring atom into which no particles can penetrate. The dimensions of the shadowing and blocking cones can be determined experimentally from scattering measurements along crystal azimuths for which the interatomic spacings are accurately known. This provides an experimental determination of the scaling parameter in the potential function and reliable cone dimensions. The cone dimensions can also be constructed theoretically[2,3,4] from the relationship of b with θ and ϕ. A universal shadow cone curve has been proposed,[5] and cone dimensions for common ion–atom combinations have been reported.[4] Since the radii of these cones are of the same order as interatomic spacings (i.e., 1 to 2 Å), the ions penetrate only into the outermost surface layers.

When an isotropic ion fluence impinges on a crystal surface at a specific incident angle α, the scattered and recoiled atom flux is anisotropic. This anisotropy is a result of the incoming ion's eye view of the surface, which depends on the specific arrangement of atoms and the shadowing and blocking cones. The arrangement of atoms controls the atomic density along the azimuths and the ability of ions to channel, that is, to penetrate into empty spaces between atomic rows. The cones determine which nuclei are screened from the impinging ion flux and which exit trajectories are blocked as depicted in Figure 5.3.[1] By measuring the ion and atom flux at specific scattering and recoiling angles as a function of ion beam incident α and azimuthal δ angles to the surface, structures are observed that can be interpreted in terms of

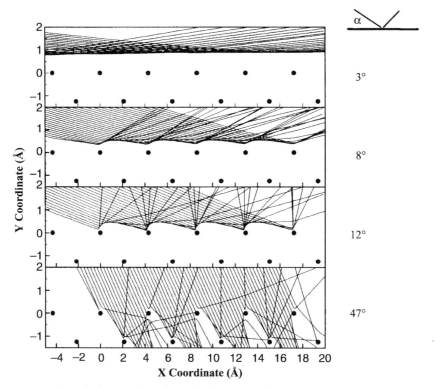

FIGURE 5.3. Classical scattering trajectories for 3-KeV Ne$^+$ impinging on a nickel (Ni) surface at different incident angles α (From Rabalais, 1994, with permission).

the interatomic spacings and shadow cones from the ion's eye view. For example, from Figure 5.2(a), it is observed that the interatomic distance d between the two atoms is

$$d = r/\sin\alpha \qquad \text{or} \qquad d = l/\cos\alpha. \tag{1}$$

Interatomic spacings determined in this manner are typically accurate to better than 0.1 Å.

5.2. STRUCTURE ANALYSIS

The atomic structure of a surface is usually not a simple termination of the bulk structure. A classification exists based on the relation of surface to bulk structure. A *bulk truncated* surface has a structure identical to that of the bulk. A *relaxed surface* has the symmetry of the bulk structure but different interatomic spacings. With respect to the first and second layers, *lateral relaxation* refers to shifts in layer registry, and *vertical relaxation* refers to shifts in layer spacings. A *reconstructed* surface has a symmetry different from that of the bulk symmetry. The methods of structural analysis using TOF-SARS will be delineated below.

5.2.1. Scattering versus Incident Angle α Scans

When an ion beam is incident on an atomically flat surface at grazing angles, each surface atom is shadowed by its neighboring atom such that only forward scattering (FS) is possible; these are large impact parameter b collisions. As α increases, a critical value $\alpha_{c,sh}{}^i$ is reached each time the ith layer of target atoms moves out of the shadow cone allowing for large angle backscattering (BS) or small b collisions as shown[1] in Figure 5.3. If the BS intensity I(BS) is monitored as a function of α, steep rises with well defined maxima are observed when the focused trajectories at the edge of the shadow cone pass close to the center of neighboring atoms as shown[1] in Figure 5.4. From the shape of the shadow cone, in other words, the radius (r) as a function of distance (l) behind the target atom (Figure 5.2), the interatomic spacing (d) can be directly determined from the I(BS) versus α plots. For example, by measuring $\alpha_{c,sh}{}^1$ along directions for which specific crystal azimuths are aligned with the projectile direction and using $d = r / \sin \alpha_{c,sh}{}^1$, one can determine interatomic spacings in the first-atomic layer. The first- to second-layer spacing can be obtained in a similar manner from $\alpha_{c,sh}{}^2$ measured along directions for which the first- and second-layer atoms are aligned, providing a measure of the vertical relaxation in the outermost layers.

5.2.2. Scattering versus Azimuthal Angle δ Scans

Fixing the incident beam angle α and rotating the crystal about the surface normal while monitoring the backscattering intensity provides a scan[7] of the crystal azimuthal angle δ. Such scans reveal the periodicity of the crystal structure. For example, one can obtain the azimuthal alignment and symmetry of the outermost layer by using a low α value such that scattering occurs from only the first-atomic layer. With higher α values, similar information can be obtained for the second-atomic layer. Shifts in the first- to second-layer registry can be detected by carefully monitoring the $\alpha_{c,sh}{}^2$ values for second-layer scattering along directions near those azimuths for which the second-layer atoms are expected, from the bulk structure, to be directly aligned with the first-layer atoms. The $\alpha_{c,sh}{}^2$ values will be maximum for those δ values where the first- and second-layer neighboring atoms are aligned.

When the scattering angle θ is decreased to a forward angle ($<90°$), both shadowing effects along the incoming trajectory and blocking effects along the outgoing trajectory contribute to the patterns. The blocking effects arise because the exit angle $\beta = \theta - \alpha$ is small at high α values. Surface periodicity can be read directly from these features as shown[1] in Figure 5.5 for Pt(110). Minima are observed at the δ positions corresponding to alignment of the beam along specific azimuths. These minima are a result of shadowing and blocking along the close-packed directions, thus providing a direct reading of the surface periodicity.

Azimuthal scans obtained for three surface phases of Ni(110) are shown[1,8] in Figure 5.6. The minima observed for the clean and hydrogen-covered surfaces are due only to nickel (Ni) atoms shadowing neighboring Ni atoms, whereas for the minima observed for the oxygen (O)-covered surface are due to both O and Ni atoms shadowing neighboring Ni atoms. Shadowing by hydrogen (H) atoms is not observed because the maximum deflection in the neon (Ne^+) trajectories caused by H atoms is $< 2.8°$.

5.2.3. Recoiling versus Incident Angle α Scans

Adsorbates can be efficiently detected by recoiling them into forward-scattering angles ϕ as shown in Figure 5.2(b). As α increases, the adsorbate atoms move out of their neighboring atom shadow cones so that direct collisions from incident ions

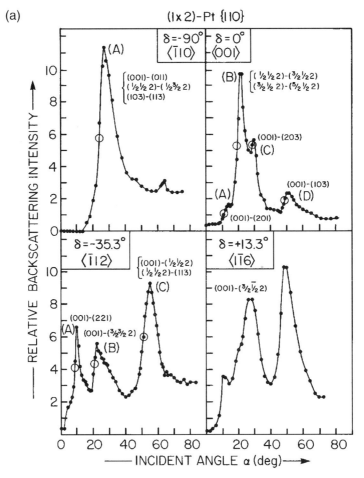

FIGURE 5.4. (a) Scattering intensity versus incident angle α scans for $Pt(110) - (1 \times 2)$ at $\theta = 149°$ along the $\langle \bar{1}10 \rangle$, $\langle 001 \rangle$, $\langle \bar{1}12 \rangle$, and $\langle 1\bar{1}6 \rangle$ azimuths. (From Rabalais, 1994 and Mason *et al.*, 1991, with permission).

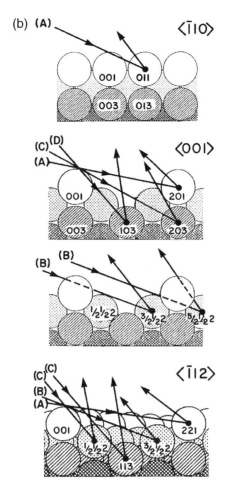

FIGURE 5.4. (*Continued*) (b) Cross-section diagrams along the three azimuths illustrating scattering trajectories for the peaks observed in the scans of Figure 4(a).

are possible. When the b value necessary for recoiling of the adsorbate atom into a specific ϕ becomes possible, adsorbate recoils are observed. Focusing at the edge of the shadow cone produces sharp rises in the recoiling intensity as a function of α. By measuring $\alpha_{c,sh}$ corresponding to the recoil event, the interatomic distance of the adsorbate atom relative to its nearest neighbors can be determined from b and the shape of the shadow cone.

Example plots of oxygen recoil intensity versus α for two different chemisorbed O atom coverages on W(211) are shown[1] in Figure 5.7. Low-dose exposure forms a p(2 × 1) structure consisting of 0.5-monolayer (ML) coverage and high-dose exposure forms a p(1 × 2) structure consisting of 1.5-ML coverage. Sharp rises appear at low α, and sharp decreases appear at high α. The rises are due to peaking of the ion flux at the edges of the shadow cones of neighboring atoms, and the decreases are due

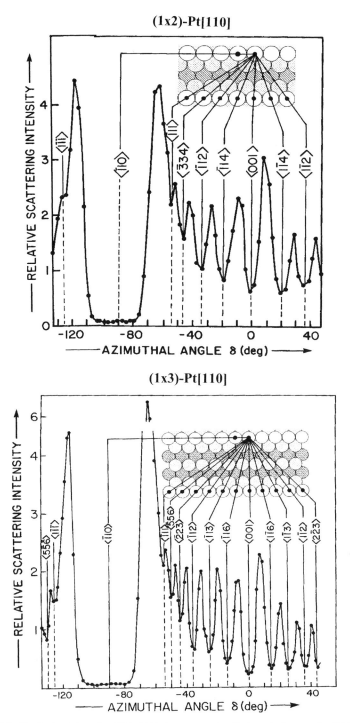

FIGURE 5.5. Scattering intensity of 2-keV Ne$^+$ versus azimuthal angle δ scans for Pt(110) in the (1 × 2) and (1 × 3) reconstructed phases. Scattering angle $\theta = 28°$ and incident angle $\alpha = 6°$ (From Rabalais, 1994 and Mason *et al.*, 1991, with permission).

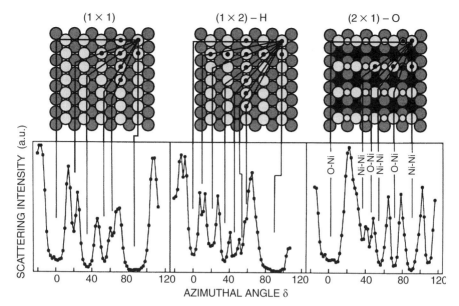

FIGURE 5.6. Scattering intensity of 4-keV Ne$^+$ versus azimuthal angle δ for a Ni(110) surface in the clean (1 × 1), (1 × 2) − H missing row, and (2 × 1) − O missing row phases. The hydrogen atoms are not shown. The oxygen atoms are shown as small open circles. O-Ni and Ni-Ni denote the directions along which O and Ni atoms, respectively, shadow the Ni scattering center (From Rabalais, 1994, and Roux *et al.*, 1991, with permission).

to blocking of recoil trajectories by neighboring atoms. The critical α values for both shadowing, $\alpha_{c,sh}$, and blocking, $\alpha_{c,bl}$, can be used for determination of interatomic spacings. At high coverage these critical values are $\alpha_{c,sh} = 24°$ and $\alpha_{c,bl} = 42°$, which is considerably higher and lower, respectively, than the values $\alpha_{c,sh} = 16°$ and $\alpha_{c,bl} = 48°$ obtained at low coverage. This indicates that as coverage increases, both the shadowing and blocking effects are enhanced due to close packing of O atoms; this results from shadowing and blocking of O atoms by their neighboring O atoms. The α_c values correspond to O atoms separated by a distance of two W lattice constants at low coverage and one W lattice constant at high coverage.

5.2.4. Recoiling versus Azimuthal Angle δ Scans

Plots of recoiling intensity versus azimuthal angle δ reveal the surface periodicity of the recoiled atoms in a manner similar to that of the scattering intensity. Azimuthal scans of silicon (Si) recoils from the Si(100) − (2 × 1) and Si(100) − (1 × 1) − H surfaces are shown[1−10] in Figure 5.8. The patterns are symmetrical about the ⟨011⟩ ($\delta = 0°$) azimuth. The positions of the minima are consistent with the structures indicated above the figures. The repetition of the symmetry features every 90° indicates that there are two domains that are rotated by 90° with respect to each other. Azimuthal scans of H recoils from the Si(100) − (2 × 1) − H and Si(100) − (1 × 1) − H surfaces are shown[1,11] in Figure 5.9. The observed minima are due to Si atoms shadowing

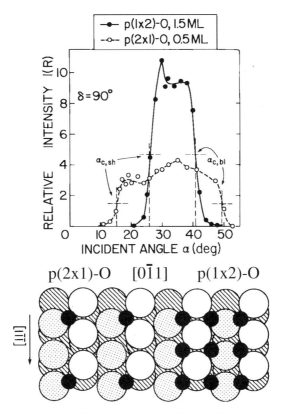

FIGURE 5.7. Plots of oxygen recoil intensity versus incident angle α for both high (1.5 ML) and low (0.5 ML) oxygen coverages on a W(211) surface. The critical shadowing $\alpha_{c,sh}$ and blocking $\alpha_{c,bl}$ angles are indicated (From Rabalais, 1994, with permission).

neighboring H atoms. The patterns are consistent with the structures indicated above the figures.

5.3. AZIMUTHAL ALIGNMENT OF THE INCIDENT ION BEAM

Sharp spatial anisotropy of scattered ion intensity can be observed when an ion beam is incident at a grazing angle on a surface that has so-called surface-semichannels. These semichannels are formed by close-packed rows in the first-atomic layer that serve as the *walls* of the channel and similar rows in the second atomic layer that serve as the *base* of the channel. Such surface semichannels are able to effectively focus scattered ions under certain conditions when the incident plane of the projectile is parallel to the channels. This focusing effect can provide an accurate method of aligning the incident ion beam with the azimuthal angles of the crystal. The W(211) surface provides a good example of surface semichanneling. It has semichannels

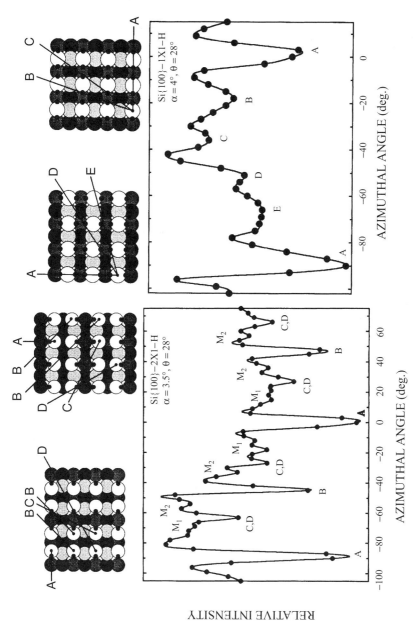

FIGURE 5.8. Azimuthal angle δ scans of the silicon recoil intensity for the clean Si(100) – (2 × 1) and Si(100) – (1 × 1) – H dihydride surfaces. The minima are identified from the structural drawings above the scans. The hydrogen atoms are not shown (From Rabalais, 1994 and Wang *et al.*, 1993, with permission).

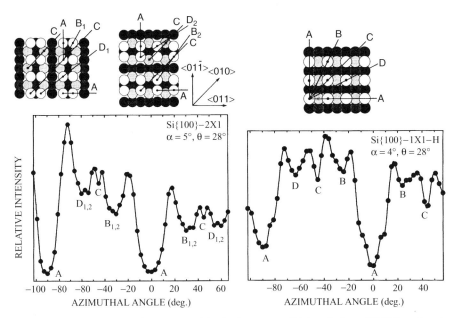

FIGURE 5.9. Azimuthal angle δ scans of the hydrogen recoil intensity for a Si(100) surface in the $(2 \times 1) - H$ monohydride and $(1 \times 1) - H$ dihydride phases. The minima are identified from the structural drawings above the scans. The hydrogen atoms are indicated as small dark circles (From Rabalais, 1994 and Wang *et al.*, 1993, with permission).

along the $[1\bar{1}\bar{1}]$ azimuthal direction. Semichanneling measurements can be made by directing the ion beam along the $[1\bar{1}\bar{1}]$ direction and collecting a series of spectra as a function of azimuthal angle in the region $\delta = \pm 25°$ about the $[1\bar{1}\bar{1}]$ axis. Specific beam incident α and exit β angles are chosen, and the scattering intensity I(S) is monitored as a function of δ. Figure 5.10 shows plots of I(S) versus δ at $\alpha = 10°$ and different β values.[10] The scans are symmetric about the $[1\bar{1}\bar{1}]$ troughs, and their structure is very sensitive to β. A sharp peak (full width at half-maximum [FWHM] $= 4–5°$) is observed at low β due to surface semichanneling. As β increases above the specular conditions (i.e., $\alpha = \beta = 10°$), the intensity of the focusing peak decreases until a minimum is finally observed.

5.4. TOF-SARS AND LEED

TOF-SARS and low-energy electron diffraction (LEED) are complementary techniques. LEED probes long-range order with a minimum domain size of 100–200 Å. It provides a measure of surface and adsorbate symmetry in reciprocal space. This is invaluable preliminary data for TOF-SARS structural determinations. Surface structure calculations from the reciprocal space LEED data are complicated, whereas

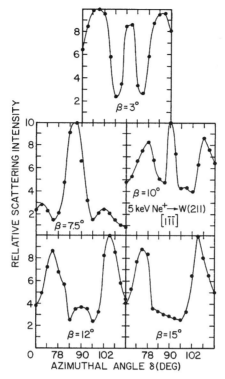

FIGURE 5.10. Surface semichanneling illustrated by plots of I(S) versus δ in the region near the W(211) $\delta = 90°$ [1$\bar{1}\bar{1}$] azimuth with $\alpha = 10°$ and different β values using 5-keV Ne$^+$ (From Wang *et al.,* 1993, with permission).

such calculations from the real-space TOF-SARS data are straightforward. Although complicated, LEED calculations are capable of analyzing structures with large, complicated unit cells. Due to the simple real-space analysis of TOF-SARS data, it cannot handle structures with large unit cells or with many inequivalent atoms in the unit cell. Coupling of LEED and TOF-SARS provides a powerful combination for surface structure investigations.

REFERENCES

1. J. W. Rabalais, *Surf. Sci.* **299/300,** 219 (1994).
2. E. S. Mashkova, and V. A. Molchanov, "Medium Energy Ion Reflection From Solids," Elsevier, North-Holland, Amsterdam, 1985.
3. J. F. Zeigler, J. P. Biersack, and U. Littmark, "The Stopping and Range of Ions in Solids," Pergamon, NY, 1985.
4. A. Kutana, I. L. Bolotin, and J. W. Rabalais, *Surf. Sci.* **495,** 77 (2001).

5. O. S. Oen, *Surf. Sci.* **131,** L407 (1983).

6. C. S. Chang, U. Knipping, and I. S. T. Tsong, *Nucl Instrum. Meth. B* **18,** 11 (1986); *ibid, idem,* **B35,** 152 (1988).

7. F. Masson and J. W. Rabalais, *Chem. Phys. Lett.* **179,** 63 (1991); *ibid, Surf. Sci.* **253,** 245 (1991); *ibid, idem,* **253,** 258 (1991).

8. C. D. Roux, H. Bu, and J. W. Rabalais, *Surf. Sci.* **259,** 253 (1991).

9. Y. Wang, M. Shi, and J. W. Rabalais, *Phys. Rev.* **B48,** 1678 (1993).

10. J. W. Rabalais, *J. Vac. Sci. Technol.* **A9,** 1293 (1991).

11. M Y. Wang, and J. W. Rabalais, *Phys. Rev.* **B48,** 1689 (1993).

6

REAL-SPACE SURFACE CRYSTALLOGRAPHY FROM SCATTERING AND RECOILING IMAGING SPECTROMETRY

Nature has provided atomic-size probes (energetic ion beams) and atomic lenses (periodic arrangements of atoms in crystals) that are capable of projecting the interatomic vectors from a surface into macroscopic real-space images with a magnification of $\sim 10^9$. Ion beams in the keV energy range impinging on crystal surfaces are atomic-size probes of these nanoscale magnifying and focusing lenses formed by the periodic structures. These lenses disperse the scattered and recoiled atom trajectories into macroscopic projections according to their velocities as a function of projectile/target atom masses and deflection angles. The resulting patterns reflect the near-surface interatomic vectors. The spatially and temporally resolved images of these patterns can be acquired by means of ion scattering spectrometry through the use of position-sensitive microchannel plate detectors (as described in Chapter 3). This provides direct, element-specific, real-space projections of surface structure. The technique is called scattering and recoiling imaging spectrometry (SARIS). It provides unique capabilities for surface composition and structure analysis and for investigating the interactions of ions with surfaces. The classical ion trajectory program called scattering and recoiling imaging code (SARIC) that simulates the spatial and time intensity distributions of the SARIS images was developed for this purpose.

6.1. AN IMAGING SPECTROMETRY FROM NATURE'S OWN ATOMIC LENSES

If a scattered ion trajectory intersects the site position of a neighboring atom, the ion trajectory is deflected by the repulsive potential of the blocking atom. This results in a

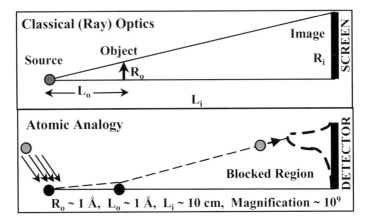

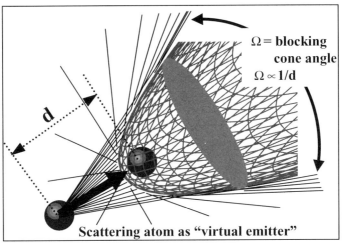

FIGURE 6.1. (Top) Schematic illustration of the magnification of a blocking cone using classical optics and atomic particle optics. For classical optics, the magnification is of the order of 10^1. For atomic particle optics, scattering trajectories from a virtual atomic emitter are blocked by a neighboring atom of radius r at distance l. The diverging nature of the blocking cone results in magnification by a factor of $\sim 10^9$ at a macroscopic distance L. (Bottom) Example of an interatomic vector and blocking cone. The contour lines define the repulsive potential surface of the blocking cone (From Rabalais, 2001, with permission).

blocking cone centered around the blocking atom, as shown in Figure 6.1.[1] Scattered or recoiled atoms do not penetrate into these cones, but instead are focused at the edges of the cones, generating a shadow much like that cast by an umbrella. When viewed by a position-sensitive detector at a large distance from the surface, the shadow cast by the blocking atom is of macroscopic size due to the diverging nature of the cone shape. This magnification of a blocking cone can be as high as 10^9 and can be compared to a slide projector: the virtual atomic source is the lamp, the blocking atoms are features in the slide, and the detector is the screen (as shown in Figure 6.1). For ion scattering, the interatomic spacings d and the "blocking atom radius" r are of

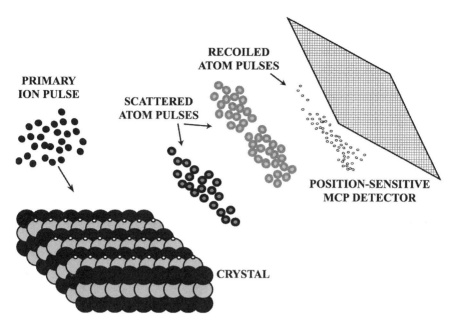

FIGURE 6.2. Schematic drawings of the SARIS experimental configuration, showing scattered and recoiled atoms from a pulsed ion beam being detected by a large-area position-sensitive microchannel plate (MCP) detector (From Rabalais, 2001, with permission).

atomic size, whereas the source-to-image distance D is macroscopic size, resulting in an image size R that is magnified by $D/d = R/r \sim 10^9$.

A schematic drawing of the SARIS experimental setup is shown[1] in Figure 6.2, and the various angles involved are defined[2] in Figure 6.3. A pulsed ion beam impinges on a surface in an ultrahigh vacuum chamber and scatters and recoils atoms in pulses. The velocities of these scattered and recoiled atoms are analyzed by measuring their flight times from the sample to a position-sensitive microchannel plate (MCP) detector. The short ion beam pulses allow selection of time windows appropriate for acquisition of selected scattered or recoiled particles. Instrumental details are provided in Chapter 3.

Examples of SARIS images are shown[1,3] in Figure 6.4 along with schematic drawings of the collision sequences that produce them. For each frame, the ordinate represents the particle exit angles (β) from the surface, and the abscissa represents the crystal azimuthal angles (δ) (i.e., an image in β,δ-space). This is most easily visualized from Figures 6.2 and 6.3 by assuming yourself sitting on the sample surface, facing the detector, and viewing the trajectories as they leave the surface. The scattered particles are the reflected (primary) atoms helium, neon, and argon (He, Ne, and Ar) and the recoiled particles are the ejected (target) atoms. Since >95% of these scattered and recoiled particles are neutral,[4] the term *atoms* is used instead of ions; the detector has equal sensitivity to both of these when their energies are > 1000 eV. The final energies (or velocities) of both scattered and recoiled atoms are determined by the masses of the colliding atom pair and the deflection angles.

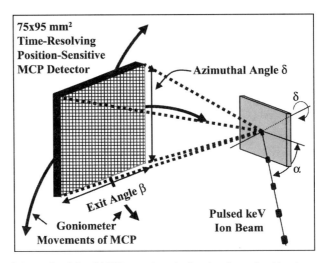

FIGURE 6.3. Schematic of the SARIS experiment, showing the pulsed ion beam, sample, and microchannel plate (MCP) detector (From Bykov *et al.,* 2000, with permission).

6.2. SHADOWING AND BLOCKING

When a beam of parallel energetic ions interacts with a target atom, the target atom core of size $\sim$0.01 nm acts as a virtual point source, scattering the atoms with a near isotropic intensity distribution. This phenomenon is due to the short de Broglie wavelengths ($\sim$0.1 pm) of keV atoms, which constrains them to classical behavior, and their large collision cross sections ($\sim$1 $\times$ 10^{-3} nm^2), which provides sensitivity to only the near-surface region. Such atomic collisions are described qualitatively as *three-dimensional pool*[5] or quantitatively by classical mechanics.[6] Scattering of ions by a target atom produces an excluded region called a shadow cone behind the target atom into which no ion can penetrate (as shown in Figure 6.1). There is zero flux density inside the cone, unit density far outside of the cone, and highly focused flux density at the boundary of the cone. This anisotropic distribution of ion flux after interaction with a target atom is the basis of the structural determinations. If a neighboring atom lies in the focusing region, enhanced scattering and/or recoiling intensity is observed. TOF-SARS measures the intensity change due to the shadowing effect on neighboring atoms as a function of incident or azimuthal angle beam direction. A shadow cone originates from the interaction of an atom with collimated atoms from a flux of parallel ions (ion source in ion beam line), resulting in a parabolid cone shape. If a scattered atom trajectory intersects the site position of a neighboring atom, the trajectory is deflected by the repulsive potential of the blocking atom. This results in a blocking cone centered around the blocking atom. It can be regarded as an interaction of a target atom (blocking atom) with diverging ions emitted from a nearby point source (scattering atom). Unlike a shadow cone, a blocking cone diverges as shown in Figure 6.1.

There are a large number of scatterer and blocker combinations in a solid. For ordered crystals, the periodicity and short interaction range ($<$1 nm) of the atoms reduces

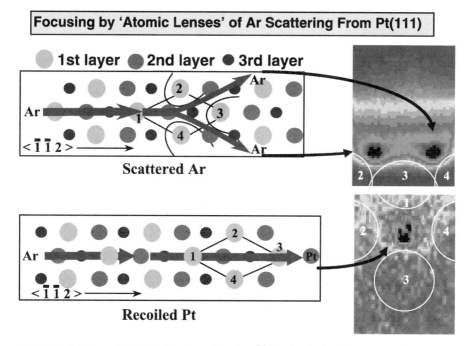

FIGURE 6.4. Views of the Pt(111) surface along the $\langle\bar{1}\bar{1}2\rangle$ azimuth showing argon (Ar$^+$) impinging from the left side. (Top) The Ar beam scatters from a first-layer platinum (Pt) atom (atom 1) and splits into two focused beams by an atomic lens formed by neighboring first-layer Pt atoms (atoms 2, 3, and 4). (Bottom) The impinging Ar recoils a second-layer Pt atom through an atomic lens formed by first-layer Pt atoms (atoms 1–4). The resulting experimental SARIS images are shown on the right side (From Rabalais, 2001 and Kim *et al.*, 1997, with permission). (See Color Plate)

this number of combinations to a tractable level. These scatterer and blocker combinations can be viewed as interatomic vectors (Figure 6.1) with density $\sim 10^{19}$ m^{-2}, resulting in reinforced focusing and blocking patterns. The projections of interatomic vectors appear as regions of low intensity in the SARIS images due to blocking of isotropically scattered ion trajectories from near-surface atoms. The resulting patterns allow direct construction of a model of the surface structure, providing the basis of SARIS.

6.3. AZIMUTHAL EQUIDISTANT MAPPING AND PROJECTION OF SARIgrams

The angular dispersion of scattered and recoiled particles is directly related to the structure of the sample under investigation. The anisotropic features observed correspond to both in- and out-of-plane scattering and recoiling processes. By varying the relative orientation of the MCP with respect to the sample using a goniometer, it is possible to collect particles leaving the target in almost 2π steradians. Different scattering and recoiling phenomena can be assessed by collecting images at different MCP positions and time intervals. In order to handle the abundance of structural information[7,8] resulting from such measurements, it is desirable to display the angular relationship between crystal lattice planes and features of the SARIS images in the form of an azimuthal equidistant mapping[9] and projection into SARIgrams. Consider an *imaginary* hemispherical detector with the sample surface at the center of the hemisphere as shown[2] in Figure 6.5. All particles leaving the sample surface in 2π sr can be captured by such a detector. The MCP can be rotated to any position on the surface of such a hemisphere. Such a position-sensitive MCP is in the shape of a planar rectangle, resulting in distortion of the angular relations expected on a hemisphere. The coordinates (x_d, y_d) of each pixel on the MCP can be transformed to the corresponding angular coordinates on a hemisphere by using the direction cosines describing the spatial orientation of the vector that connects the center of the hemisphere to the detector pixel. Details are shown in Figure 6.5, where **D** is the target to detector distance, **H** and **W** are the detector height and width, and β and δ are the particle exit and azimuthal angles, respectively. When projecting the vector (x, y, z) onto the base plane (x, y) of the hemisphere, it is desirable to keep the same projected azimuth and to have the distance from the center to projection point be a linear measure of the vector exit angle β. Such a transformation provides a portion of an azimuthal equidistant projection[9] called a SARIgram on which the angular relations between crystallographic planes and poles can be related to the SARIS image as shown in Figure 6.5. By rotating the MCP to different positions on the hemisphere, it is possible to obtain a complete projection. Having a three-dimensional spatial intensity distribution mapped onto the two-dimensional plot, it is possible to directly associate the features observed on the map with crystallographic directions characteristic for a given type of crystal bulk unit cell. The advantage of SARIgrams is as follows. The image areas of low intensity, so-called blocking cones, appear in the plot as slightly distorted circles reflecting the fact that a normal cross section of such a cone is indeed a circle. The diameter of such a circle measured along the radial coordinate is directly related to the angular size of the blocking cone. In mapping individual SARIS images onto the hemispherical surface, the intensity measured by a pixel of the MCP is transformed to an equivalent solid angle increment on the hemispherical surface and then projected onto the SARIgram.

In order to determine the major crystallographic directions along which the characteristic features of the intensity distributions of the SARIgram are observed, consider the general case of a crystal bulk unit cell specified by three non-coplanar vectors $\vec{v}_1, \vec{v}_2, \vec{v}_3$ defined by their lengths v_1, v_2, v_3 and pairwise angles $\gamma_{12}, \gamma_{23}, \gamma_{13}$. The spatial orientation of the unit cell is determined by the Miller indices of a facet of interest $(k, l, m,)$ where the $v = [k, l, m]$ vector is parallel to the Z-axis (surface normal), and the angle between v_1 and the (x, z)-plane (i.e., the surface azimuthal angle) is δ. Next we find the components of all three unit cell vectors and then decompose any arbitrary direction vector ω into $\{v_1, v_2, v_3\}$ space. From the definition of the bulk unit cell, it can be written as

$$(v_i v_j) = v_i v_j \cos \gamma_{ij}; \; i, j = 1, 2, 3. \tag{1}$$

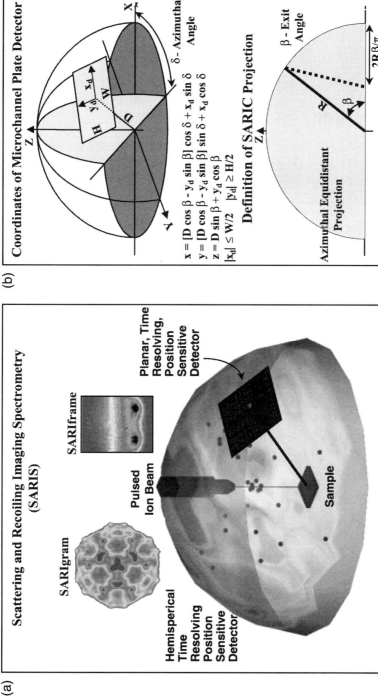

(a)

Scattering and Recoiling Imaging Spectrometry (SARIS)

SARIgram

SARIframe

Hemispherical Time Resolving Position Sensitive Detector

Pulsed Ion Beam

Planar, Time Resolving, Position Sensitive Detector

Sample

(b)

Coordinates of Microchannel Plate Detector

X

δ - Azimuthal Angle

Z

Y

$x = [D \cos \beta - y_d \sin \beta] \cos \delta + x_d \sin \delta$
$y = [D \cos \beta - y_d \sin \beta] \sin \delta + x_d \cos \delta$
$z = D \sin \beta + y_d \cos \beta$
$|x_d| \leq W/2 \quad |y_d| \geq H/2$

Definition of SARIC Projection

Z

β - Exit Angle

$2R\beta/\pi$

R

β

Azimuthal Equidistant Projection

FIGURE 6.5. (a) Hypothetical hemispherical detector, simulated SARIgram, and experimental SARIframe for 4-keV Ne$^+$ scattering from Pt(111). (b) Coordinates of the microchannel plate (MCP) detector and definition of the SARIC azimuthal equidistant projection (From Rabalais, 2001, and Bykov et al., 2000, with permission).

The condition $\upsilon||Z$ is equivalent to $(\upsilon \varepsilon_z) = |\upsilon|$, where $\{\varepsilon_x,\ \varepsilon_y,\ \varepsilon_z\}$ is an orthonormal basis of the Cartesian system of coordinates. Knowing all nine components of $\upsilon_1,\ \upsilon_2,\ \upsilon_3$, any vector ω given by the (x,y)-coordinates of its SARI-projection can be decomposed into its $\omega_1, \omega_2, \omega_3$ components. The integer parts of the $\omega_1/\upsilon_1,\ \omega_2/\upsilon_2,$ ω_3/υ_3 ratios are the Miller indices of an arbitrary direction ω in $\{\upsilon_1,\ \upsilon_2,\ \upsilon_3\}$ space.

6.4. SIMULATED AND EXPERIMENTAL SARIgrams

Examples of azimuthal equidistant projections are shown[1,2,10] for 4-keV He$^+$ scattering from a Ni(110) surface in Figure 6.6 and from Pt(111) and Si(100) surfaces in Figure 6.7 along with the standard stereographic projections for these surfaces. The *simulated* SARIC images of an azimuthal equidistant projection of He scattering from these surfaces was obtained by collecting the scattered He atoms on a virtual hemispherical detector centered over the sample. The contour plot is an azimuthal equidistant projection of the spatial intensity distribution onto the plane of the hemisphere base with angular coordinate equal to the crystal azimuthal angle δ and radial coordinate equal to the polar exit angle β. The *experimental* SARIS scattering intensity distributions collected for different MCP orientations were mapped to the imaginary hemisphere and then also projected onto the base, corresponding to one quadrant of the azimuthal equidistant projection. The features observed on the maps represent experimental azimuthal equidistant projections that can be directly associated with crystallographic directions characteristic for *fcc* (110), *fcc* (111), and diamond (100) surfaces. The image areas of low intensity correspond to blocking cones. The sizes of these areas (i.e., the diameters of such circles measured along the radial coordinate) are directly related to the angular sizes of the cones. The agreement of the experimental SARIS images with the standard *fcc* (110), *fcc* (111), and diamond (100) stereographic projections confirms the identity of the termination planes of these surfaces.

6.5. TERMINATION LAYER OF CdS(0001)

Examples of SARIS images are shown[1,2,11] in Figure 6.8 for 4-keV Kr$^+$ incident along the $\langle \bar{1}010 \rangle$ azimuth of cadium sulfide CdS(0001). Time frames corresponding to specific scattering and recoiling events can be identified from the binary collision approximation. For each frame, the ordinate represents the particle exit angles (β) from the surface, and the abscissa represents the crystal azimuthal angles (δ) (i.e., an image in β,δ-space). Each frame of Figure 6.8 represents the velocity-resolved spatial distribution of particles arriving at the MCP in consecutive 16.7-ns windows. These consecutive frames resolve the different spatial distributions for scattered krypton (Kr) and recoiled cadmium (Cd), sulfur (S), H, and O atoms.

    The images are symmetrical about the $\langle \bar{1}010 \rangle$ azimuth, as is the crystal structure. The spatial distribution of the scattered Kr signal appears as two intense spots at the

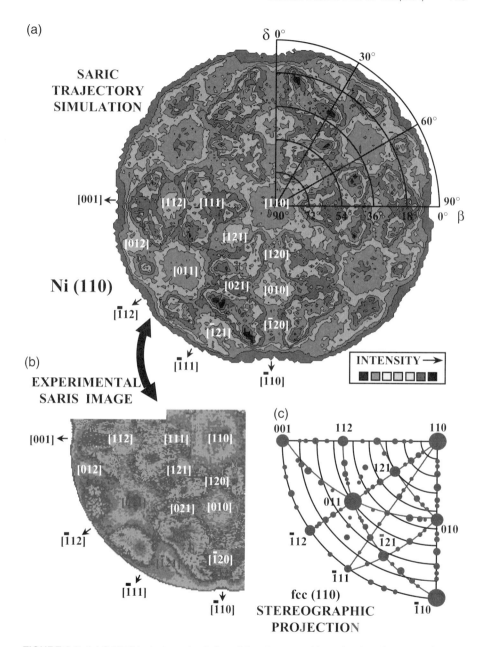

FIGURE 6.6. (a) SARIC trajectory simulation of the stereographic projection of 4-keV He$^+$ scattering from a Ni(110) surface, (b) experimental SARIS image corresponding to one quadrant of the simulation, and (c) a standard *fcc* (110) stereographic projection (From Rabalais, 2001, Bykov *et al.*, 2000, and Bolotin *et al.*, 2000, with permission). (See Color Plate)

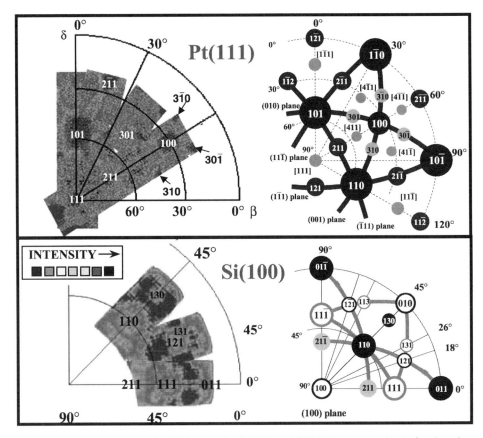

FIGURE 6.7. Experimental SARIS images for Pt(111) and Si(100) in one quadrant of an imaginary hemispherical detector along with the standard stereographic projections for these surfaces. The experimental scattering intensity distributions were collected for different microchannel plate orientations, then mapped to the imaginary hemisphere, and then projected onto the base of the hemisphere. (From Rabalais, 2001, with permission). (See Color Plate)

edges of the 2.05-μs frame at a low exit angle of $\beta \sim 9°$ and azimuthal positions of $\delta \sim \pm 8°$; this frame corresponds to single scattering (SS) of Kr from Cd. The 2.18 and 2.31-μs frames correspond to slower particles and exhibit additional features at higher exit angles. These images exhibit a horizontal streak of intensity in the $\beta \sim 17$–$26°$ range, which is considerably higher than that of SS Kr at 2.05 μs; this streak corresponds to multiply scattered (MS) Kr. The vertical streak of intensity observed in both of these frames that is centered along the symmetrical azimuth at high β is due to recoiling of Cd atoms. This focusing arises from recoiling of Cd atoms from the second-bilayer and focusing along the outgoing trajectory by first-bilayer Cd and S atoms. Note that this vertical streak is $\sim 2°$ off the symmetrical position due to a slight misalignment of the azimuthal angle.

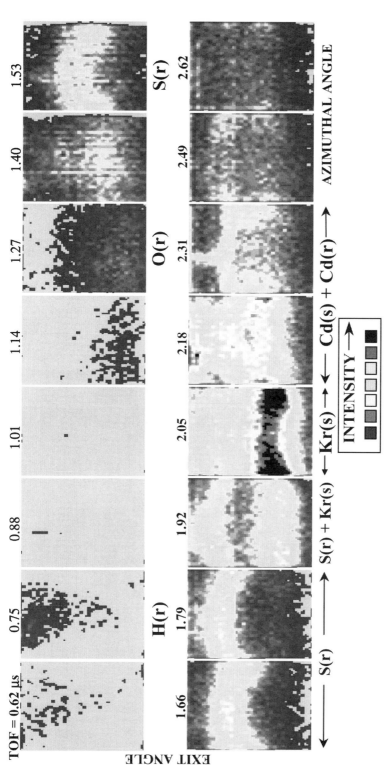

FIGURE 6.8. Selected consecutive 16.7-ns SARIS frames for 4-keV Kr⁺ scattering from CdS(0001) with the beam aligned along the ⟨1̄010⟩ azimuth. The time-of-flight (TOF) times corresponding to each consecutive image are shown above the frames. Labels corresponding to the dominant scattering, Kr(s), and recoiling, H(r), O(r), and S(r), events are listed below their respective frames (From Rabalais, 2001, Bykov *et al.*, 2000, and Kim *et al.*, 1998, with permission). (See Color Plate)

Structure analyses can be obtained from these images by using SARIC simulations as a function of the surface structure parameters. Examples of SARIC images for both the Cd- and S-terminated surfaces of CdS using bulk interatomic spacings and bond angles are shown[1,2,11] in Figure 6.9. Both scattering from or recoiling of first-layer atoms occurs at lower exit angles β than similar processes from second-layer atoms. Trajectories from second-layer atoms require higher β values to escape from the surface. Comparison of the images of Figures 6.8 and 6.9 shows that there is agreement only with the simulated images of the Cd-terminated surface and not with those of the S-terminated surface. For the simulated images of the Cd-terminated surface, the two low β peaks for scattered Kr(s)/Cd are at $\beta \sim 9°$ (Figure 6.9), in agreement with the 2.05-μs frame of Figure 6.8. The arch for recoiled S(r) atoms exhibits a maximum near $\beta \sim 30°$ (Figure 6.9), as observed in the 1.66- and 1.79-μs frames of Figure 6.8. Simulations for the S-terminated surface in Figure 6.9 yield the two scattered Kr(s)/Cd peaks at $\beta \sim 13°$ and the recoiled S(r) arch maximum near $\beta \sim 14°$, in disagreement with the respective frames of Figure 6.8. This shows that

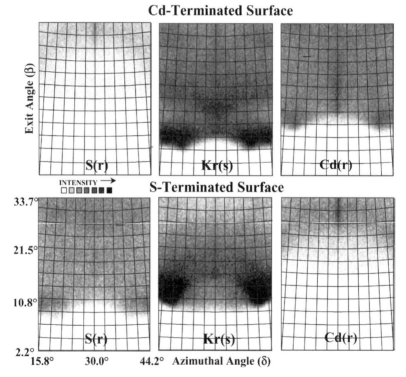

FIGURE 6.9. Simulated images from the SARIC program for both the cadmium (Cd)- and sulfur (S)-terminated CdS(0001) surfaces using the same conditions as the experimental images of Figure 6.7. Separate images are shown for krypton (Kr) scattered from Cd atoms [Kr(s)/Cd] and for recoiled Cd atoms [Cd(r)] and S atoms [S(r)]. These images are not time resolved; therefore, the Kr(s)/Cd image contains both quasi-single scattering and multiple scattering Kr atoms (From Rabalais, 2001, Bykov *et al.*, 2000, and Kim *et al.*, 1998, with permission).

SARIS is sensitive to and is capable of identifying the outermost termination layer of a surface.

6.6. CHEMISORPTION SITE OF CHLORINE ON Ni(110)

SARIS images of 4-keV Ar^+ scattering from clean Ni(110), and this surface after exposure to Cl_2 gas, are shown[2] in Figure 6.10(a). Each frame represents the velocity-resolved spatial distributions of Ar atoms scattering along the $\langle \bar{1}10 \rangle$ azimuth and arriving at the MCP in windows of 16.7-ns duration. The selected window at 1.4 μs corresponds to the flight time of quasi-SS Ar atoms from the Ni sample to the MCP. The images of Figure 6.10(a) are symmetrical about the $\langle \bar{1}10 \rangle$ azimuth, as is the crystal structure. This is in accord with low-energy electron diffraction (LEED) measurements that exhibit a (1 × 1) pattern for clean Ni and a (3 × 1) pattern for Cl/Ni. The differences in these images, i.e., [(Cl/Ni) − (clean Ni)], clearly reveal regions of altered intensities due to the presence of Cl atoms. Combining the difference images along $\langle \bar{1}10 \rangle$ and 15° off the azimuth, in order to span a broader azimuthal range, reveals intensification (focusing) of trajectories along $\langle \bar{1}10 \rangle$ and blocking for angles on either side of this azimuthal direction.

There are three chemically active symmetrical sites on the Ni(110) − (3 × 1) surface as shown[2] in Figure 6.10(b): SARIC simulations were performed with Cl at each of these sites and at different heights above the surface. A two-dimensional reliability,[3] or R factor, based on the differences between the experimental and simulated images, was calculated as a function of Cl atom height above the surface. A minimum in R was obtained for the short-bridge (SB) position with the Cl atoms 1.9 ± 0.1 Å above the Ni rows, corresponding to a Cl–Ni bond length of 2.3 ± 0.1 Å; the sensitivity of R to changes in the Cl height is ∼5% for a 0.1 Å change. The bond length for Cl bridge-bonded to two Ni atoms is 2.45 Å, confirming dissociative chemisorption and bonding of Cl atoms to Ni atoms. Previous scanning tunneling microscopy (STM) work[12] on this system is in agreement with this assignment, showing that Cl_2 chemisorbs dissociatively on Ni(110) and forms atom pairs oriented along the [001] direction, with the Cl–Cl spacings comparable to the bulk Ni lattice constant.

The (3 × 1) LEED pattern can be obtained from two different types of structures: (1) single rows of Cl atoms or (2) single missing rows of Cl atoms along $\langle 001 \rangle$ that repeat every third-SB position. Recoil of Ni atoms along the $\langle 001 \rangle$ azimuth is highly sensitive to these two different coverages due to deflection of Ni atoms by Cl atoms along this direction. Ni atoms recoiled by Kr projectiles are resolved in the 2.7-μs window shown[2] in Figure 6.10(c). Excellent agreement is obtained between experimental and simulated images for single rows of Cl atoms at the SB site; the other sites provide poor agreement, in accord with the scattered Ar results of Figures 6.10(a) and 6.10(b). The perturbations by the Cl atoms on the recoiled Kr trajectories can be deciphered from simulated trajectories. Along $\langle 001 \rangle$, the Cl atoms are not situated along adjacent sides of a single deep trough where they can enhance focusing of trajectories. Instead, they deflect the Ni recoil trajectories, resulting in a decrease in intensity along $\langle 001 \rangle$ as shown in Figure 6.10(c).

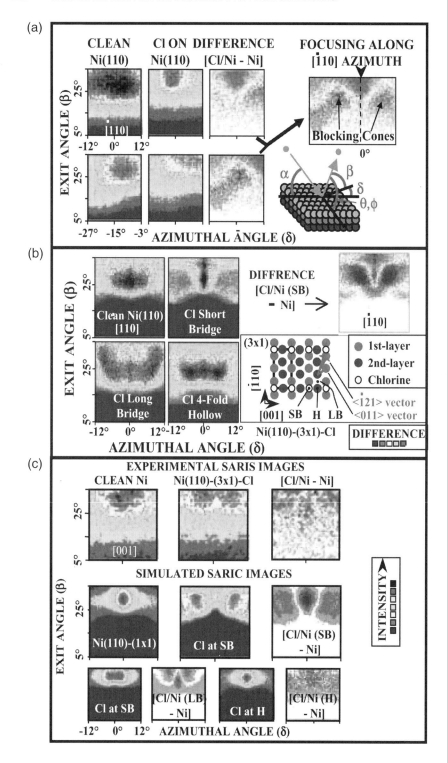

6.7. QUANTITATIVE ANALYSIS OF THE Pt(111) SURFACE

A series of time-resolved He scattering images taken as a function of azimuthal angle δ are shown in Figure 6.11.[1,2,13] The crystal was rotated about its surface normal by $3°$ for each image. Each image is taken from a 16.7-ns frame corresponding to the SS TOF. The observed images are rich in features that change in position and intensity as a function of azimuthal angle. The regions of low intensity correspond to the positions of the centers of the blocking cones; these regions have circular or oval shapes with distortions caused by other overlapping blocking cones. The regions of high intensity correspond to the positions of intersection or near overlap of blocking cones; atom trajectories are highly focused along the edges of the cones.

The images at $\delta = 0°$ and $60°$ along the $\langle \bar{2}11 \rangle$ and $\langle \bar{1}\bar{1}2 \rangle$ azimuths, respectively, are symmetrical about a vertical line through the center of the frame, as is the crystal structure along these azimuths as shown[1,2,13] in Figure 6.12(a). The shifts in the positions and sizes of the blocking cones can be monitored as δ is rotated away from the $0°$ or $60°$ directions. There are large variations in the intensities as a function of δ, with the highest intensities being observed along the directions $\delta = 22 \sim 32°$ and $56 \sim 60°$. These features result from focusing of ions onto second-layer atoms by the shadow cones of first layer atoms. The second-layer atoms are in sites that are asymmetrical with respect to the first-layer, resulting in nonplanar scattering trajectories. Intense features in asymmetrical positions are observed at higher exit angles. These features correspond to semichanneling[7] in asymmetrical channels. Semichannels are so-called valleys in surfaces through which scattered ions are guided. Along the $\langle \bar{1}01 \rangle$ direction, the first-layer atoms form the walls, and the second-layer atoms form the floor of the semichannel. However, the second-layer rows are not centered in the bottom of the channel, resulting in an asymmetrical channel. As a result, the scattered atom trajectories are bent and focused along directions determined by the asymmetry of the channel.

The frames along the $0°$ $\langle \bar{2}11 \rangle$ and $60°$ $\langle \bar{1}\bar{1}2 \rangle$ azimuths in Figure 6.12 were selected to compare with those of blocking cone analyses and ion trajectory simulations.[2] The arrangement of the first-layer atoms is identical along both of these azimuths; however, the second- and third-layer atoms have a different arrangement with respect to the first-layer atoms as shown in Figure 6.12(a). He atoms scattered from first- and third-layer atoms experience a different arrangement of blocking cones on their exit from the surface. The positions of the blocking cones were calculated from the projections of the interatomic vectors onto the MCP detector (Figure 6.12a). The critical blocking

FIGURE 6.10. (a) SARIS images of 4-keV Ar^+ scattering from Ni(110)-(1 × 1) and Ni(110)-(3 × 1)-Cl along $\langle \bar{1}10 \rangle$ (top) and $15°$ off this azimuth (bottom). The difference image reveals focusing of scattered Ar trajectories in the centers of the $\langle \bar{1}10 \rangle$ troughs. (b) SARIC simulations of 4 keV Ar^+ scattering from Ni(110)-(1 × 1) and Ni(110)-(3 × 1)-Cl with Cl in the short bridge (SB), long bridge (LB), and four-fold hollow (H) sites. The SB site provides the best agreement with the experimental images of (a). (c) SARIS images and SARIC simulations of Ni atoms recoiled by 4-keV Kr^+ along the $\langle 001 \rangle$ azimuth from Ni(110)-(1 × 1) and Ni(110)-(3 × 1)-Cl (From Bykov *et al.*, 2000, with permission). (See Color Plate)

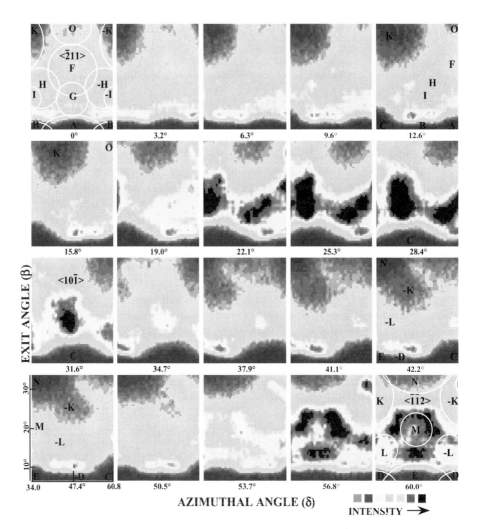

FIGURE 6.11. A series of 20 time-resolved SARIS frames for 4-keV He[+] scattering from Pt(111)-(1 × 1) taken every 3° of rotation about the azimuthal angle δ, starting with the $\langle \bar{2}11 \rangle$ azimuth at $\delta = 0°$ and ending with the $\langle \bar{1}\bar{1}2 \rangle$ azimuth at $\delta = 60°$. Each frame represents a 16.7-ns window centered at the time-of-flight (TOF) corresponding to quasi-single scattering as predicted by the binary collision approximation. The white circles on the $\delta = 0°$ and 60° frames represent the positions and relative sizes of calculated blocking cones (From Rabalais, 2001, Bykov *et al.*, 2000, and Kim *et al.*, 1998, with permission). (See Color Plate)

angles or sizes of the cones were calculated from SARIC. The results are shown in Figure 6.13.[2,3] The blocking of scattering trajectories from nth-layer atoms by their neighboring nth-layer atoms is observed at low β since these atoms are all in the same plane. This first-layer-atom-first-layer-atom blocking contributes most of the intensity at low β. The arcs corresponding to the edges of the blocking cones (Figure 6.13)

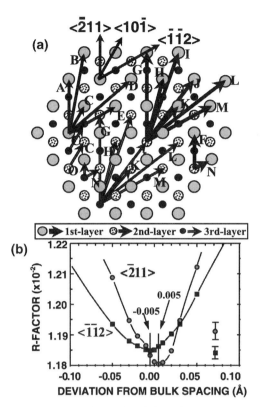

FIGURE 6.12. (a) Schematic drawing of the Pt{111}-(1 × 1) surface. Arrows indicate the nearest neighbor first–first-, second–first-, and third–first-layer interatomic vectors. (b) Two-dimensional *R*-factors as a function of the deviation (*d*) of the first–second-interlayer spacing from the bulk value. The experimental and simulated images along the $\langle\bar{2}11\rangle$ and $\langle\bar{1}\bar{1}2\rangle$ azimuths were used in the comparison (From Bykov *et al.,* 2000 and Kim *et al.,* 1997, with permission).

resulting from the vectors **A, B, D,** and **E** in Figure 6.12 occur at $\beta \sim 10°$. The features at higher β correspond to scattering trajectories from second- and third-layer atoms that are blocked and focused by first-layer atoms. The cones resulting from the vectors **F, G,** and **O** along the azimuthal direction $\langle\bar{2}11\rangle$ and **M** and **N** along $\langle\bar{1}\bar{1}2\rangle$ are due to scattering trajectories from second- and third-layer atoms that are blocked by first-layer atoms along these symmetrical directions. These are centered along the azimuths and are directed to higher β values for shorter interatomic spacings. Blocking cones due to the vectors **H, I, K,** and **L** result from second- and third-layer scattering and are observed at δ values off of the 0° and 60° directions due to nonplanar scattering trajectories.

Quantitative analyses can be achieved by using the SARIC simulation and minimization of the *R*-factor[3] between the experimental and simulated images as a function of the structural parameters. SARIC was used to generate simulated images of 4-keV He⁺ scattering from bulk-terminated Pt{111} as a function of the

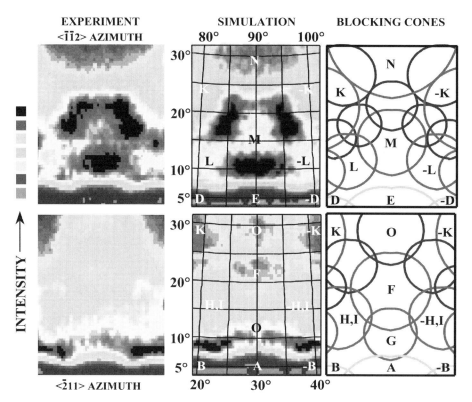

FIGURE 6.13. Experimental SARIS images (left), simulated SARIC images (center), and blocking cone analysis (right) for He$^+$ scattering along the $\langle \bar{2}11 \rangle$ and $\langle \bar{1}\bar{1}2 \rangle$ azimuths of Pt(111). For the calculated blocking cones, first–first, first–second, and first–third layer interactions are identified by green, red, and blue lines, respectively. The letters correspond to the interatomic vectors identified in Figure 6.11(a). (From Bykov *et al.*, 2000, Kim *et al.*, 1997 and Kim *et al.*, 1997 with permission). (See Color Plate)

first–second-interlayer spacing d. Isotropic thermal vibrations with an amplitude of 0.1 Å were included in the model. A two-dimensional reliability, or R factor, based on the differences between the experimental and simulated patterns, was calculated as a function of the deviation d of the first–second-interlayer spacing from the bulk value. The plots shown in Figure 6.12(b) exhibit minima at $\Delta d_{\min} = -0.005$ and $+0.005$ Å for the $\langle \bar{1}\bar{1}2 \rangle$ and $\langle \bar{2}11 \rangle$ azimuths, respectively. The optimized simulated images corresponding to $d_{\min}$ are shown in Figure 6.13; there is good agreement between these simulated and experimental images. The R-factors are sensitive to changes in the interlayer spacing at the level of 0.01 Å. Based on these data, it is concluded that the Pt{111} surface is bulk-terminated with the first–second-layer spacing within ± 0.01 Å, or 0.4%, of the 2.265 Å bulk spacing. This sensitivity is less than the uncertainty due to the thermal vibrations because SARIS samples the average positions of lattice atoms.

6.8. DIRECT DETECTION OF HYDROGEN ATOMS ON Pt(111)

Although the presence of hydrogen atoms on a surface is generally considered to present only a minor perturbation on the scattering trajectories of heavy atoms, recent low-keV channeling measurements[2,8,14,15] using TOF-SARS have shown that the hydrogen atoms can perturb the trajectories of heavier projectiles such as Ne to a detectable level. Such low-energy ion channeling measurements are capable of probing the positions of light elements on heavy substrates with analysis by simple geometrical constructs. Examples of SARIgrams for 5-keV Ne$^+$ scattered from the clean Pt(111) and Pt(111)-(1 × 1)-H surfaces are shown[15] in Figure 6.14.

6.8.1. Clean Pt(111)

For the clean Pt surface, sharp intense features with an apparent 60° periodicity that correspond to focusing of the scattered projectiles by atomic lenses of the crystal surface are observed. These intense features are separated from one another by low-intensity regions that originate from blocking cones situated along the exit trajectories of the scattered Ne atoms. Specifically, for this angular configuration, the low-intensity regions are due to superposition of blocking cones centered at $\beta = 0°$ formed by scattered trajectories from first-layer atoms that are blocked by neighboring first-layer atoms as well as cones centered at $\beta > 0°$ resulting from scattering from subsurface atoms and blocking by first-layer atoms. Each of the crystallographic directions

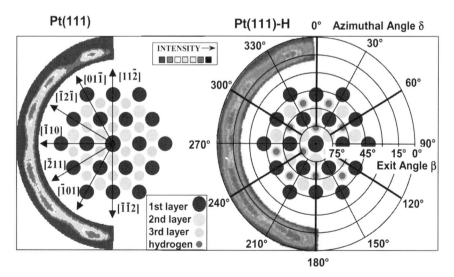

FIGURE 6.14. SARIgrams for 5-keV Ne$^+$ scattering from Pt(111) and Pt(111)-(1 × 1)-H surfaces. The SARIgrams were constructed by using an azimuthal equidistant projection of 32 individual SARIS images. Incident angle $\alpha = 19°$; scattering angle to detector normal $\theta = 24°$; 17 cm flight path to detector center (From Bykov *et al.,* 2000, and Lui *et al.,* 1999, with permission).

indicated on the SARIgram in conjunction with the [111] crystal direction defines a plane of symmetry that bisects the corresponding blocking feature (i.e., the feature is symmetrical about these azimuths). It is observed that the recorded resultant blocking is most pronounced at the [01$\bar{1}$], [$\bar{1}$10], and [$\bar{1}$01] azimuths and appears as only small indentations at the [$\bar{1}$2$\bar{1}$] and [$\bar{2}$11] azimuths. The angular positions of these three dominant blocking features suggest a sixfold symmetry because they result from only first-layer scattering. Scattering with a higher incidence angle allows sampling of subsurface layers and reveals the threefold symmetry of the (111) surface.

6.8.2. Pt(111)-(1 × 1)-H

The SARIgram of the hydrogen-covered Pt(111) surface exhibits similar basic features as those of the clean surface, although severely broadened. It is clear that the forementioned blocking features are attenuated to different extents. After hydrogen adsorption, the intensity gradients, i.e., $[\partial I(\delta,\beta)/\partial\delta]_{7°<\beta<15°}$, at the azimuthal directions 240° and 300° almost vanish. At the azimuthal directions 210°, 270°, and 330°, these gradients remain large and recognizable, although the features at 270° do not remain as clear as those at 210° and 330°. The reason for the different features at 270° is still not clear. This could be caused by H atoms occupying both *fcc* and *hcp* sites at this coverage or by small displacements of the Pt atoms as a result of chemisorption. These observations are in accord with the fact that the H atoms on the surface are capable of deflecting the Ne trajectories by a maximum of ~3°, which amounts to a deflection of ~9 mm on the MCP under the experimental configuration. Accordingly, the two small indentations at the [$\bar{1}$2$\bar{1}$] and [$\bar{2}$11] azimuths vanish completely.

The scattering of Ne atoms by H atoms can be better understood by considering the actual trajectories of the colliding atoms. Figure 6.15 shows[2,15] a flux of

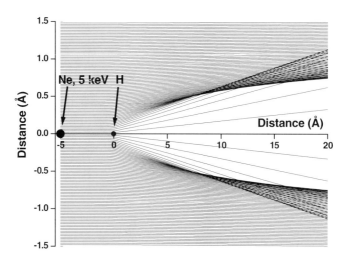

FIGURE 6.15. A flux of parallel 5-keV Ne atom trajectories impinging on a H atom with impact parameters *b* that are incremented by 0.02 Å (From Bykov *et al.*, 2000, and Lui *et al.*, 1999).

parallel 5-keV Ne atom trajectories impinging on a H atom. For the H/Ne case where $M_t/M_p < 1$, some small b collisions result in penetration of the cone by the projectile atom M_p, although a distinct focusing of large b trajectories exists, resulting in a semi-transparent cone. Such semitransparent cones have also been observed for hydrogen chemisorbed on a tungsten surface.[15,16] For $b = 0$, the projectile atoms M_p follow the target atoms M_t through the center of the cone.

The importance of the SARIgram of the hydrogen-exposed surface is threefold. (1) It demonstrates that H atoms are capable of deflecting Ne atom trajectories and that this deflection can be directly observed. (2) The delocalization of intensities on the SARIgram after hydrogen adsorption infers an increase in the number of scatters per unit cell. (3) The sensitivity to local atomic arrangements provides a direct probe of the chemically active sites for hydrogen adsorption.

6.9. INTERPRETATION OF SARIgrams

The changes in the shapes and intensities of the blocking features in the SARIgrams of Figure 6.14 as a result of exposure to hydrogen are a manifestation of (1) the different geometrical intersections of the respective blocking cones on the detector surface and (2) the proximity of the H atoms to the Ne trajectories. Hydrogen atoms adsorbed at the *fcc* sites are contained in the planes defined by the [111] and the [$\bar{1}2\bar{1}$] or [$\bar{2}11$] crystal directions and situated 0.9 Å above the surface plane in the *fcc* sites as found previously.[17] Therefore, they can directly perturb the incoming and outgoing Ne trajectories along these azimuths. This defocusing by the H atoms results in the large changes observed in Figure 6.14 along 240° and 300° in the SARIgram of the hydrogen-exposed surface. The features along 210°, 270°, and 330° remain salient after exposure to hydrogen because most of the Ne trajectories scattering along these directions do not make direct collisions with H atoms.

The large b collisions result in significant deflections of the incoming Ne beam such that the scattered Ne trajectories exhibit large deviations from the focused beams of the clean surface. Since these large b collisions occur with the highest probability, these types of trajectories are the dominant contributors to the broadening observed as a result of hydrogen chemisorption. Similar features can be observed along the [$\bar{2}11$] azimuth. These specific data, however, do not distinguish between chemisorption at the *fcc* or *hcp* sites. Low-energy ion channeling experiments provide this distinction.[7,14,18]

6.10. QUANTITATIVE ANALYSIS OF SARIS IMAGES

6.10.1. SARIS Frames

Surface crystallography is deducted from experimental SARIS images by comparison with simulated SARIC images. Example experimental SARIS frames and simulated SARIC frames for 20-keV He$^+$ scattering from a clean Pt(111) surface are shown[10] in Figures 6.16 and 6.17, respectively. Each frame represents the velocity-resolved

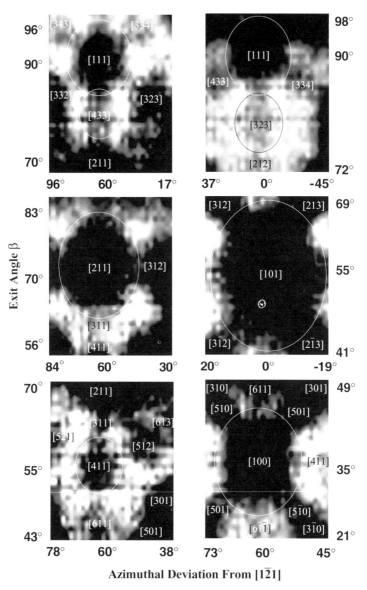

Azimuthal Deviation From [1$\bar{2}$1]

FIGURE 6.16. Experimental SARIS frames in six different positions of β,δ-space for scattering of 20-keV He$^+$ from Pt(111). The δ angles are defined with respect to the [1$\bar{2}$1] azimuth. Planar backgrounds have been subtracted from the images, and different intensity scales are used for each image in order to best emphasize the blocking phenomena (From Bolotin *et al.,* 2000, with permission).

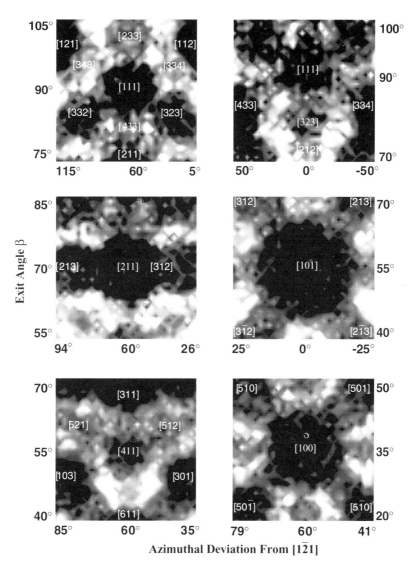

FIGURE 6.17. Simulated SARIC frames in six different positions of β,δ-space for scattering of 20-keV He$^+$. The δ angles are defined with respect to the [1$\bar{2}$1] azimuth (From Bolotin *et al.*, 2000, with permission).

spatial distributions of He atoms scattering along different angles and arriving at the MCP in windows of 16.7-ns duration. The selected window at $\sim$0.15 μs corresponds to the flight time of quasi-SS He atoms from Pt. The ordinate of each frame is the particle exit angle (β) from the surface and the abscissa is the crystal azimuthal angle (δ). The δ angles are defined with respect to the [1$\bar{2}$1] azimuth. Since the MCP is planar rather than spherical, the lines of constant azimuthal angle and constant exit angle are not linear and the values of β and δ in the figures serve only as guides. When the incident ions interact with individual atoms, blocking cones or "rings" are observable in the figures. When they interact simultaneously with a group of atoms that are arranged as an atomic lens, intense continuous streaks can be produced. Such streaks can be observed in Figures 6.6 and 6.7.

The simulated frames[10] of Figure 6.17 are in good agreement with the experimental frames of Figure 6.16. They differ in two areas. First, some of the cones that originate from deeper layers, e.g. the $\langle 433 \rangle$ cone in Figure 6.16 results from sixth-layer scattering, are barely detectable in the simulation of Figure 6.17 due to poor statistics from the small number of scattered trajectories that reach the surface. Second, the experimental cones have ellipsoidal shapes whereas the simulated cones have circular shapes, although in both cases their centers are at the same β,δ position. The ellipsoidal shapes of the experimental cones are a result of a small misalignment of the MCP and sample surface. This problem is readily corrected as described elsewhere.[10]

The images of Figures 6.16 and 6.17 can be separated into groups based on the atomic layer from which the primary scattering event takes place, as shown[10] in Figure 6.18. Second-layer scattering produces the largest blocking cones along the $\langle 110 \rangle$ and $\langle 001 \rangle$ directions due to the short interatomic spacings. Intense focusing spots are observed at the edges of these cones; there are six spots for the $\langle 110 \rangle$ azimuth and eight spots for the $\langle 001 \rangle$ azimuth, corresponding to the number of nearest-neighbor projections (Figure 6.7) surrounding these azimuths. For example, the nearest

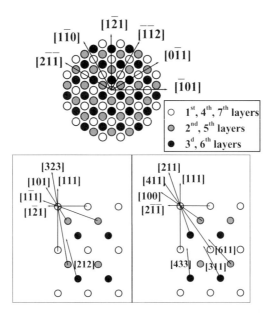

FIGURE 6.18. Plan view of the (111) surface and cross sections perpendicular to the (111) surface along the [$\bar{1}2\bar{1}$] and [$2\bar{1}\bar{1}$] directions showing the interatomic vectors responsible for the blocking cones of Figures 6.16 and 6.17. (From Bolotin *et al.*, 2000, with permission).

neighbor projections of the [110] azimuth are the [211], [310], [21$\bar{1}$], [12$\bar{1}$], [130] and [121] and those of the [100] azimuth are [3$\bar{1}$0], [4$\bar{1}\bar{1}$], [30$\bar{1}$], [41$\bar{1}$], [310], [411], [301], and [4$\bar{1}$1]. Third-layer scattering produces the next largest blocking cones along the $\langle 112 \rangle$ and $\langle 103 \rangle$ directions. The dark gray curved lines connecting the blocking cones of Figure 6.7 are due to overlapping second- and third-layer cones produced by lines of blocking surface atoms lying in different $\{100\}$, $\{110\}$, and $\{111\}$ planes. Fourth-layer scattering produces blocking cones along the $\langle 114 \rangle$ and $\langle 111 \rangle$ directions. Fifth- and sixth-layer scattering can be observed as minor cones along the [323], [611], [433], [311] and other directions.

6.10.2. Blocking Cones

Since blocking cones represent diverging trajectories, the sizes of the cones cannot be described by their radii.[10] The sizes can be defined as the angular aperture corresponding to the position of 70% of the maximum scattered intensity at the cone edges. The sizes of the cones are inversely proportional to the length of the interatomic vectors d. The largest cone is observed for the $\langle 110 \rangle$ direction corresponding to second- to first-layer vectors in close-packed rows.

The relative sizes of the experimental blocking cones from Figure 6.16 as a function of the interatomic distance d for different He$^+$ energies are shown in Figure 6.19(a). The relative sizes are plotted as $\tan \psi / \tan \psi_o$ which is equal[10] to $(b^2/a)/(b_0^2/a_0)$, where a and b are semimajor and semiminor axes, respectively, of the best fitting ellipses to the observed cones. These ellipses are shown on Figure 6.16. The normalization (b_0^2/a_0) corresponds to the size of the blocking cone in the [112] direction. This cone was chosen for normalization because it is completely contained inside the MCP area for all He$^+$ energy values used. The points of Figure 6.19(a) follow a d^{-k} dependence, where $k = 0.74 \pm 0.02$.

The relative sizes of the experimental blocking cones as a function of the He$^+$ energy E are shown in Figure 6.19(b). Results for the $\langle 110 \rangle$, $\langle 112 \rangle$, and $\langle 100 \rangle$ cones exhibit a E^{-l} dependence. For the specific azimuths, the values are $l_{\langle 110 \rangle} = l_{\langle 100 \rangle} = 0.26 \pm 0.04$ and $l_{\langle 112 \rangle} = 0.21 \pm 0.04$. The average value is $l = 0.24 \pm 0.03$. Only three or four different energies are available for each cone, therefore this result has a greater uncertainty than that of the d dependence.

6.10.3. Two-Atom Model for Blocking Cones

A simple SARIS-based two-atom model for blocking cone analysis is presented here. Note that this is different from the two-atom model used in Rutherford backscattering (RBS) and medium-energy ion scattering (MEIS)[18] where a point emission approximation is made for the backscattering atom. In the SARIS-based model, only two atoms are used in the target; however, a complete three-dimensional trajectory simulation is performed by using random arrival trajectories and, hence, a range of impact parameters. This simulation produces a blocking cone that is unperturbed by ion trajectories resulting from scattering or recoiling from neighboring atoms.

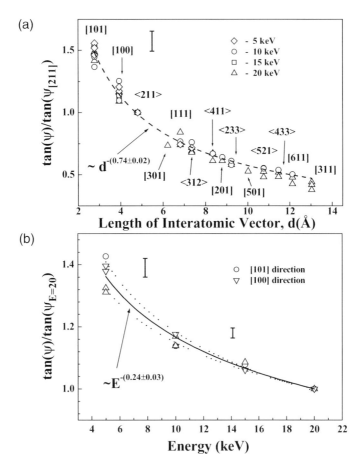

FIGURE 6.19. (a) (points) Relative sizes of experimental blocking cones from the experimental images of Figure 6.7 as a function of the interatomic distance d for different He$^+$ energies (dashed line). Best-fit line to the experimental points. (b) (points) Relative sizes of experimental blocking cones from the images of Figure 6.7 as a function of the He$^+$ kinetic energy E. The solid line is the best fit to the experimental points for all directions, and the dashed lines are the best fits to the [101] and [211] directions. Error bars corresponding to different parts of the measurements are shown on the plots (From Bolotin *et al.*, 2000, with permission).

Consider a system of two atoms with interatomic distance d in which the incident He$^+$ is scattering from atom 1. It is necessary to determine the critical blocking angle ψ_c when atom 2 begins to block the scattered He$^+$ trajectories, where ψ is the emission angle with respect to the interatomic axis. The value of ψ_c is determined by the size of the blocking cone. The critical blocking angle ψ_c is defined as the angle at which the scattered intensity curve I_s attains 70% of its maximum intensity. The value of 70% was chosen after comparison of the influence of different sample temperatures on the sizes of the blocking cones using SARIC. In SARIC, the temperature is described by

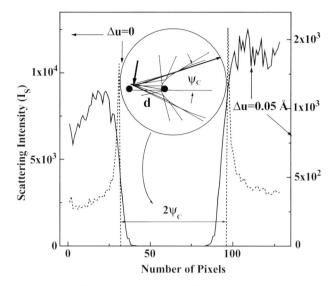

FIGURE 6.20. Results of a SARIC calculation using different root-mean-square (rms) vibrational amplitudes $\Delta u \cdot \Delta u = 0.0$ Å (corresponds to 0 K and the two fixed-atom model shown in the insert) and $\Delta u = 0.05$ Å (corresponds to the amplitude of a Pt vibration at room temperature). The edges of the blocking cones from both calculations intersect at $I_s \sim 70\%$ (From Bolotin *et al.*, 2000, with permission).

the amplitude of thermal vibrations, which is simulated by the distribution of average rms shifts (Δu) of atoms from their equilibrium positions. The results of a SARIC calculation with $\Delta u = 0.0$ Å (corresponding to 0 K and the two fixed-atom model) and $\Delta u = 0.05$ Å (corresponding to the amplitude of a Pt vibration at room temperature) are shown[10] in Figure 6.20. The edges of the blocking cones from both calculations intersect at $I_s \sim 70\%$.

The two-atom model for scattering can be tested by performing SARIC calculations for an isolated system of two atoms using three different He$^+$ energies as shown[10] in Figure 6.21. The results for each energy value in Figure 6.21(a) exhibit a $d^{-(0.72 \pm 0.01)}$ dependence on the interatomic spacing. For the two cones along the $\langle 110 \rangle$ and $\langle 100 \rangle$ azimuths, the energy dependence in Figure 6.21(b) can be described as $E^{-(0.26 \pm 0.01)}$. Both of these expressions are in excellent agreement with the experimental results of Figure 6.19.

6.11. ADVANTAGES OF SARIS

The major advantages of SARIS are as follows:

1. Both planar and nonplanar scattering and recoiling trajectories are detected simultaneously by means of a position-sensitive MCP detector, thereby providing more detailed data in a shorter time.

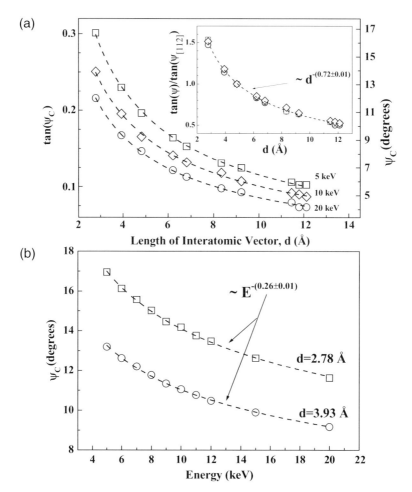

FIGURE 6.21. (a) (points) SARIC calculations of the relative sizes of the blocking cones as a function of interatomic distance from the two-atom model of scattering using three different He$^+$ energies. (lines) Best-fit lines to the calculated points. The inset shows the relative dependence, similar to Figure 6.19. (b) (points) SARIC calculations of the relative sizes of the blocking cones as a function of energy from the two-atom model of scattering. (lines) Best-fit lines to the calculated points (From Bolotin *et al.*, 2000, with permission).

2. The size of detector acceptance angle can be selected. While TOF-SARS uses a small-area detector with acceptance angle of $\sim 10^{-4}$ sr, the MCP detector in SARIS can accept a large solid angle, corresponding to a large range of exit (β) and azimuthal (δ) angular space.

3. The data acquisition time is reduced from >2 hr for a TOF-SARS structural analysis data set to <10 min for a SARIS structural analysis data set.

4. The two-dimensional scattering and recoiling data, coupled with high sensitivity to surface structure, enables measurement of interatomic spacings to an accuracy of < 0.01 Å by R-factor analysis and classical ion trajectory simulations.

5. Surface kinetic and dynamic processes can be monitored on a time scale of seconds.

6. Atomic scale microscopy and spatial averaging are combined due to the sensitivity to short-range order (i.e., individual bond lengths [<1 nm], and the macroscopic size of the primary beam).

8. Analysis of the spatial distributions of scattered and recoiled ion fractions are possible, which together with trajectory simulations, can provide site-specific ion–surface electron tunneling probabilities.

REFERENCES

1. J. W. Rabalais, *Analytical Chem.* 73, **206A** (2001).

2. V. Bykov, L. Houssiau, and J. W. Rabalais, *J. Phys. Chem.* **104,** 6340 (2000).

3. C. Kim, C. Hoefner, and J. W. Rabalais, *Surface Sci.* **388,** L1085 (1997).

4. L. Houssiau, J. Wolfgang, P. Nordlander, and J. W. Rabalais, *Phys. Rev. Lett.* **81,** 5153 (1998).

5. R. Poole, *Science* **246,** 995 (1989).

6. H. Goldstein, "Classical Mechanics," Addison Wesley Pub. Co., 2001, Reading, MA.

7. C. Hoefner, and J. W. Rabalais, *Phys. Rev.* **B 58,** 9990 (1998); *ibid, Surface Sci.* **400,** 189 (1998); C. Hoefner, V. Bykov, and J. W. Rabalais, *Surface Sci.* **393,** 184 (1997).

8. K. L. Lui, Y. Kim, W. M. Lau, and J. W. Rabalais, *J. Appl. Phys.* **86,** 5256 (1999).

9. J. J. Rousseau, "Basic Crystallography," Wiley Chichester, 1998; A. H. Robinson, R. D. Sale, "Elements of Cartography," (3rd-ed.), Wiley, NY, 1969; E. J. Raisz, "Principles of Cartography," McGraw-Hill, New York, 1962.

10. I. L. Bolotin, L. Houssiau, and J. W. Rabalais, *J. Chem. Phys.* **112,** 7181 (2000).

11. C. Kim, J. Ahn, V. Bykov, and J. W. Rabalais, Intern. *J. Mass Spectro. Ion Proc.* **174,** 305 (1998); J. Ahn and J .W. Rabalais, *J. Phys. Chem.* **102,** 223 (1998).

12. T. W. Fishlock, J. B. Pethica, F. H. Jones, R. G. Egdell, J. S. Foord, *Surface Sci.* **377,** 629 (1997).

13. C. Kim, and J. W. Rabalais, *Surface Sci.* **385,** L938 (1997); C. Kim, C. Hoefner, V. Bykov, J. W. Rabalais, *Nucl. Instrum. Meth. Phys. Res.* **B 125,** 315 (1997).

14. K. M. Lui, Y. Kim, W. M. Lau, and J. W. Rabalais, *Appl. Phys. Lett.* **75,** 587 (1999).

15. K. M. Lui, I. Bolotin, A. Kutana, V. Bykov, W. M. Lau, and J. W. Rabalais, *J. Chem. Phys.* **111,** 11095 (1999).

16. O. Grizzi, M. Shi, H. Bu, J. W. Rabalais, and P. Hochman, *Phys. Rev.* **B 40,** 10127 (1989); M. Shi, O. Grizzi, H. Bu, J. W. Rabalais, and P. Nordlander, *Phys. Rev.* **B 40,** 10163 (1989).

17. K. Umezawa, T. Ito, M. Asada, S. Nakanishi, P. Ding, W. A. Lanford, B. Hjorvarsson, *Surface Sci.* **387,** 320 (1997); P. J. Feibelman, D. R. Hamman, *Surf. Sci.* **182,** 411 (1987); B. J. J. Koeleman, S. T. de Zwart, A. L. Boers, B. Poelsema, L. K. Verhey, *Phys. Rev. Lett.* **56,** 1152 (1986).

18. J. R. Tesmer and M. Nastasi (Eds.), "Handbook of Modern Ion Beam Materials Analysis," J. C. Barbour, C. J. Maggiore, and J. W. Mayer (Contrib. Eds.), *Materials Res. Society,* Pittsburgh, 1995.

7

APPLICATIONS OF TOF-SARS AND SARIS TO SURFACE STRUCTURE ANALYSIS

Low-energy ion scattering spectrometry has developed into a major surface structure analysis technique due to its extreme surface sensitivity and its ability to obtain so-called real-space structural details. Azimuthal angle scans using grazing incidence angle bombardment of a single crystal surface and observation of the scattered and/or recoiled particles under specular conditions can probe the principal azimuths of a surface and hence its periodicity. Incident angle scans using either a backscattering or forward-scattering condition can probe subsurface layers and interatomic spacings. This chapter provides selected examples of surface structure analysis from a variety of surfaces using time-of-flight scattering and recoiling spectrometry (TOF-SARS) and scattering and recoiling imaging spectrometry (SARIS). There are many more examples of structural analysis by ion scattering techniques in the Compilation of Ion Scattering Publications at the end of this book.

7.1. CLEAN SURFACES: RECONSTRUCTION AND RELAXATION

7.1.1. Surface Periodicity from TOF-SARS Azimuthal Angle Scans

When the angles between the direction of motion of scattering particles and the atomic rows are small enough, the scattering potential can be approximated as a multiatom continuous potential due to the chains of atoms along the principal azimuths. For such a case, most of the trajectories arriving along the atomic rows are scattered specularly (i.e., $\alpha = \theta/2 = \beta$, where α, β, and θ are the incident, exit, and scattering angles, respectively). Lindhard[1] defined a critical incident angle α_L above which the continuum approximation breaks down; for $\alpha > \alpha_L$, focusing is lost due to scattering from individual atomic centers rather than continuous multiatom potentials. If the surface has semichannels, for example, the (100) and (110) faces of fcc metals, this specular focusing effect along the crystallographic axes is enhanced. Semichannels are a group of three (or more) adjoint parallel atomic rows, two of which are in the first-layer and the third is in between in the second layer. With the first-layer rows serving as the walls and the second-layer row serving as the base, the semichannel acts as a convergent lens, and focusing occurs when the particles enter and leave the semichannel. The combination of axial focusing (i.e., along the atomic rows) reveals the crystallographic axes of the surface. Indeed, scans of azimuthal angle under specular conditions produce intensity maxima along the principal azimuths.

In the *shadowing mode,* the primary ion beam is directed at a grazing incident angle α to the surface ($\alpha < \alpha_L$) and, as the crystal is rotated azimuthally, the scattered flux I(s) is monitored in off-specular conditions (i.e., at a scattering angle θ considerably higher than 2α). Consequently, the crystallographic axes, and hence the periodicity, are determined entirely by shadowing effects along the incoming trajectory. Under these off-specular conditions, the scattered ion flux is minimum along the crystallographic axes as the crystal is rotated azimuthally. The reasons for this are twofold. First, the interatomic spacings are relatively short along the principal azimuths so that, for grazing incidence ions, the first-layer scattering centers lie within the shadow cones of their neighbors as shown[2] in Figure 7.1. As a result, the impact parameter

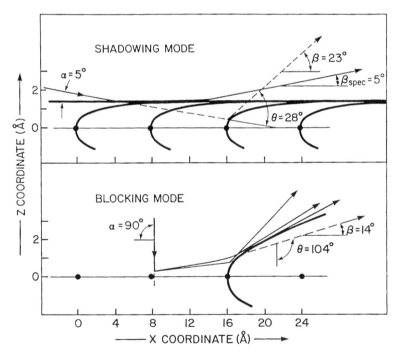

FIGURE 7.1. Schematic drawings of scattering trajectories along a surface atomic row, in both the shadowing and blocking modes. The solid arrowed lines represent actual trajectories, and the dashed lines represent hypothetical trajectories that would be observed if there were no shadowing and blocking effects (i.e., isolated binary collisions). β_{spec} is the exit angle corresponding to specular scattering (top frame). Example numerical values of the angles are indicated for each mode (From Masson et al., 1991, with permission).

required for off-specular scattering (dashed line in the figure) is not attained, and most of the scattered flux is detected at the specular θ (solid arrowed line). Second, semichanneling between the atomic rows, which occurs at such low-incidence angles, is observed in near-specular conditions. In the intermediate azimuthal directions, where the layer spacings are longer, incident trajectories can reach the target atoms at the proper impact parameter for off-specular scattering, and maxima are correspondingly observed. Those incident trajectories may even be focused by pairs of first-layer neighboring atoms between which they have to travel before reaching the scattering centers. Such focusing further increases the intensity observed between the principal azimuths.

In the *blocking mode,* the primary ion beam is directed along the surface normal, and the scattered flux is monitored at a fixed scattering angle θ that corresponds to a grazing exit angle β. Under these conditions, the crystallographic axes, and hence the periodicity, are determined entirely by blocking effects along the outgoing trajectory. As the crystal is rotated through its azimuthal angles δ, intensity minima are observed along the principal azimuths. The reason for this is that, for grazing exit, the trajectories of scattered particles intersect the blocking cones of neighboring atoms along these principal azimuths, as shown in Figure 7.1 (dashed line); a larger β would be required to escape blocking effects (solid arrowed lines). In intermediate azimuthal directions, where first-layer atoms are farther apart, scattered particles are able to escape at the low exit angle β employed and may even be focused by pairs of neighboring atoms on their way to the detector.

I(s) versus δ scans in the shadowing mode are shown[2] in Figure 5.5 for both (1×2) and (1×3) reconstructed Pt(110) surfaces. Sharp variations in I(s) are observed as a function of δ. The minima are located at the δ values corresponding to alignment of first-layer neighboring atoms (i.e., the crystallographic axes). These minima converge upon rotation from $\langle 001 \rangle$ to $\langle \bar{1}11 \rangle$ until azimuthal resolution is lost beyond the $\langle \bar{1}11 \rangle$ direction, culminating in a steep slope. This convergence is due to the successively smaller angular rotations that are required to align first-layer atoms across the long transtrough distances. The wide, deep minima along the $\langle \bar{1}10 \rangle$ direction is due to the close-packed rows; a large δ rotation is required for the first-layer atoms to move out of the shadow cones cast by their neighbors.

I versus δ scans in the blocking mode are shown[2] in Figure 7.2. Similar structures are observed as in the shadowing mode. The β values that provide the optimum structure in Figure 7.2 are much larger than the α values of Figure 5.5. This is due to the fact that the blocking cones of scattered particles that have suffered energy losses are considerably larger than the shadowing cones of incident particles that have not yet suffered energy losses. This work demonstrates that low-energy ion scattering is capable of real-space assessment of the periodicities of single crystal surfaces. This is an excellent complement to low-energy electron diffraction (LEED) experiments that provide information in reciprocal space.

7.1.2. Interlayer Spacings from TOF-SARS Incident Angle Scans

A common feature of incident angle α scans is low backscattering (BS) intensity at grazing incidence as shown[3] in Figure 7.3. This low intensity is due to the shadowing

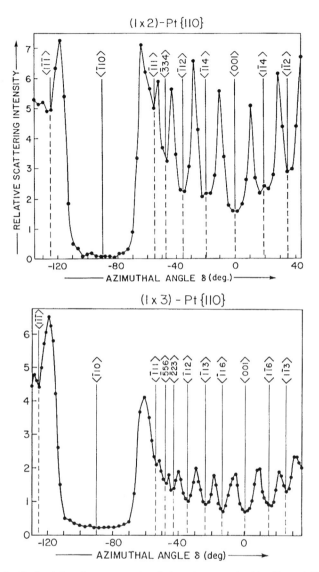

FIGURE 7.2. Scattering intensity versus azimuthal angle scans for (1×2)- and (1×3)-Pt(110) in the blocking mode. $\beta = 13°$ for (1×2) and $\beta = 12°$ for (1×3) (From Masson *et al.*, 1991, with permission).

of each surface atom by its preceding neighbor. An interpretative illustration of the observed peaks is provided[3] in Figure 7.4. The peaks are labeled $(XYZ)–(X'Y'Z')$, indicating that the incident trajectories are focused by atom (XYZ) onto the scattering center $(X'Y'Z')$. X is the coordinate along $\langle 001 \rangle$, Y is that along $\langle \bar{1}10 \rangle$, both in terms of the (1×1) unit mesh, and Z denotes the layer in which the atom is located. The α-values at which the sharp rises occur are called the critical incident angles α_c. They

COLOR PLATE 1

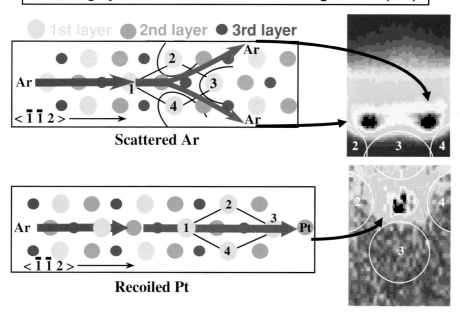

COLOR PLATE 6.4. Views of the Pt(111) surface along the $\langle\bar{1}\bar{1}2\rangle$ azimuth showing argon (Ar$^+$) impinging from the left side. (Top) The Ar beam scatters from a first-layer platinum (Pt) atom (atom 1) and splits into two focused beams by an atomic lens formed by neighboring first-layer Pt atoms (atoms 2, 3, and 4). (Bottom) The impinging Ar recoils a second-layer Pt atom through an atomic lens formed by first-layer Pt atoms (atoms 1–4). The resulting experimental SARIS images are shown on the right side (From Rabalais, 2001 and Kim *et al.* 1997, with permission).

COLOR PLATE 2

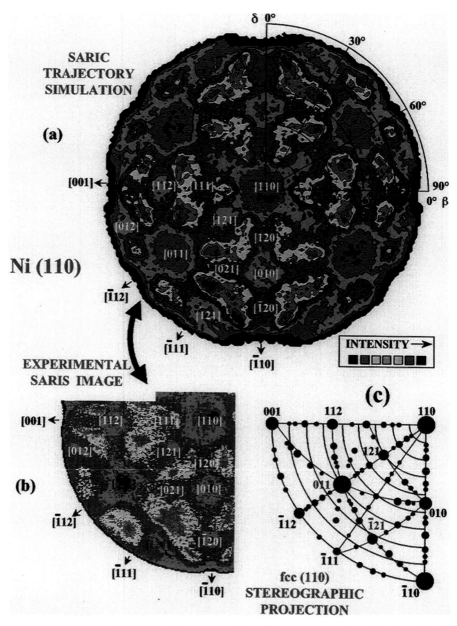

COLOR PLATE 6.6. (a) SARIC trajectory simulation of the stereographic projection of 4-keV He$^+$ scattering from a Ni(110) surface, (b) experimental SARIS image corresponding to one quadrant of the simulation, and (c) a standard *fcc* (110) stereographic projection (From Rabalais, 2001, Bykov *et al.* 2000, and Bolotin *et al.* 2000, with permission).

COLOR PLATE 3

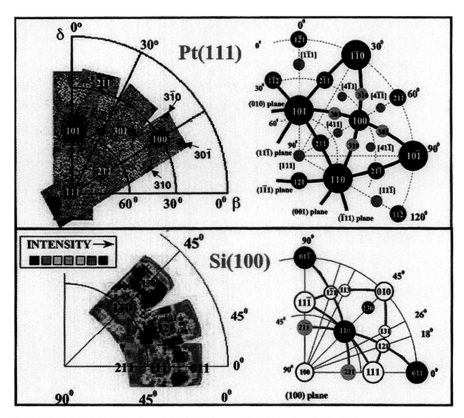

COLOR PLATE 6.7. Experimental SARIS images for Pt(111) and Si(100) in one quadrant of an imaginary hemispherical detector along with the standard stereographic projections for these surfaces. The experimental scattering intensity distributions were collected for different microchannel plate orientations, then mapped to the imaginary hemisphere, and then projected onto the base of the hemisphere. The intensity scale is as in Figure 6.4 (From Rabalais, 2001, with permission).

COLOR PLATE 4

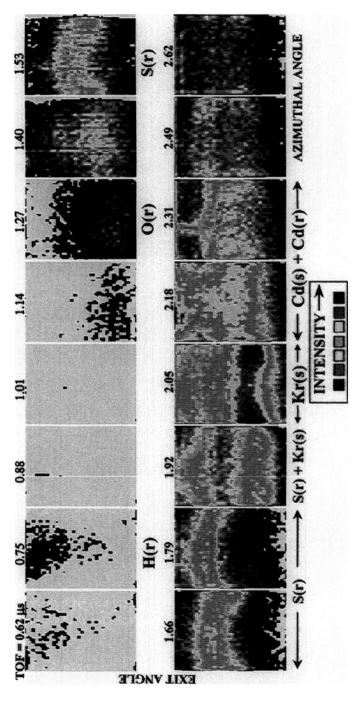

COLOR PLATE 6.8. Selected consecutive 16.7-ns SARIS frames for 4-keV Kr^+ scattering from CdS(0001) with the beam aligned along the $(\bar{1}010)$ azimuth. The time-of-flight (TOF) times corresponding to each consecutive image are shown above the frames. Labels corresponding to the dominant scattering, Kr(s), and recoiling, H(r), O(r), and S(r), events are listed below their respective frames (From Rabalais, 2001, Bykov *et al.* 2000, and Kim *et al.* 1998, with permission).

COLOR PLATE 5

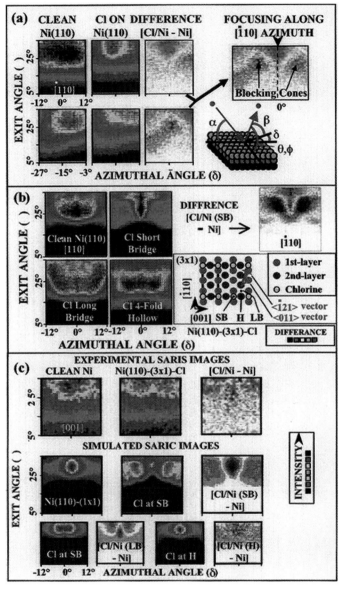

COLOR PLATE 6.10. (a) SARIS images of 4-keV Ar$^+$ scattering from Ni(110)-(1 × 1) and Ni(110)-(3 × 1)-Cl along ⟨$\bar{1}$10⟩ (top) and 15° off this azimuth (bottom). The difference image reveals focusing of scattered Ar trajectories in the centers of the ⟨$\bar{1}$10⟩ troughs. (b) SARIC simulations of 4 keV Ar$^+$ scattering from Ni(110)-(1 × 1) and Ni(110)-(3 × 1)-Cl with Cl in the short bridge (SB), long bridge (LB), and four-fold hollow (H) sites. The SB site provides the best agreement with the experimental images of (a). (c) SARIS images and SARIC simulations of Ni atoms recoiled by 4-keV Kr$^+$ along the ⟨001⟩ azimuth from Ni(110)-(1 × 1) and Ni(110)-(3 × 1)-Cl (From Bykov *et al.* 2000, with permission).

COLOR PLATE 6

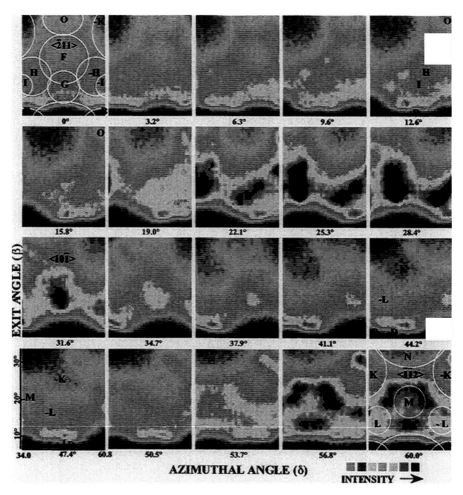

COLOR PLATE 6.11. A series of 20 time-resolved SARIS frames for 4-keV He$^+$ scattering from Pt(111)-(1 × 1) taken every 3° of rotation about the azimuthal angle δ, starting with the $\langle \bar{2}11 \rangle$ azimuth at $\delta = 0°$ and ending with the $\langle \bar{1}\bar{1}2 \rangle$ azimuth at $\delta = 60°$. Each frame represents a 16.7-ns window centered at the time-of-flight (TOF) corresponding to quasi-single scattering as predicted by the binary collision approximation. The white circles on the $\delta = 0°$ and 60° frames represent the positions and relative sizes of calculated blocking cones (From Rabalais, 2001, Bykov *et al.* 2000, and Kim *et al.* 1998, with permission).

COLOR PLATE 7

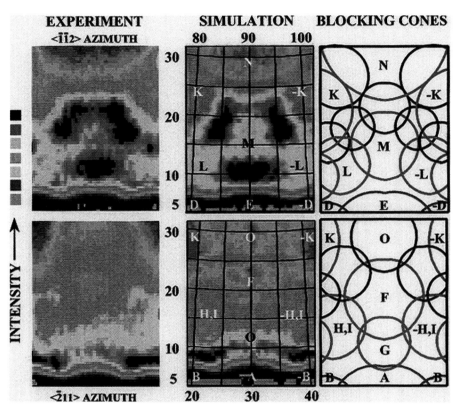

COLOR PLATE 6.13. Experimental SARIS images (left), simulated SARIC images (center), and blocking cone analysis (right) for He$^+$ scattering along the $\langle\bar{2}11\rangle$ and $\langle\bar{1}\bar{1}2\rangle$ azimuths of Pt(111). For the calculated blocking cones, first–first, first–second, and first–third layer interactions are identified by green, red, and blue lines, respectively. The letters correspond to the interatomic vectors identified in Figure 6.11(a). (From Bykov *et al.* 2000, Kim *et al.* 1997 and Kim *et al.* 1997 with permission).

COLOR PLATE 8

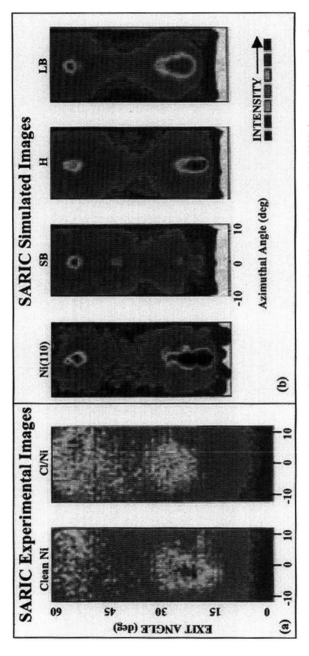

COLOR PLATE 7.28. (a) Time-integrated composite SARIS images and (b) SARIC simulations for 4-keV Ne$^+$ scattering on the clean Ni(110)-(1 × 1) and Ni(110)-(3 × 1)-Cl surfaces along the $\delta = 0°$ [100] azimuth. Incident angle $\alpha = 40°$; scattering angle $\theta = 85°$; exit angle $\beta = 45$; SB–Cl, in short-bridge site; H–Cl, in long-bridge site; LB–Cl, in fourfold hollow site (From Houssiau et al. 1999, with permission).

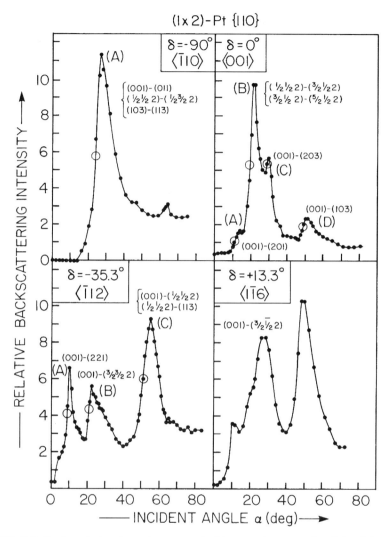

FIGURE 7.3. Backscattering intensity versus incident angle scans for Pt(110)-(1 × 2) at $\theta = 149°$ along different azimuths (From Masson *et al.*, 1991, with permission).

are measured at one-half the peak height minus the background and are indicated by open circles in Figure 7.3. An interpretative diagram is not shown for the $\langle 1\bar{1}6 \rangle$ azimuth because off-planar contributions are involved in this case that require three-dimensional illustrations.

For quantitative structural analysis (i.e., "precise" determination of the atomic positions in the outermost layers), one has to select those α_c's that correspond to a (focusing atom)–(scattering atom) pair for which the interatomic spacing is to be determined. For example, in order to evaluate the first–second-layer spacing D_{12}, the

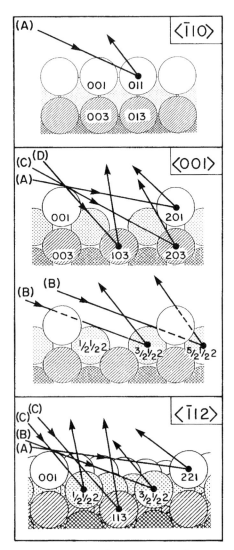

FIGURE 7.4. Cross section diagrams along $\langle \bar{1}10 \rangle$, $\langle 001 \rangle$, and $\langle \bar{1}12 \rangle$ azimuths, illustrating scattering trajectories for the peaks observed in Figure 7.3 (From Masson *et al.*, 1991, with permission).

peak (001)-$(\frac{1}{2}\frac{3}{2}2)$ along $\langle \bar{1}12 \rangle$ may be used. However, peak (C) at $\alpha_c = 51°$ along $\langle \bar{1}12 \rangle$ may not serve that purpose because it also involves a contribution from third-layer atoms. The objective is therefore to *selectively block* this third-layer contribution by reducing the scattering angle θ. Calculations, using the Ziegler–Biersack–Littmark interaction potential predict that a maximum $\theta \sim 105°$ is required to block the $(\frac{1}{2}\frac{1}{2}2)$-$(113)$ contribution and that a minimum $\theta \sim 75°$ must be chosen so as not to block the (001)-$(\frac{1}{2}\frac{1}{2}2)$ contribution.

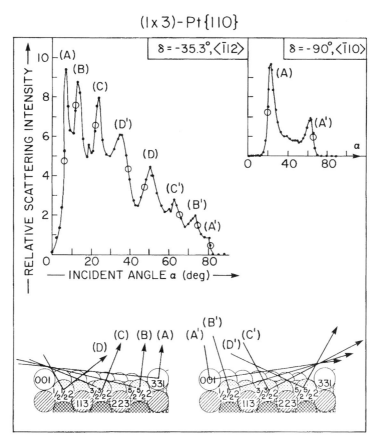

FIGURE 7.5. Scattering intensity versus incident angle scans for Pt(110)-(1 × 2) at $\theta = 93°$ along the $\langle \bar{1}10 \rangle$ and $\langle \bar{1}12 \rangle$ azimuths (From Masson *et al.,* 1991, with permission).

A smaller scattering angle can be used in order to achieve more restrictive blocking of deep layer contributions. Such α scans at $\theta = 93°$ along the $\langle \bar{1}10 \rangle$ and $\langle \bar{1}12 \rangle$ azimuths are shown[3] in Figure 7.5. There are now eight peaks along this azimuth instead of the three peaks observed at $\theta = 149°$ in Figure 7.3. An α scan along the simple $\langle \bar{1}10 \rangle$ azimuth is also shown in Figure 7.5 for $\theta = 93°$. Along $\langle \bar{1}10 \rangle$, two peaks (A) and (A′) are observed. The (A) peak at $\alpha_c = 20.5°$ is analogous to the peak at $\alpha_c \sim 24°$ at $\theta = 149°$ in Figure 7.3. The $3.5°$ difference is due to the fact that the impact parameter b required for scattering at $\theta = 93°$ ($b = 0.244 \,\text{Å}$) is larger than that for $\theta = 149°$ ($b = 0.08 \,\text{Å}$). As α is increased, a second peak (A′), which is well defined and centered at $\alpha \sim 62°$, appears. This peak cannot be due to scattering from atom (013) below (011), since this contribution is clearly blocked by atom (021). Instead, this peak arises from trajectories scattered by atom (001) at $\theta = 93°$ that are focused at the edge of the blocking cone cast by neighboring atom (011) towards the detector. The intensity then drops to zero at $\alpha > 70°$ because (011) now totally blocks those scattered trajectories. In the range $35° < \alpha < 55°$, a relatively low intensity is observed since scattering in nonfocusing conditions is taking place. If one defines an exit angle scale as $\theta = 93°$ minus the incident angle, peak (A′) is quantitatively characterized by the critical exit angle β_c as indicated by the open circle in Figure 7.5. The value $\beta_c = 26.5°$ is in good agreement with the expected value from blocking cone calculations. It is to be noted that, as for peak (A), peak (A′) also involves contributions from the second- and third-layers in the trough.

Along the $\langle \bar{1}12 \rangle$ azimuth, peaks (A), (B), and (C) at $\alpha_c \sim 8°$, $19°$, and $47°$, respectively, are analogous to the three peaks observed at $\theta = 149°$, with the restriction that for peak (C), the contribution from atom (113), observed at $\theta = 149°$, is blocked by atom $(\frac{3}{2}\frac{3}{2}2)$ at $\theta = 93°$, as desired. These peaks are observed at lower α_c than at $\theta = 149°$, reflecting the larger impact parameter required at $\theta = 93°$. The three other peaks (A′), (B′), and (C′) are due to the same effect that gives rise to peak (A′) along $\langle \bar{1}10 \rangle$, in other words, the trajectories that are sequentially scattered at $\theta = 93°$ by atoms (001), $(\frac{1}{2}\frac{1}{2}2)$, and $(\frac{3}{2}\frac{3}{2}2)$, are focused at the edge of the blocking cone cast by atom (221), as illustrated on the cross section diagram.

7.1.3. Surface Structure from SARIS Images

SARIS images can be interpreted qualitatively by using a simple blocking cone analysis as described in Chapter 6. Quantitative interpretation is achieved by using a reliability or R-factor comparison of the experimental and simulated patterns as a function of the surface structural parameters.

Figure 7.6(a)[4] shows experimental SARIS data for He^+, Ne^+, and Ar^+ scattering from the Au(110) surface. Each of the four time-resolved images are 16.7-ns frames taken at the flight times corresponding to quasi-single scattering (QSS); these frames are at 0.355 μs for He^+, 0.816 μs for Ne^+, and 1.18 μs for Ar^+. These QSS frames correspond to the most intense images in the time-resolved series. QSS denotes a projectile atom that undergoes a single small-impact parameter collision that may be preceded and/or followed by several large-impact parameter collisions. The energy of a QSS atom can be comparable with that of a single-scattered atom. The images for $\delta = 90°$ along the $\langle 001 \rangle$ azimuth are symmetrical about a vertical line through the center of the frame, as is the crystal structure along this azimuth. The images for $\delta = 78°$ and $81°$ are asymmetrical, as is the crystal structure along these directions. The regions of low intensity correspond to blocking of scattered trajectories by neighboring atoms. The positions and sizes of these blocked regions are weakly dependent of the scattering cross sections and scattering angles, but strongly dependent on the atomic positions. The regions of high intensity correspond to focusing of atom trajectories at the edges of blocking cones and at the positions of intersection or near overlap of blocking cones. The qualitative positions and sizes of the blocking cones can be calculated from the directions and lengths of the interatomic vectors determined from the crystal structure. These qualitative blocking positions taken as a group define the zero-order pattern of the image. For example, the blocking of scattering trajectories from nth-layer atoms by their neighboring nth-layer atoms is observed at low exit angles β, as these atoms are all in the same plane. The axes of these blocking cones

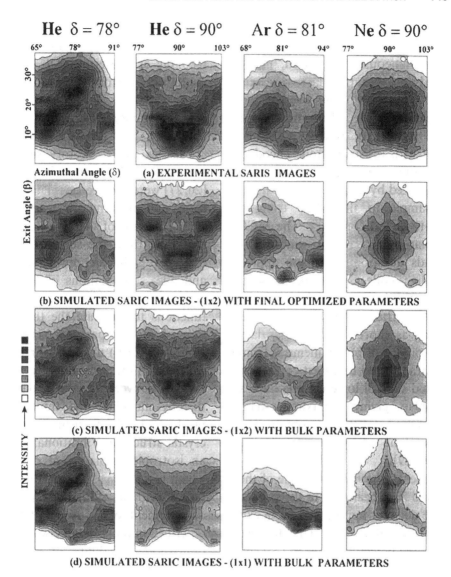

FIGURE 7.6. (a) Spatial- and time-resolved experimental SARIS images for 4-keV He^+, Ne^+, and Ar^+ scattering from a Au(110)-(1 × 2) surface. The primary ion and azimuthal alignment of the beam are indicated above the figures. Each image corresponds to a 16.7-ns window centered at the TOF corresponding to quasi-single scattering; these are also the most intense frames in the TOF series. (b) Simulated SARIC images for a (1 × 2) reconstructed surface with all structural parameters optimized according to the values that yield the minimum R-factor. (c) Simulated SARIC images for a (1 × 2) reconstructed surface with bulk interatomic spacings. (d) Simulated SARIC images for a (1 × 1) bulk-terminated surface (From Hoefner *et al.*, 1998, with permission).

are at $\beta = 0°$, and the arcs corresponding to the edges of the cones are observed at $\beta < 10°$. The features at higher β correspond to the scattering trajectories from subsurface nth-layer atoms that are blocked and focused by atoms in $(n–m)$th-layers nearer to the surface, where $n > m$. The observed images are complicated due to overlapping of blocking cones and focusing at the edges of the cones.

Figures 7.6(b–d) show simulated SARIC images for He^+, Ne^+, and Ar^+ scattering from the Au(110) surface. The results of three different simulations are shown for each azimuth and primary ion corresponding to: (b) the final optimized missing-row (1×2) surface structure that provides the minimum R-factor; (c) a missing-row (1×2) surface with all interatomic spacings equal to the bulk spacings; and (d) a (1×1) bulk-terminated surface. The images of Figure 7.6(b) are in excellent agreement with the experimental images, exhibiting minimum R-factors. The R-factors for the structures of Figures 7.6(c and d) are significantly larger. The minimum R-factor found for Figure 7.6(a) shows that experimental SARIS images coupled with simulated SARIC images are sensitive to structural changes in the surface and subsurface layers of Au(110) down to five atomic layers for changes at a level of ± 0.025 Å. The structural parameters included in the analysis[4] were interlayer spacings, row pairing, and buckling.

Some fcc(110) surfaces undergo order–disorder or roughening transitions at temperatures well below the bulk melting point. SARIS has been applied[5] to the temperature and ion bombardment-induced disordering of the Au(110) surface. The images are highly sensitive to the deconstruction or disordering in the outer surface layers. The basic features of the (1×2) missing-row structure are still observed near 700 K, where three-dimensional roughening begins; however, they are completely obliterated on the sputtered amorphousized surface.

7.1.4. Other Surface Structure Examples from Clean Surfaces

TOF-SARS has been applied[6] to determination of surface relaxation, both changes in first–second-layer spacing and first–second-layer registry, for the W(211) surface. The results show that there is a vertical contraction of 9.3% and a lateral shift of 3.6% along the $\langle 1\bar{1}\bar{1} \rangle$ azimuth from the bulk-truncated surface. Structures associated with scattering from the first- to fourth-atomic layers were delineated.

The structure of the reconstructed Ir(110) surface has been investigated[7] by TOF-SARS and LEED. The results indicate that the reconstructed surface consists of primary domains of faceted (1×3) structures (with two missing first-layer rows and one missing second-layer row) along with secondary domains of (1×1) structures (with no missing rows). Estimates of the interatomic spacings in the (1×3) domains indicate that the second-layer atoms are shifted from the bulk values laterally by $\sim 6\%$ towards the center of the trough and that the first–second-layer spacing is contracted by $\sim 8\%$.

The CdS(0001) surface has been studied by both TOF-SARS[8] and SARIS.[9] The lack of inversion symmetry in the $\langle 0001 \rangle$ direction of this wurtzite structure gives rise to a crystallographic polarity (i.e., the surface is terminated in either a cadmium or sulfur layer). TOF-SARS spectra of the Cd- and S-terminated surfaces are shown in Figure 5.1, SARIS images of the Cd-terminated surface are shown in Figure 6.8, and SARIC simulations for both the Cd- and S-terminated surfaces are shown in Figure 6.9. The results show that both surfaces are bulk-terminated with no detectable reconstruction within the uncertainty of the measurements (i.e., ± 0.3 Å). Both surfaces have two structural domains that are rotated by $60°$ from each other and there are steps on the surface. The Cd-terminated surface exhibits oxygen and hydrogen with the same azimuthal periodicities as the Cd atoms, indicating that they are bound as OH groups to the first-layer Cd atoms. The S-terminated surface also exhibits these impurities, although the absence of any azimuthal periodicity for recoiled O and H indicate that although these impurities occupy specific sites, they do so with random occupancy.

7.2. HYDROGEN ON SURFACES

Although chemisorption of hydrogen can dramatically alter the electronic and chemical properties of surfaces of most materials, specific details about adsorption sites of hydrogen are elusive. Hydrogen is notorious for being difficult to study, for it has small cross sections for electron scattering, X-ray scattering, and photoionization by X-rays; it has no Auger transition and is ubiquitous in secondary ion mass spectra. Elegant techniques utilizing MeV ion beam analysis, such as elastic recoil detection[10] and nuclear reaction analysis,[11] provide depth profiles of the concentration of hydrogen in materials. The high energies involved inevitably lead to depth resolutions >1 nm in most circumstances, and hence fail in delivering surface atomic site-specific information.

Ion channeling at low keV energies is a sensitive probe for determining the adsorption site of hydrogen on surfaces. Using TOF-SARS in a backscattering geometry where the angle between the incident ions and the channel axis is less than the Lindhard critical angle, the Lindhard criteria are satisfied and ion channeling is observed. Under these conditions, the atomic potentials of the individual atoms making up the walls of the channel overlap to form a continuous potential, and strongly correlated small angle collisions prevail. When the angle between the incident ions and the channel axis is increased beyond the critical angle, large angle deflections result from the more corrugated potential surfaces. This gives rise to a flux of backscattered particles that have experienced only single or quasi-single collisions with energies predictable by the binary collision approximation (BCA). This outgoing flux acts as a probe of foreign atoms adsorbed in an ordered array on the crystal surface. By performing the channeling along different crystallographic directions, the position of adsorbed atoms can be derived unambiguously by a triangulation procedure.

7.2.1. Hydrogen on Pt(111)

Examples of SARIgrams for 5-keV Ne$^+$ scattered from the clean Pt(111)-(1 × 1) and Pt(111)-(1 × 1)-H surfaces[12] are presented in Figure 6.14.

TOF-SARS in the channeling mode is capable of quantitatively probing the positions of hydrogen atoms on heavy substrates. Figure 7.7(a–c) depicts typical

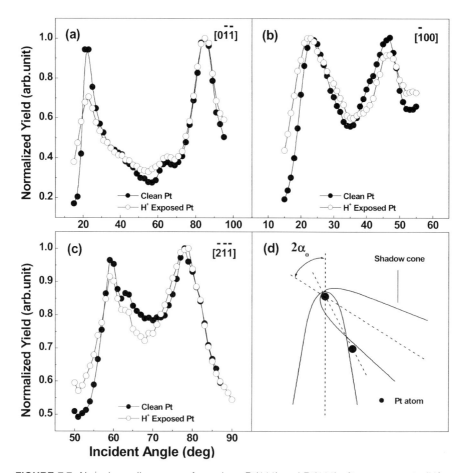

FIGURE 7.7. Ne$^+$ channeling curves from clean Pt(111) and Pt(111) after exposure to (H$^+$ + H$_2^+$) ions for channels along the (a) [011], (b) [$\bar{1}$22], and (c) [211] directions. (d) Positions of the shadow cones and the angle (2α_0) that create the maxima in (a–c) (From Lui *et al.*, 1999 and Lui *et al.*, 1999, with permission).

results[13,14] for Ne$^+$ backscattering along three different channels, namely [0$\bar{1}\bar{1}$], [$\bar{1}$00], and [$2\bar{1}\bar{1}$], which theoretically make angles of 54.7°, 35.3°, and 70.5°, respectively, with the Pt(111) surface. In general, for a clean surface, upon departure from the minimum, the backscattering yield increases symmetrically on both sides of the curve until maxima are attained. These maxima signify the two angular positions at which the shadow cones created by the first-layer atoms move across the subsurface neighboring atoms lying parallel to the channel directions (Figure 7.7d); the incident trajectories are directed onto the second- or third-layer atoms that then serve as point scatters to steer the outgoing particles. These outgoing particles act as a probe of the positions of the surface atoms. The relative interatomic distances along different crystallographic directions can be qualitatively inferred from the angular widths (2α_0)

of the channeling curves; short (long) distances provide larger (smaller) widths. The interatomic distances for Pt in the $[0\bar{1}\bar{1}]$, $[\bar{1}00]$, and $[\bar{2}\bar{1}\bar{1}]$ directions are 2.8, 3.9, and 4.8 Å, respectively, and the corresponding $(2\alpha_o)$ values are 60°, 24°, and 19°. This reveals the self-consistency of the channeling measurements.

The channeling curves became noticeably asymmetric when hydrogen is absorbed on the surface, as shown in Figure 7.7(a–c). A quantity Δ, the difference in normalized yield, can be defined as the normalized yield before hydrogen exposure minus the normalized yield after hydrogen exposure. The physical meaning of such a quantity is that for $\Delta > 0$ ($\Delta < 0$) in the angular range from α to $\alpha + d\alpha$, there is a decrease (increase) of Ne atoms reaching the detector after the surface is exposed to hydrogen. The magnitude of this decrease (increase) is proportional to $\Delta(\alpha)d\alpha$. The legitimacy of this treatment lies in the fact that the channeling curves, after normalization, should be shape-invariant if there is no change in the pattern of the surface atoms and $\Delta(\alpha)$ should be zero in the region of interest. Conversely, any change in the atomic pattern of the surface due to absorption of hydrogen atoms will be revealed as nonzero Δ. Integrating Δ in the relevant angular region should give zero if no backscattered Ne particle is lost due to deflections by hydrogen. This is a statement of conservation of the number of backscattered Ne atoms. The physical origin for a positive Δ arises from the fact that part of the backscattered Ne atoms that could have reached the detector are deflected out of the detection cone as a result of being scattered by the absorbed hydrogen atoms. These deflected atoms are forced into nearby α angles leading to the corresponding negative Δ. The differences in normalized yields for the above three channels are shown in Figure 7.8. Each of these graphs exhibits regions of positive and negative Δ. The α angles corresponding to the maximum Δ were found for the $[0\bar{1}\bar{1}]$, $[\bar{1}00]$, and $[\bar{2}\bar{1}\bar{1}]$ channels, resulting in the values $21° \pm 1°$, $41° \pm 1°$, $66° \pm 2°$, respectively. Using $\theta = 156°$ and the above angles of the maximum Δ, it is possible to construct the outgoing paths along which the Ne atoms experience the closest approach to the H atoms. Such a construction is shown in Figure 7.9, where it is observed that the three outgoing paths intersect at a single point. This point represents the position of the H atoms responsible for the deflections and shows unambiguously that the majority of the hydrogen atoms populate the fcc threefold sites. Solving for the intersection of the three lines leads to a H atom height of 0.9 ± 0.1 Å above the first-layer Pt atoms and a corresponding Pt-H bond length of 1.9 ± 0.1 Å. The result of a preferential occupation of the fcc site on the Pt surface is consistent with theoretical calculations[15] showing that the energy of the fcc site is 0.2 eV lower than that of the hcp site.

7.2.2. Hydrogen on Other Surfaces

The structure of the hydrogen adsorption sites on the Si(100) surface in the (2×1)-H monohydride, (1×1)-H dihydride, and c(4×4)-H phases has been studied[16,17] by TOF-SARS. The data provide a direct determination of the interatomic spacings in the outermost layer, including the hydrogen atom coordinates relative to the outermost layer of silicon atoms. Si-H bond lengths and bond angles are calculated from data such as in Figures 5.8 and 5.9. It has been demonstrated[18] that both surface and

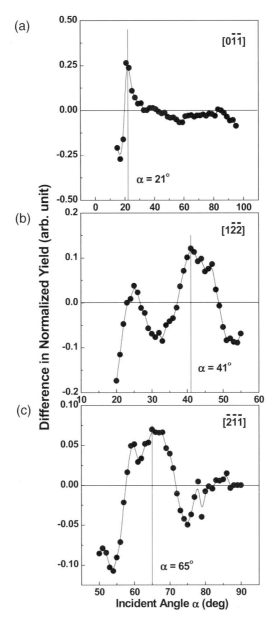

FIGURE 7.8. Differences in the scattering yields of Figure 7(a–c) versus incident angle α (From Lui *et al.,* 1999 and Lui *et al.,* 1999, with permission).

subsurface structural information can be obtained from Si(100)-(2 × 1) and Si(100)-(1 × 1)-H by using coaxial TOF-SARS. Using 2-keV He$^+$ and $\theta = 180°$ coupled with three-dimensional trajectory simulations, it has been shown that scattering features can be observed and delineated from as many as 14 atomic layers ($\sim$18 Å) below the surface. The sensitivity of the α- and δ-scans to the amplitudes of both surface

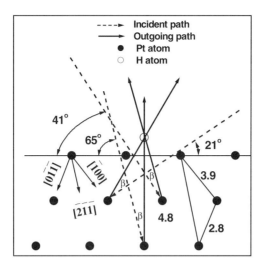

FIGURE 7.9. Side view of the Pt(111) surface showing incoming (dashed) neon (Ne) trajectories focused by first-layer platinum (Pt) atoms onto second- and third-layer Pt atoms and scattered outgoing Ne trajectories (solid lines) at a scattering angle $\theta = 156°$. The three outgoing trajectories corresponding to the α angles from Figure 7.8(a–c) that exhibit the maximum positive Δ intensity deviation intersect at a single point (From Lui *et al.*, 1999 and Lui *et al.*, 1999, with permission).

and bulk vibrations, the interlayer spacings, the intradimer spacing [in the (2×1) reconstructed phase], and the presence of hydrogen atoms [in the (1×1) phase] is demonstrated.

The structure of the Ni(110)-p(2×1)-H surface and the adsorption site of hydrogen on this surface have been studied[19] by TOF-SARS. The results for the Ni(110)-p(2×1)-H surface are contrasted to those for the clean Ni(110)-p(1×1) surface and the Ni(110)-p(2×1)-O oxygen-covered surface in Figure 5.6. Hydrogen induces a missing-row structure in which every other $\langle 1\bar{1}0 \rangle$ first-layer Ni row is missing. The hydrogen atoms are localized at the pseudo-threefold sites on this missing row structure.

Hydrogen adsorption at saturation coverage on the W(211) surface has been studied by TOF-SARS and effective-medium-theory.[20] The results show that hydrogen resides in a band that is located above the troughs along the $\langle 1\bar{1}\bar{1} \rangle$ direction with the center of this band located 0.57 ± 0.20 Å above the plane of the first-layer W atoms. The average spacing between the H atoms along the $\langle 1\bar{1}\bar{1} \rangle$ azimuth is 2.7 Å, which is equivalent to the W lattice spacing along this direction. The H atom probability distribution is maximum at the positions of the short-bridge and threefold through site positions.

7.3. OXYGEN ON SURFACES

7.3.1. Oxygen on Nickel Surfaces

The kinetics of O_2 chemisorption at low dose on Ni(111) and the nature of the chemisorption site have been studied[21] at 300 and 500 K using TOF-SARS, LEED, and SARIC simulations. Variations in the TOF-SARS spectra with different crystal alignments during O_2 dosing provide direct information on the location of oxygen atoms on the Ni(111) surface at very low coverages as well as site-specific occupation rates (S_{fcc} and S_{hcp}) and occupancies (θ_{fcc} and θ_{hcp}). A system of equations was developed that relate the slopes of the scattering and recoiling intensities versus O_2 exposure dose to these probabilities and occupancies. The results identify three chemisorption stages as a function of oxygen exposure, each with its own specific occupation rates and occupancies. The *first-stage* is observed up to $\theta_i = \theta_{fcc} \sim 0.21$ ML with $\theta_{hcp} = 0$, constant $S = S_{fcc} \sim 0.18 \pm 0.01$, and coverage ratio $w = \theta_{hcp}/\theta_{fcc} \sim 0$ for both 300 and 500 K. The *second-stage* is observed at coverages between $\theta_i \sim 0.21$ and $\theta_2 \sim 0.32$ ML, with constant $S_{fcc} = -(0.05 \pm 0.01)$ and $S_{hcp} = (0.16 \pm 0.02)$ at 300 K and $S_{fcc} = (0.005 \pm 0.003)$, and $S_{hcp} = (0.007 \pm 0.003)$ at 500 K, and coverage ratios $w = \theta_{hcp}/\theta_{fcc} \sim 1$ at 300 K and $w = \theta_{hcp}/\theta_{fcc} \sim 0.10$ at 500 K. The *third-stage,* observed for $\theta > 0.32$ ML, involves saturation coverage of the adsorption sites. SARIC simulations were used to interpret the spectra and the influence of oxygen chemisorption and vibrational effects. A method for determining the "effective Debye temperature Θ_D^*" that uses the experimental TOF-SARS intensity variations as a function of temperature, and the simulated SARIC signals as a function of the mean square vibrational amplitude $\langle u^2 \rangle$ was also developed. The result for this system is $\Theta_D^* = 314 \pm 10$ K.

TOF-SARS[22] has been used to study the O_2- and H_2O-induced reconstructed phases of the Ni(110) surface. By monitoring features in the TOF-SARS scans that are unique to specific phases, it is possible to follow the migration of the first-layer Ni atoms as a function of O_2 exposure. The results show that upon increasing exposures of the clean Ni(110)-(1 × 1) surface to O_2, a series of LEED patterns [initial p(3 × 1), p(2 × 1), and final p(3 × 1)] is produced, corresponding to three surface phases that differ only in the density of the first-layer Ni (001) rows, as shown[22] in Figure 7.10. These nascent "added rows" are stabilized by bonding to oxygen atoms that reside in the long-bridge positions along the ⟨001⟩ rows. The minima observed in the scans correspond to the azimuthal directions along which first-layer atoms are aligned. The structure of the Ni(110)-p(2 × 1)-OH surface resulting from exposure of the O_2-predosed Ni(110)-p(3 × 1)-O surface to H_2O vapor has also been investigated.[22] The results show that the reaction between H_2O and preadsorbed O removes the Ni substrate (3 × 1) reconstruction induced by O_2, producing an unreconstructed (1 × 1) Ni substrate. The resulting OH overlayer is arranged in a (2 × 1) rectangular lattice corresponding to a 0.5-ML coverage. Hydroxyl is bonded to Ni through the O atom at the short-bridge site. A study of the inhibition of the oxygen-induced (2 × 1) reconstruction of the Ni(110) surface by sulfur impurities has been made.[23] The data indicate that sulfur inhibits the diffusion of Ni atoms in the added-row reconstruction, presenting a kinetic barrier to the thermodynamically stable Ni(110)-(2 × 1)-O phase.

7.3.2. Oxygen on Other Surfaces

Oxygen chemisorption on the W(211) surface has been studied by TOF-SARS and LEED.[24] The results show that O_2 is dissociatively chemisorbed within the troughs

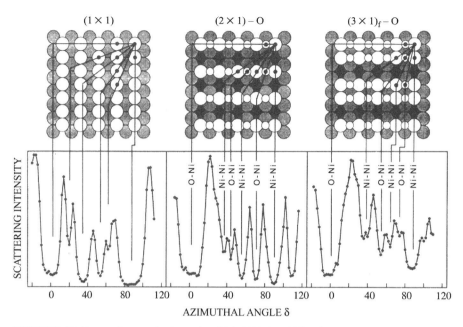

FIGURE 7.10. Forward-scattering intensity of 4-keV Ne$^+$ on Ni(110) at $\theta = 30°$ versus azimuthal angle δ for the (1 × 1) phase using $\alpha = 6°$ and the (2 × 1) and (3 × 1) phases using $\alpha = 4.5°$ (From Bu *et al.*, 1992, with permission).

along the $\langle 1\bar{1}\bar{1} \rangle$ direction. For the saturation p(1 × 2) structure, the oxygen is in threefold sites formed by two first- and one second-layer W atoms. These trough sites are occupied from low coverages up to more than 1.5-ML coverage. The data are consistent with shifting of the relaxed clean W surface structure to the bulk-truncated structure upon O_2 chemisorption.

Structural analysis of O_2 and H_2O chemisorption on a Si(100) surface has been performed by TOF-SARS.[25] The qualitative features of both the substrate and adsorbate structures were elucidated from the angular anisotropies of H, O, and Si recoils. The initial exposures to O_2 result in dissociative chemisorption with destruction of the Si(100)-(2 × 1) dimer structure; the resulting structure has no long-range or short-range order after ~1 ML O atom coverage. The Si subsurface structure is not disturbed by this surface disorder. The O atoms are bound to the Si dimer dangling bonds at disordered high positions (>0.5Å) with respect to the first-layer Si atoms at low coverage (<0.5 ML). At a higher coverage, chemisorption of O atoms occurs at subsurface positions. The data are consistent with these subsurface sites being bridging positions between the first- and second-Si layers. At all coverages, no short-range order was observed. Chemisorption of H_2O produces ordered structures in which a (2 × 1) pattern is maintained even at saturation doses. The H_2O is dissociatively chemisorbed on Si(100), resulting in H + OH on the surface. The data indicate that both H and OH species are bound to Si atoms through the dimer dangling bonds.

7.4. METAL OXIDE SURFACES

Most metal oxides can be studied by TOF-SARS without charging problems as long as a low primary ion beam current is maintained. The surface composition, termination, and structure of the $Al_2O_3(0001)$-(1×1) surface has been determined[26] through the use of TOF-SARS, LEED, and SARIC. A typical spectrum along a random direction from a surface that was unannealed and exhibited a diffuse (1×1) LEED pattern is shown in Figure 7.11. Distinct peaks due to recoiled hydrogen H(r) and oxygen O(r) atoms are observed along with a broad low-intensity structure at longer time due to scattering from aluminum Al(s). Upon annealing to 1100°C, the Al(s) intensity increased, the H(r) intensity decreased, and the Al(r) peak became visible. The H(r)

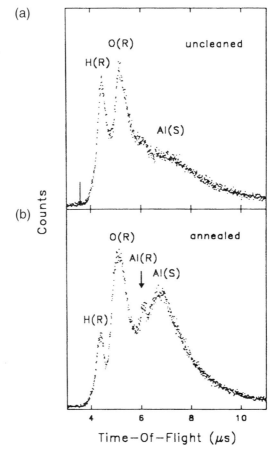

FIGURE 7.11. TOF-SARS spectra of 4-keV Ar^+ scattering and recoiling from an $Al_2O_3(0001)$-(1×1) surface with the beam aligned along a random azimuthal δ direction for (a) an uncleaned surface and (b) a surface that was annealed to 1100°C. Incident angle $\alpha = 6°$. Scattering angle $\theta = 40°$ (From Ahn *et al.*, 1997, with permission).

peak remained, even after several annealings to 1100°C. The results of the structure determination show that the surface is terminated in an Al layer that relaxes downward so that the Al first-layer to O second-layer spacing is 0.3 ± 0.1 Å. This relaxation of 63% from the bulk value is made possible through a bond length conserving buckling of the O atoms.

The oxides with ABO_3 composition and perovskite structures have (100) surfaces that have the possibility of being terminated in either AO_x or BO_x structures. These structures are highly sensitive to the method of surface preparation and treatment. The surface termination, structure, and morphology of the $LaAlO_3(100)$[27] and $SrTiO_3(100)$[28] surfaces have been studied as a function of temperature by means of TOF-SARS, atomic force microscopy (AFM), LEED, Auger electron spectroscopy (AES), and X-ray photon spectroscopy (XPS). For $LaAlO_3(100)$, the results show that the surface is terminated in a Al-O layer from room temperature up to ~ 150°C and a La-O layer at temperatures above ~ 250°C. The surfaces are terminated exclusively in either Al-O or La-O layers, with mixed terminations observed only in the intermediate region of 150–250°C. A mechanism is proposed for this low-temperature surface stoichiometry change that is linked to the observation of the creation of surface oxygen deficiencies upon heating. The $SrTiO_3(100)$ exhibits a well-ordered TiO_2-terminated surface at low temperature that converts to a disordered surface with a more intense Sr signal following prolonged annealing in O_2.

7.5. ORGANIC MOLECULES ON SURFACES

7.5.1. Methanethiol on Pt(111)

The chemisorption site of the simplest prototypical model alkanethiol compound, methanethiol [CH_3SH], on a Pt(111) surface in the temperature range 298–1073 K has been investigated[29] by TOF-SARS and LEED. Typical TOF spectra from the clean Pt surface, the surface after exposure to $\sim 7 \times 10^2$ L of CH_3SH at 298 K, and the exposed surface after heating to 373 and 1073 K are shown in Figure 7.12. Upon exposure to a saturation dose of CH_3SH, the Pt(s) peak broadens due to multiple scattering contributions and additional peaks appear due to recoil of H, C, and S and scattering of Ar from S atoms. Heating to the higher temperatures results in loss of the H(r) peak and better definition of the C(r), S(r), and S(s) peaks.

A schematic drawing of the Pt(111) surface along with the azimuthal assignments is shown in Figure 7.13. Examples of α scans along three azimuthal directions for Ar scattering from Pt and C and S recoiling from the chemisorbed surface after heating to 1073 K are shown in Figure 7.14. The α_c's for Pt(s) are as expected for the azimuthal directions shown in Figure 7.13.

These results provide the following clues that are useful in determination of the ultimate positions of the S atoms: (1) The accessibility of the S atoms to the beam along the $\delta = 30$° and 90° azimuths is the same and different from that along the $\delta = 60$° azimuth. (2) The S atoms are severely shadowed or blocked along the 30° and 90° azimuths (i.e., they are only accessible at high α). Along the 60° azimuth, the S atoms are accessible at low α. The α scans for C(r) exhibit a similar behavior

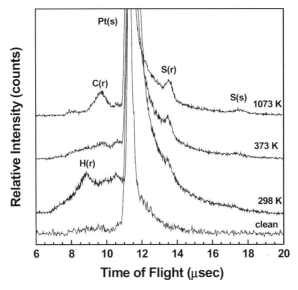

FIGURE 7.12. TOF-SARS spectra of 4-keV Ar$^+$ scattering and recoiling from a clean Pt(111) surface, the surface after exposure to $\sim$7 $\times 10^2$ L of CH$_3$SH at 298 K, and after heating the exposed surface to 373 and 1073 K. Incident angle $\alpha = 23°$; scattering angle $\beta = 45°$; azimuthal angle $\delta = 100°$ (From Kim *et al.,* 1998, with permission).

for all three δ directions, with a slowly increasing intensity over the broad range of $\alpha = 15$–$25°$ and centered at $\alpha_c \sim 19°$. This indicates that the C atoms are being shadowed and blocked by more than one neighboring atom and that these neighboring atoms are at different distances from the C atoms.

Azimuthal angle δ scans of the Ar scattering intensity Pt(s) from the clean Pt(111) surface and the sulfur and carbon recoiling intensities S(r) and C(r) from the surface after exposure to CH$_3$SH at 298 K and after heating to 373 and 1073 K are shown in Figure 7.15. The δ scan for the clean Pt surface exhibits well-defined 60° periodicity with deep, wide minima along the $\delta = 0°$, 60°, and 120° directions due to the short interatomic spacings along these azimuths. The minima along $\delta = 30°$ and 90° are shallow and narrow due to the long interatomic spacings along these azimuths. Chemisorption at 298 K results in a drastic reduction of the intensities and lifting of the 60° periodicity. The shapes of the minima at $\delta = 30°$ and 90° are different after chemisorption, indicating that these two directions are no longer equivalent. This difference becomes more apparent after heating to 373 K and 1073 K, where three sharp maxima are clearly observed in the region near $\delta = 30°$. This modification of the δ scan of scattered Ar upon chemisorption provides important clues to the nature of the chemisorption site of CH$_3$SH:

1. The reduction in scattering intensity indicates that the CH$_3$SH molecule perturbs the scattered ion trajectories (i.e., the molecules reside either on top of or very near the surface).

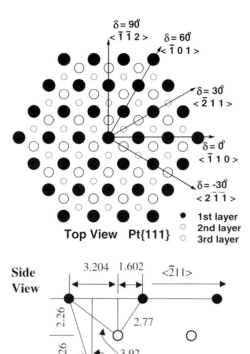

FIGURE 7.13. (Upper) Plan view of the ideal bulk-terminated Pt(111) surface illustrating the azimuthal angle δ assignments. (Lower) Vertical slice through the surface along the $\langle \bar{2}11 \rangle$ azimuth showing the ideal interatomic spacings and angles (From Kim *et al.*, 1998, with permission).

2. The lifting of the 60° periodicity shows that the chemisorption pattern reduces the surface periodicity to 120°.
3. The changes in the scattering features along the $\delta = 30°$ and 90° directions and the absence of change along the $\delta = 0°$, 60°, and 120° directions upon chemisorption indicates that the molecules reside at sites along the $\delta = 30°$ and 90° azimuths where they can directly perturb scattering trajectories.
4. The different scattering features along the $\delta = 30°$ and 90° azimuths indicate that the molecules are chemisorbed in specific sites that are not equivalent when viewed along these two different azimuthal directions.

Only minor periodic variations are observed in the δ scans of recoiled S and C upon chemisorption at 298 K. No periodicity was observed in the H recoil scans. These periodic variations increase upon heating to 373 K and become very distinct at 1073 K. The lack of a distinct periodicity at 298 K can be due to at least two phenomena. First, the molecules may not be in well-ordered sites upon chemisorption at room temperature. In this situation, migration of the sulfur moieties on the surface is also possible. Annealing may serve to organize these sulfur moieties into stable,

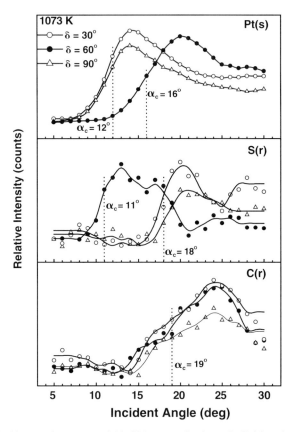

FIGURE 7.14. Incident angle α scans of 4-keV Ar scattering intensity Pt(s) and sulfur and carbon recoiling intensities S(r) and C(r) along the three azimuthal directions $\delta = 0°$, $30°$, and $90°$ from the Pt(111) surface after exposure to $\sim 7 \times 10^2$ L of CH_3SH at 298 K and after heating the exposed surface to 1073 K. The positions of the critical incident angles α_c are denoted by dashed lines (From Kim *et al.*, 1998, with permission).

well-ordered sites. Second, the many vibrational and rotational degrees of freedom of a molecule such as CH_3SH on a surface along with the large vibrational amplitudes of such light atoms results in a range of possible target atom positions that tends to obliterate the azimuthal features. The small periodic variations observed at 298 K imply that the molecules are in well-ordered sites, but the periodicities are unclear due to the vibrational and rotational excursions of the atoms. Well-defined $120°$ periodicities are observed for the recoiled S and C upon annealing, particularly for the 1073 K scan. The widths of the minima at $\delta = 30°$ and $90°$ are clearly different, in agreement with the observation from Ar scattered off Pt atoms. (i.e., the $60°$ periodicity becomes a $120°$ periodicity) This can only occur if the adsorbed molecule occupies a distinct threefold site and the dissociated constituent atoms also occupy distinct threefold sites.

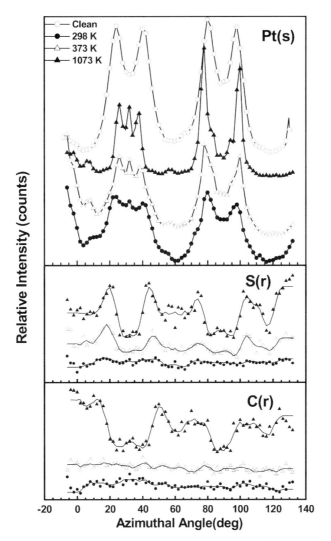

FIGURE 7.15. Azimuthal angle δ scans of Ar scattering intensity Pt(s) and sulfur and carbon recoiling intensities S(r) and C(r) for the clean Pt(111) surface, the surface after exposure to $\sim 7 \times 10^2$ L of CH_3SH at 298 K, and after heating the exposed surface to 373 and 1073 K. Incident angle $\alpha = 13°$; Scattering angle $\theta = 45°$ (From Kim *et al.*, 1998, with permission).

For identification of the specific threefold site(s) of the adsorbates, it is necessary to resort to the use of classical ion trajectory simulations on models of the chemisorbed surface. Examples of calculated shadow cones for 4-keV Ar collisions with Pt, S, and C atoms are shown in Figure 7.16. The accuracy of the Ar/Pt shadow cone shape was verified by using the known Pt interatomic spacings (d's) along specific azimuths and comparison to the experimental critical incident angles (α_c). For

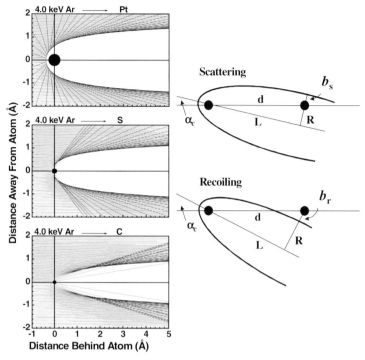

FIGURE 7.16. (Left) Examples of calculated shadow cones for 4-keV Ar collisions with Pt, S, and C atoms. (Right) Schematic diagram showing the use of the experimental critical incident angles (α_c's), the calculated shadow cone shapes (i.e., R versus L), and the calculated impact parameters (b) to determine the positions of the S and C atoms on the Pt surface. The p_s is the impact parameter for scattering of the projectile atom into angle θ, and the p_r is the impact parameter for recoiling of a target atom into angle ϕ. R is the radius of the shadow cone at a distance L behind the target atom (From Kim *et al.*, 1998, with permission).

scattering, the radius R of the shadow cone at a distance L behind a target atom is related to α_c, d, and b_s as

$$R = (d \sin \alpha_c) - b_s \quad \text{and} \quad L = (d \cos \alpha_c), \tag{1}$$

where b_s is the impact parameter for scattering into angle θ. For recoiling, the radius is modified as

$$R = (d \sin \alpha_c) + b_r, \tag{2}$$

where b_r is the impact parameter for recoiling into angle ϕ. Note that for heavy projectiles colliding with light atoms, some of the scattered trajectories at small p values penetrate into the repulsive potential of the cone, resulting in a poorly defined shadow cone radius. This is particularly apparent for the light C atom. The consequence of this indistinct cone radius and penetration of Ar trajectories is that the C atoms are extremely poor shadowers of neighboring atoms. The S atom provides a somewhat better defined cone radius with little penetration of the Ar trajectories.

The chemisorption site of methanethiol and its subsequent dehydrogenation products on Pt(111) have been determined[29] from the LEED pattern, the calculated shadow cones, and the experimental α and δ scans. The data from the 1073 K measurements was used in this analysis since it provided the most distinct features. There are many possible chemisorption sites and atomic positions that can be selected for the model calculations. The structural information obtained from the LEED data and the incident angle α and azimuthal angle δ scans were used to reduce the number of possible chemisorption sites to a tractable level. Using this evidence and a triangulation approach, the only chemisorption structure that was found to be consistent with all of the experimental data is one in which the S and C atoms are above the surface in fcc and hcp three-fold sites, respectively, as shown in Figure 7.17. These data are consistent with the S and C atoms residing at sites that are $\sim$1.6 $\pm$0.2 and $\sim$1.5 $\pm$0.4 Å, respectively, above the surface and Pt-S, Pt-C, and S-C bond distances of $\sim$2.3 $\pm$ 0.1, 1.7 $\pm$ 0.1, and 2.2 $\pm$ 0.2 Å, respectively, as shown in Fig. 7.17(c). The S-C moieties are shown in Figure 7.17(d) arranged in a $(\sqrt{3} \times \sqrt{3})R$ 30° pattern with the S atoms in fcc sites and the C atoms in hcp sites aligned along the $\langle \bar{2}11 \rangle$ azimuth

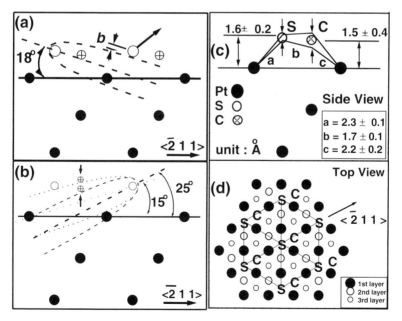

FIGURE 7.17. Illustration of the use of experimental critical incident angles α_c and calculated shadow cones to determine the interatomic spacings and relative heights of (a) the sulfur (S) atoms and (b) the carbon (C) atoms on the Pt(111) surface. (c) Final positions of the S and C atoms above the Pt(111) surface as determined from the TOF-SARS and LEED results. (d) Illustration of an S–C adsorbate structure that would produce a $(\sqrt{3} \times \sqrt{3})R$ 30° LEED pattern (From Kim et al., 1998, with permission).

at $\delta = 30°$. These C atoms could also occupy the equivalent hcp sites aligned with the $\langle \bar{1}\bar{1}2 \rangle$ azimuth at $\delta = 90°$ or the $\langle 2\bar{1}\bar{1} \rangle$ azimuth at $\delta = -30°$. Random occupancy of these equivalent hcp sites by the C atoms would still give a $(\sqrt{3} \times \sqrt{3})R\,30°$ pattern.

The above analysis is based on the elevated temperature data. The room temperature data are qualitatively similar. The 298 K LEED pattern has the same symmetry $[(\sqrt{3} \times \sqrt{3})R30°]$ as the 993 K pattern, although it is faint and has an increased background intensity; the original (1×1) structure of the Pt surface remains. The 298 K incident and azimuthal angle scans have the same features as the 1073 K data, although they are broadened and less distinct. These results indicate that the room temperature chemisorption site is similar to that at the higher temperatures, although the heights of the S and C atoms above the surface may be different, and the adsorbates may not be as well ordered as they are at higher temperatures.

7.5.2. Other Organic Molecules on Surfaces

Determining the absolute coverage θ_{max} of a chemisorbed species on a solid surface is a delicate experiment in surface chemistry. A knowledge of θ_{max} is necessary for determination of the absolute sticking coefficient and may permit a better understanding of the adsorption process and the behavior of the adsorbed species under given conditions. The saturation coverage θ_{max} of ethylene C_2H_6 on Pt(111) has been determined[30] by using chemisorbed CO on Pt(111) as a standard system for calibration. The measurements were made with a pulsed 4-keV K^+ beam, allowing analysis of the directly recoiled hydrogen, carbon, and oxygen atoms. The results support a coverage of $\theta_{max} = 0.5$ ML of ethylidyne units on the surface.

TOF-SARS has been applied to obtain the molecular orientation at the surface of organic polymers. A model system consisting of monocrystalline thin layers of paraffin (hexatriacontane) with the lamellae oriented parallel to the substrate surface (polymer chains perpendicular to the surface) and with the lamellae oriented perpendicular to the substrate surface (polymer chain parallel to the surface) has been studied.[31] The fluxes of H and C atoms recoiling from 4-keV Ar^+ have been measured as a function of the beam incident angle and crystal azimuthal angle. The anisotropy observed in the azimuthal angle scans and the relative intensities of the H and O atom recoils in the incident angle scans provide information on the surface structure. It is possible to detect the backbone direction of a monocrystalline paraffin layer deposited with the molecular axes parallel to the substrate. When the lamellae are lying flat on the substrate surface, the presence of the methyl groups directed outward from the surface can be detected. The method could also be applied to well-organized organic monolayer films deposited by the Langmuir–Blodget technique. Precise quantitative structural information on the molecular configurations requires simulation of recoiling trajectories for the complex polymer structure.

Chemisorption of benzene and phenol on a clean Ni(110)-(1×1) surface and an oxygen-predosed Ni(110)-(3×1)-O surface has been investigated[32] by TOF-SARS. The Ne scattering and H, C, and O recoiling fluxes exhibited strong angular anisotropies as a function of beam incident and azimuthal angles. These anisotropies are due to C and O atoms shadowing their neighboring atoms within the benzene molecules and resulting phenoxide species. The results show that both benzene and phenoxide are chemisorbed as molecules that have very good short-range order despite the absence of long-range order observable by LEED. Both benzene and phenoxide are oriented nearly parallel to the surface, with a maximum inclination angle of 15°. The C atoms in the para positions of benzene and C—O bond in phenoxide are oriented along the ⟨001⟩ azimuth. The C—H bond is bent out of the plane of the hexagonal ring so that the H atoms are above the C atom plane.

Surface elemental and structural characterization of hexadecanthiol and heptade-canethiol (C_{16} and C_{17} for short) and 16,16,16-trifluorohexadecanethiol (FC_{16}) self-assembled monolayers (SAMs) on a Au(111) surface has been obtained from TOF-SARS.[33] The clean Au surface was also characterized by TOF-SARS and LEED in order to identify the azimuthal orientation of the SAMs with respect to the substrate. Classical ion trajectory simulations were used to relate the experimental scattering and recoiling data to the surface structure. The scattered and recoiled atoms originate from the outermost five to six atomic layers; azimuthal anisotropy was observed in the measurements. The results provide a model for the SAMs in which the alkyl chains chemisorb with the S atoms situated above the fcc threefold sites of the Au(111) surface to form a continuous film with a $(\sqrt{3} \times \sqrt{3})R30$ structure that fully covers the Au surface. The orientation of the molecular axis azimuth of the SAMs relative to the Au azimuthal directions was determined. The data indicate that the molecular chains have specific tilt and twist angles relative to the Au surface and six coexisting domains resulting from the six equivalent tilt directions of the molecular axis. Dramatic changes in the anisotropic patterns of the TOF-SARS azimuthal scans from the surfaces of the SAMs with different terminations were observed. These phenomena result from the different tilt angles of the CH_3 and CF_3 groups. The data are consistent with free rotation of both the CH_3 and CF_3 groups. The C_{16} SAM exhibited the clearest azimuthal features and was more resistant to radiation damage from the incident Ar^+ scattering beam than the other films. Due to the tilt angle of the SAMs, an "ion's eye view" of the structure (i.e., the positions of the atomic cores as experienced by the incoming keV ions) reveals a regular array of sloping cavities within each unit cell.

7.6. SEMICONDUCTOR SURFACES

7.6.1. Gallium Nitride (GaN) Epilayers on Sapphire

GaN epilayers are typically grown by metalorganic chemical vapor deposition (MOCVD) in the direction normal to the (0001) basal plane. This basal plane is similar to the (111) planes of zinc-blended structures that have a polar configuration (i.e., they can be terminated in either of the two atomic subplanes). The polarity of these surfaces can have important effects in semiconductor interfaces. The surface composition and structure of such GaN films have been studied by TOF-SARS.[34] The clean surfaces were prepared by cycles of sputtering (1-keV N_2^+ ions, 0.5 μA/cm², 10 min) and annealing (10 min) using a dynamic N_2 backfill. The measurements were carried out on several different GaN samples that were all prepared in the same manner, providing a random sampling of the terminations. Schematic drawings of the unreconstructed GaN($000\bar{1}$) N-terminated and (0001) Ga-terminated surfaces are shown in Figure 7.18; the $\langle 1000 \rangle$ azimuth is defined as $\delta = 0°$.

The termination layer was determined by using grazing incidence ($\alpha = 6°$) TOF-SARS at random azimuthal angles and a scattering–recoiling angle of $\theta = \phi = 40°$. Using random azimuthal angles and a low incident angle avoids the anisotropic effects of scattering along the principal low-index azimuths and provides scattering intensities that are from the first-atomic layer and exposed atoms in the second-atomic layer. Typical spectra from both clean GaN($000\bar{1}$) and (0001) surfaces taken along a random azimuthal direction (not aligned along a high symmetry azimuth) are shown in Figure 7.19. The Ga(s)/N(r) ratio is 1.82 on the (0001) surface and 0.52 on the ($000\bar{1}$) surface, indicating termination in Ga and N layers, respectively. The Ga-terminated surface exhibits a low H(r) intensity after cleaning, whereas the N-terminated surface exhibits a high H(r) intensity. The intense H(r) peak on the N-terminated surface was present under all conditions that gave rise to the clear (1×1) LEED pattern, indicating that the H concentration on the (1×1) surface is comparable to the N concentration.

The bulk crystallographic directions can be obtained by measuring the intensities along the $0°$ and $30°$ azimuths because there are two domains on the surfaces, as will be shown later. Along the $0°$ azimuth, the second-layer atoms are partially shadowed by the first-layer atoms because the lateral spacing between them is only $0.9\,\text{Å}$, and the radii of the shadow cones are of the order of $1\,\text{Å}$. Along the $30°$ azimuth, it is possible for the first-layer atoms to totally shadow second-layer atoms, making them inaccessible for scattering. The degree of shadowing is different for projectiles approaching along the $30°$ or $90°$ directions due to the arrangement of the first- and second-layer atoms. However, the I_{Ga} and I_N intensity ratios along the $30°$ and $90°$ directions are the same for both the ($000\bar{1}$) and (0001) surfaces; this indicates that both of the surfaces have two spatial domains that are rotated by $60°$. Example spectra used in obtaining these intensities are shown in Figure 7.20. For the ($000\bar{1}$) N-terminated surface, the second-layer Ga accessibility is extremely sensitive to azimuthal direction; specifically, it is greatly reduced along the $30°$ and $90°$ directions relative to the $0°$ and $60°$ directions. This is due to shadowing and blocking by first-layer N atoms. This confirms that these directions correspond to the $\langle 1\bar{1}00 \rangle$ azimuth or similar azimuths that are rotated by $60°$, such as $\langle 0\bar{1}10 \rangle$, $\langle \bar{1}010 \rangle$, $\langle \bar{1}100 \rangle$, $\langle 01\bar{1}0 \rangle$, or $\langle 10\bar{1}0 \rangle$ (see Figure 7.18). For the (0001)-(1×1) Ga-terminated surface, the first-layer Ga accessibility is greatly reduced along the $0°$ and $60°$ directions relative to the $30°$ and $90°$ directions. This is due to the enhanced self-shadowing and -blocking of Ga atoms by neighboring Ga atoms along the $0°$ and $60°$ directions as a result of the short interatomic spacings. Self-shadowing and blocking are reduced along the $30°$ and $90°$ azimuths due to the long interatomic spacings. This confirms that the $0°$ and $60°$ directions correspond to the $\langle 1000 \rangle$ azimuth or similar azimuths that are rotated by $60°$, such as $\langle 0\bar{1}00 \rangle$, $\langle 0010 \rangle$, $\langle \bar{1}000 \rangle$, $\langle 0100 \rangle$, or $\langle 00\bar{1}0 \rangle$ (see Figure 7.18).

The surface periodicities of the H, N, and Ga atoms can be determined by monitoring I_{Ga} for Ar scattering from Ga atoms and I_H and I_N for Ar recoiling of H and

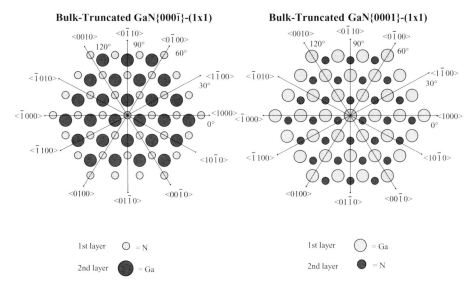

FIGURE 7.18. Plan view of the ideal bulk-terminated, GaN(000$\bar{1}$)-(1 × 1) N-terminated, and (0001)-(1 × 1) Ga-terminated surfaces illustrating the azimuthal angle δ assignments. Another domain is obtained by 60° rotation of this surface about the surface normal (From Ahn *et al.*, 1997, with permission).

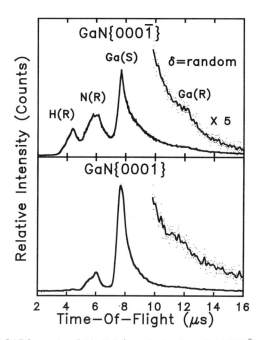

FIGURE 7.19. TOF-SARS spectra of 4-keV Ar$^+$ scattering from GaN (000$\bar{1}$) and (0001) surfaces with the ion beam aligned along a random azimuthal direction. Incidence angle $\alpha = 6°$; Scattering angle $\theta = 40°$ (From Ahn *et al.*, 1997, with permission).

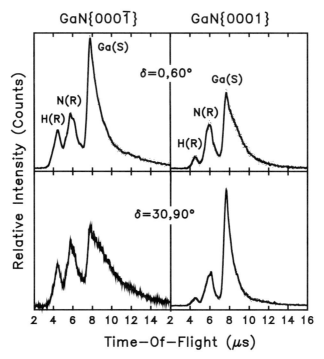

FIGURE 7.20. Typical TOF-SARS spectra of 4-keV Ar^+ scattering from GaN $(000\bar{1})$ and (0001) surfaces with the ion beam directed along the $\langle 1000 \rangle$ $(\delta = 0°)$ and $\langle 1\bar{1}00 \rangle$ $(\delta = 30°)$ azimuths. Spectra from $\delta = 60°$ are identical to those from $\delta = 0°$, and spectra from $\delta = 90°$ are identical to those from $\delta = 30°$. Incident angle $\alpha = 12°$; scattering angle $\theta = 40°$ (From Ahn *et al.*, 1997, with permission).

N atoms as a function of crystal azimuthal angle δ. Figures 7.21 and 7.22 show the azimuthal anisotropy of scattering and recoiling intensities for both the $(000\bar{1})$ and (0001) surfaces, respectively. The minima are coincident with low-index azimuths where the surface atoms are inside of the shadowing or blocking cones cast by their aligned, closely spaced nearest neighbors, resulting in low intensities. Wide, deep minima are expected from short interatomic spacings because of the larger degree of rotation about δ required for atoms to emerge from neighboring shadows. Schematic diagrams of the surface structures illustrating the scattering and blocking directions are presented along with the δ-scans.

The I_N δ-scans were performed in the shadowing mode (i.e., a grazing incident angle of $\alpha = 10°$ was used). For the $(000\bar{1})$ surface (Figure 7.21), the data exhibit minima every 30°, with the minima labeled 0° (C) and 60° (E) being deeper than those at 30° (D) and 90° (F). The 0° and 60° directions correspond to the $\langle 1000 \rangle$ and $\langle 0\bar{1}00 \rangle$ azimuths, respectively, where the N–N interatomic spacings are shortest and self-shadowing of first-layer N atoms by neighboring first-layer N atoms is most efficient. The 30° and 90° minima correspond to the $\langle 1\bar{1}00 \rangle$ and $\langle 0\bar{1}10 \rangle$ azimuths, respectively, where the N–N interatomic spacings are longer and self-shadowing is less efficient.

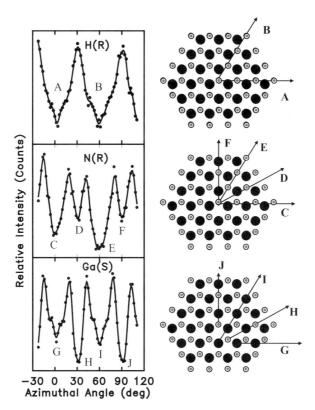

FIGURE 7.21. Azimuthal angle δ-scans of hydrogen (H) atom recoil H(r), nitrogen (N) atom recoil N(r), and argon (Ar$^+$) scattering from gallium (Ga) atoms Ga(s) on the GaN(000$\bar{1}$) surface. The projectile is 4-keV Ar$^+$, and the conditions are $\alpha = 20°$, $\theta = 30°$ for H(r), $\alpha = 10°$, $\theta = 40°$ for N(r), and $\alpha = 12°$, $\theta = 90°$ for Ga(s) (From Ahn *et al.,* 1997, with permission).

The (0001) surface (Figure 7.22) has a very similar I_N azimuthal pattern, except that there is a 30° shift in the position of each corresponding minimum with respect to that of the (000$\bar{1}$) surface; the minima along 30° (B) and 90° (D) are deeper than those at 0° (A) and 60° (C). The second-layer N atoms are efficiently shadowed at 30° and 90° by the first-layer Ga atoms, although they are only partially shadowed and blocked at 0° and 60°. As shown in the model (Figure 7.18), the degree of shadowing and blocking along the 30° and 90° azimuths is not identical due to the relationship of the first- and second-layer atoms. The minima observed at these two angles have identical widths and depths, suggesting that the surface exists in two different domains; if there would be only one domain, different widths and depths for these minima would be expected.

For the (000$\bar{1}$) surface (Figure 7.21), the data also exhibit minima every 30°, but in contrast to the I_N minima, the minima labeled 30° (H) and 90° (J) are much deeper than those at 0° (G) and 60° (I). Shadowing and blocking of second-layer Ga by first-layer N is most efficient along the 30° and 90° directions due to the aligned overlayer

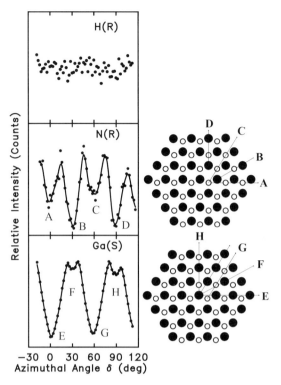

FIGURE 7.22. Azimuthal angle δ-scans of hydrogen (H) atom recoil H(r), nitrogen (N) atom recoil N(r), and gallium atom scattering Ga(s) on the GaN(0001) surface. The scattering conditions are identical to those of Figure 7.21 (From Ahn *et al.*, 1997, with permission).

N atoms. Along the $0°$ and $60°$ directions, the overlayer N atoms are not aligned with Ga atoms, resulting in less severe shadowing and blocking. The minima observed at $30°$ and $90°$ have identical widths and depths, also suggesting that the surface exists in two different domains. The periodicity of the I_{Ga} δ-scan of the (0001) surface (Figure 7.22) shows a $30°$ shift, in terms of the positions of minima, with respect to that of $(000\bar{1})$ surface. The shallow minima along the $30°$ (F) and $90°$ (H) azimuths result from the long interatomic spacings along these directions, and the deep minima at $0°$ (E) and $60°$ (G) result from the short interatomic spacings along these directions.

The I_H δ-scans of Figures 7.21 and 7.22 were performed in the blocking mode (i.e., a small exit angle $\beta = 10°$ was used). For the $(000\bar{1})$ surface (Figure 7.21), the data exhibit $60°$ periodicity with deep minima along the $0°$ and $60°$ directions, unlike the I_N and I_{Ga} scans. The H atoms are too light to cause significant shadowing of the incoming Ar projectiles. However, recoiled H atoms can be blocked by first-layer neighboring N atoms. The radii of the shadowing and blocking cones for H atom collisions with N and Ga atoms are only $\sim 0.3\,\text{Å}$ compared to $>1\,\text{Å}$ for Ne^+ and Ar^+ collisions with N and Ga atoms. By using a small exit angle β, recoiled H atoms can be blocked by nearest neighbor N atoms. The observed periodicity arises if the H atoms are bound to the outerlayer N atoms and protruding outward from the surface plane. The short H–N interatomic spacings along $0°$ and $60°$ result in overlayer recoiled H atoms being blocked by their first-layer N neighbors. Along the $30°$ and $90°$ directions, the H–N interatomic spacings are longer, and the recoiling H trajectories are outside of the neighboring N atom blocking cones. In contrast to the I_H δ-scan of the $(000\bar{1})$ surface, the (0001) surface (Figure 7.22) shows no appreciable periodicity in the I_H-δ-scan. This indicates the H atoms are most likely randomly distributed on the surface.

These results show that it is possible to identify and characterize both N-terminated $(000\bar{1})$-(1×1) and Ga-terminated (0001)-(1×1) surfaces GaN(0001) films. No evidence for the presence of mixed terminations was observed. For the $(000\bar{1})$ surface, H atoms are bound to the N atoms of the outmost layer with a coverage of $\sim 3/4$ ML and protrude outward from the surface. The (0001) surface is highly reactive towards adsorption of C and O from residual gases; however, unlike the $(000\bar{1})$ surface, it adsorbs very little hydrogen. This Ga-terminated surface is stabilized and has a more ordered structure as a result of the contamination. Both surfaces exhibit two structural domains.

7.6.2. Other Semiconductor Surfaces

The composition and structure of the (1×1) and (4×2) phases of InP(100)[35] surfaces, InAs(100)-(4×2)[36] epilayers on InP(100), and GaP(100)-(4×2)[37] have been characterized by TOF-SARS and LEED. For the InP(100)[35] surface, the results show that the unreconstructed (1×1) phase is terminated in an In layer. Because this ordered (1×1) structure is only obtained in the presence of surface impurities (hydrogen, carbon, and oxygen) and excess phosphorus, it is not a stable intrinsic phase of clean InP(100). The reconstructed (4×2) phase is terminated in an In layer with second-layer P atoms exposed in the In missing row troughs. This reconstruction is an In missing-row-trimer P dimer (MRTD) structure in which every fourth In $\langle 0\bar{1}1 \rangle$ row is missing, the In atoms are trimerized along the $\langle 011 \rangle$ azimuth, and the second-layer P atoms exposed in the $\langle 0\bar{1}1 \rangle$ troughs are dimerized. For the InAs(100)-(4×2) epilayers on InP, the data[36] are consistent with In-termination and a missing-row-dimer (MRD) reconstruction. For the GaP(100)-(4×2)[37] surface, the data indicate termination in a Ga layer with second-layer P atoms exposed in the Ga missing row troughs. This is a Ga MRTD structure in which every fourth Ga $\langle 0\bar{1}1 \rangle$ row is missing, the Ga atoms are trimerized along the $\langle 011 \rangle$ azimuth, and the second-layer P atoms exposed in the $\langle 0\bar{1}1 \rangle$ troughs are dimerized. These mixed semiconductor surfaces are compatible with the autocompensation-electron counting rules. The basic assumption of autocompensation is that for reconstructed semiconductor surfaces to be stable, they must have either completely filled or completely empty dangling bonds that yield semiconducting behavior. According to the electron counting model, the dangling bonds on the more electronegative atom must be filled, and the dangling bonds on the more electropositive atom must be empty.

TOF-SARS spectra[38] of Ge epilayers grown on a Si(100)-(2×1) substrate have been compared to those of a clean Ge(100) surface. The Ge epilayer has a well-ordered buckled dimer, consistent with a (2×1) LEED pattern. The epilayer grows by an islanding mode in which deposited Ge atoms migrate together to form two-dimensional islands with (2×1) symmetry. The buckling of the dimers is not observed on the Ge(100) crystal, implying that the strain in the overlayer results in an increased barrier height for dimer interconversion.

Si epilayers grown on a Ni(100)-(1×1) substrate that exhibit a c(2×2) structure have been investigated[39] by TOF-SARS. The alignment of the azimuthal patterns of the Si and Ni recoils shows that the Si atoms occupy fourfold hollow sites on the Ni surface. The incident angle α scans are highly sensitive to defects in the c(2×2)-Si epilayer.

7.7. EPILAYERS ON NICKEL

7.7.1. Silver on Ni(100)

Monolayer thick Ag films were grown[40] on a Ni(100) surface using direct ion beam deposition of $^{107}Ag^+$ ions with 20-eV kinetic energy and the substrate at 25°C. Deposition was performed by means of a mass-selected, low-energy, ultrahigh vacuum ion beam system with a well-defined ion energy. Typical TOF spectra from the clean Ni(100) surface and this surface after deposition of $\sim 2 \times 10^{15}$ Ag atoms/cm^2 are shown in Figure 7.23. The spectrum of Ni exhibits a single sharp, intense peak centered at 6.2 μs due to scattering of Ar from Ni, [i.e., Ni(s)] and a broad, low-intensity

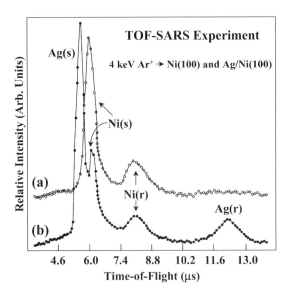

FIGURE 7.23. TOF-SARS spectra of 4-keV Ar$^+$ scattering from (a) a clean nickel Ni(100) surface and (b) a silver (Ag) film on the Ni(100) surface. Ni(s) and Ag(s) are quasi-single scattering peaks of argon (Ar) from Ni and Ag atoms, respectively. Ni(r) and Ag(r) peaks are due to direct recoiling of Ni and Ag atoms, respectively, from Ar atoms. Incident angle $\alpha = 15°$; scattering and recoiling angle $\theta = 30°$; azimuthal angle $\delta = 30°$ (From Todorov *et al.,* 1999, with permission).

peak centered at 7.9 μs due to recoiling of Ni [i.e. Ni(r)]. Two additional peaks are observed upon deposition of Ag; a sharp, intense peak centered at 5.6 μs due to scattering of Ar from Ag [i.e., Ag(s)] and a broad, low-intensity peak centered at 12.3 μs due to recoiling of Ag [i.e., Ag(r)]. Although the Ag/Ni sample has only approximately one monolayer of Ag on the surface, the Ag(s) peak is ~2.5 times more intense than the Ni(s) peak. This is a result of two factors: first, the Ag layer is above the Ni, causing severe shadowing and blocking of the Ni atoms; second, the differential scattering cross section of Ar from Ag is higher than that from Ni; the cross section ratio is Ω_{Ag}/Ω_{Ni} ~1.4. The low intensity of the recoil peaks compared to the scattering peaks is typical and is due to the lower recoiling cross sections. These results are consistent with a monolayer Ag film covering the Ni substrate. If the film would be thicker than a single monolayer, Ni(s) and Ni(r) would not be observed because of the shallow sampling depth of TOF-SARS.

TOF-SARS spectra simulated by means of the SARIC program are shown in Figure 7.24. The incident and scattering angles used in the simulation were identical to those of Figure 7.23, and the amplitude of the thermal vibrations was 0.1 Å. The simulation of the clean Ni spectrum exhibits the Ni(s) and Ni(r) peaks in good agreement with Figure 7.23. The simulation of the Ag/Ni film was obtained by constructing a sample with a single monolayer of Ag arranged in a fcc(111) structure on the Ni(100) surface. The simulated spectrum exhibits the additional Ag(s) and Ag(r) peaks as well as the Ni peaks with intensities in good agreement with the experimental spectrum. These simulated spectra serve two purposes. First, they provide a positive identification for the peaks observed in the experimental spectra. Second, they confirm that the Ag overlayer is approximately one monolayer thick.

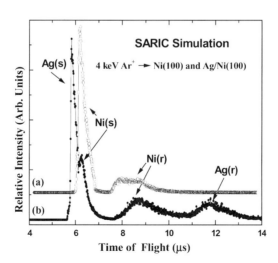

FIGURE 7.24. Simulated TOF-SARS spectra of 4-keV Ar$^+$ scattering from (a) a clean Ni(100) surface and (b) a single monolayer silver (Ag) film with a fcc (111) structure on the Ni(100) surface. Incident angle $\alpha = 15°$; scattering and recoiling angle $\theta = 30°$; azimuthal angle $\delta = 30°$ (From Todorov *et al.*, 1999, with permission).

Note that the simulated Ni(r) peak of Figure 7.24 is broad and appears to be composed of two overlapping components centered near 7.8 and 8.6 μs. Analysis of the associated trajectories shows that the short and long time components result from Ni recoils that originate from the first- and second-layers, respectively. These components are not resolved in the experimental spectrum. The simulated Ni(r) peak of Figure 7.24(b) is composed of only one broad component whose peak is centered at ~8.6 μs. This position is closest to the long time component of Figure 24(a). The trajectories indicate that this peak is broad due to the fact that they originate from the second layer and suffer multiple collision energy losses. The position of the experimental peak in Figure 7.23(b) indicates that the deposited film has a small amount of Ni atoms in the uppermost layer due to mixing processes induced by the energetic Ag atoms.

The surface periodicity of the structures can be determined by monitoring the scattering or recoiling intensities as a function of the azimuthal angle δ. Ni and Ag recoil δ scans are shown in Figure 7.25. The recoil peaks were used for this applica-

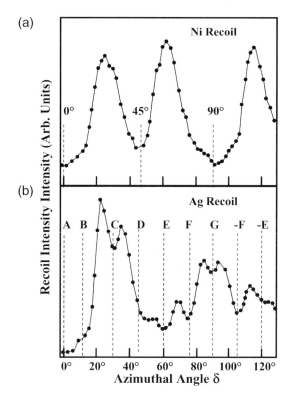

FIGURE 7.25. Azimuthal angle δ scans of the nickel (Ni) and silver (Ag) atom recoil intensities from 4-keV Ar$^+$ ions on (a) the clean Ni(100) surface and (b) a Ag film on the Ni(100) surface. Incident angle $\alpha = 15°$; scattering and recoiling angle $\theta = 30°$. The azimuthal positions labeled by letters A to G on (b) correspond to the arrows on the structural diagram of Figure 7.26 (From Todorov et al., 1999, with permission).

tion because they are well separated from each other; the sloping background from the Ni(s) peak is easily subtracted from the Ni(r) peak. The Ag(s) and Ni(s) peaks have significant overlap and are more difficult to deconvolute with quantitative accuracy. The δ scan for the clean Ni(100) surface exhibits well-defined fourfold or 90° periodicity. Deep, wide minima are observed along the $\langle 001 \rangle$ and $\langle 010 \rangle$ directions due to the short first-layer interatomic spacings along these azimuths. The full widths of these minima at half-height (FWMH) are $\Delta\delta \sim 37°$. The narrow minimum observed along the $\langle 011 \rangle$ direction with FWMH of $\Delta\delta \sim 21°$ is due to the long interatomic spacings along this azimuth. These azimuthal alignments are illustrated in Figure 7.26.

The δ scan for the Ag overlayer in Figure 7.25 exhibits an entirely different pattern from that of the clean Ni(100) surface. Wide, deep minima are observed only along the $\delta = 0°$ and 60° directions (A and E), indicative of close-packed azimuths. Several overlapping minima are observed at other δ positions. These complex δ features can all be interpreted if the Ag overlayer is in a fcc (111) structure, as shown in Figure 7.26. Some of the indicated directions in Figure 7.26, such as B_1 and B_2, D_1 and D_2, as well as F_1 and F_2, are not resolved in the experimental δ scan of Figure 7.25 because of the small angular differences between them. These results clearly show that the Ag overlayer has a fcc (111) structure and not the fcc (100) structure of the Ni substrate.

7.7.2. Chlorine on Ni(110)

Chlorine chemisorbed on the Ni(110) surface has been investigated[41] by SARIS in order to determine the Cl site on the Ni substrate. SARIC simulations with Cl at different sites on the Ni surface were used for comparison to the experimental images.

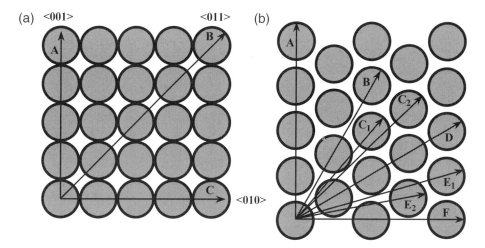

FIGURE 7.26. Plan views of (a) an ideal fcc Ni(100) surface and (b) an ideal fcc Ag(111) surface. The arrows correspond to the nearest-neighbor directions of the structures (From Todorov *et al.*, 1999, with permission).

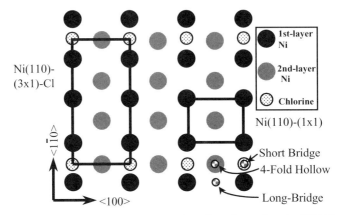

FIGURE 7.27. Schematic diagram of the Ni(110) surface illustrating the Ni(110)-(1 × 1) and Ni(110)-(3 × 1)-Cl structures and the positions of the various Cl chemisorption sites considered in the simulations (From Houssiau *et al.,* 1999, with permission).

Three different chemisorption sites were examined: the short-bridge sites (SB) above the $[1\bar{1}0]$ rows, the long-bridge sites (LB) above the [100] rows, and the four-fold hollow site (H) located in the center of the unit cell. These sites are shown in Figure 7.27. For each site, a Ni-Cl distance of 2.24 Å was chosen, corresponding to the sum of the average covalent radii of Ni (0.99 Å) and Cl (1.25 Å).

Figure 7.28(a) represents a composite SARIS image, obtained by merging three time-integrated experimental frames, taken along the $\delta = 0°$ [100] azimuth. The exit angles cover a range from 0–60°. The image displays the two focusing spots. Two

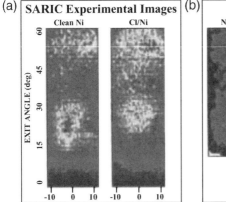

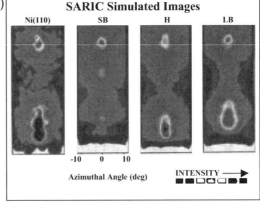

FIGURE 7.28. (a) Time-integrated composite SARIS images and (b) SARIC simulations for 4-keV Ne$^+$ scattering on the clean Ni(110)-(1 × 1) and Ni(110)-(3 × 1)-Cl surfaces along the $\delta = 0°$ [100] azimuth. Incident angle $\alpha = 40°$; scattering angle $\theta = 85°$; exit angle $\beta = 45$; SB–Cl, in short-bridge site; LB–Cl, in long-bridge site; H–Cl, in fourfold hollow site (From Houssiau *et al.,* 1999, with permission).

distinct differences are observed upon chemisorption: (1) The intensity of the low β spot is decreased relative to that of the high β spot. (2) The low β spot is shifted to higher exit angles by about $5°$ relative to its position on the clean surface. On clean Ni, this spot has an elongated shape and a high intensity centered at $\beta = 20°$. After exposure to Cl_2, the spot has a weaker intensity, centered at $\beta = 26°$, and the upper spot is now the most intense one. An increased intensity at $\sim\beta = 40°$ is also observed.

SARIC simulations for 4-keV Ne^+ on the clean Ni(110) surface provide good agreement with the experimental data, as shown in Figure 7.28(b). The simulations reproduce the two spots of high intensity at the correct β angles. Figure 7.28(b) shows the images simulated for the short bridge (SB) site, the long bridge (LB) site, and the fourfold hollow (H) site. For this particular scattering geometry, only the SB chemisorption site induces a strong variation in the intensities of the focusing spots. The intensity of the upper β spot increases relative to that of the lower β spot, as is observed in the SARIS frames of Figure 7.28(a). The lower spot evolves from a high-intensity elongated spot centered at $\sim20°$ to a low-intensity spot centered at $\sim26°$, which agrees with the experimental data. The simulation also suggests the appearance of a third area of moderately high intensity, located between the two other spots, that reflects the increased intensity at $\sim40°$ in Figure 7.28(a) after chemisorption. The simulated images for Cl in the LB and H sites exhibit only a minor perturbation, in poor agreement with the experimental images.

Based on these results, it is concluded that the SB position is the chemisorption site. Trajectory analyses show that the low exit angle spot is due to focusing of Ne ions into the [100] troughs. The presence of Cl atoms situated above these troughs in the SB sites defocuses the trajectories and dramatically perturbs the focusing, unlike the other sites. The center of the focusing spot is also raised to higher exit angles by about $5°$, which again is consistent with the experimental data and is not explainable with the LB or H sites.

REFERENCES

1. J. Lindhard, *Mat. Fys. Medd. Dan. Vid. Selsk.* **34,** 14 (1965).

2. F. Masson, and J. W. Rabalais, *Chem. Phys. Lett.* **179,** 63 (1991).

3. F. Masson, and J. W. Rabalais, *Surf. Sci.* **253,** 245 (1991).

4. C. Hoefner, and J. W. Rabalais, *Surf. Sci.* **400,** 189 (1998).

5. C. Hoefner, and J. W. Rabalais, *Phys. Rev.* **B 58,** 9990 (1998).

6. O. Grizzi, M. Shi, H. Bu, J. W. Rabalais, and P. Hochman, *Phys. Rev.* **B 40,** 10127 (1989).

7. M. Shi, H. Bu, and J. W. Rabalais, *Phys. Rev.* **B 42,** 2852 (1990).

8. J. Ahn, and J. W. Rabalais, *J. Phys. Chem.* **B 102,** 223 (1998).

9. C. Kim, J. Ahn, V. Bykov, and J. W. Rabalais, *Intern. J. Mass Spectrom. Ion Phys.* **174,** 305 (1998).

10. J. L'Ecuyer, C. Brassard, C. Cardinal, J. Chabbal, L. Deschenes, J. P. Labrie, B. Terrault, J. G. Martel, and R. St.-Jacques, *J. Appl. Phys.* **47,** 881 (1976).

11. G. Amsel, and W. A. Lanford, *Ann. Rev. Nucl. Part.Sci.* **34,** 435 (1984) and references therein.

12. K. M. Lui, I. Bolotin, A. Kutana, W. M. Lau, and J. W. Rabalais, *J. Chem. Phys.* **111,** 11095 (1999).

13. K. M. Lui, Y. Kim, W. M. Lau, and J. W. Rabalais, *Appl. Phys. Lett.* **75,** 587 (1999).

14. K. M. Lui, Y. Kim, W. M. Lau, and J. W. Rabalais, *J. Appl. Phys.* **86,** 5256 (1999).

15. P. J. Feibelman, and D. R. Hamman, *Surf. Sci.* **182,** 411 (1987).

16. Y. Wang, M. Shi, and J. W. Rabalais, *Phys. Rev.* **B 48,** 1678 (1993).

17. M. Shi, Y. Wang, and J. W. Rabalais, *Phys. Rev.* **B 48,** 1689 (1993).

18. Y. Wang, S. V. Teplov, O. S. Zaporozchenko, V. Bykov, and J. W. Rabalais, *Surf. Sci.* **296,** 213 (1993).

19. C. Roux, H. Bu, and J. W. Rabalais, *Surf. Sci.* 259, 253 (1991); H. Bu, C. D. Roux, and J. W. Rabalais, *Surf. Sci.* **271,** 68 (1992).

20. O. Grizzi, M. Shi, H. Bu, J. W. Rabalais, R. R. Rye, and P. Nordlander, *Phys. Rev. Lett.* **63,** 1408 (1989); M. Shi, O. Grizzi, H. Bu, J. W. Rabalais, R. R. Rye, and P. Nordlander, *Phys. Rev.* **B 40,** 10163 (1989).

21. I. L. Bolotin, A. Kutana, B. Makarenko, and J. W. Rabalais, *Surf. Sci.* **472,** 205 (2001).

22. H. Bu, C. D. Roux, and J. W. Rabalais, *J. Chem. Phys.* **97,** 1465 (1992); C. D. Roux, H. Bu, and J. W. Rabalais, *Surf. Sci.* **279,** 1 (1992).

23. C. D. Roux, H. Bu, and J. W. Rabalais, *Chem. Phys. Lett.* **200,** 60 (1992).

24. H. Bu, O. Grizzi, M. Shi, and J. W. Rabalais, *Phys. Rev.* **B 40,** 10147 (1989).

25. H. Bu, and J. W. Rabalais, *Surf. Sci.* **301,** 285 (1994).

26. J. Ahn, and J. W. Rabalais, *Surf. Sci.* **388,** 121 (1997).

27. J. Yao, P. B. Merrill, S. S. Perry, D. Marton, and J. W. Rabalais, *J. Chem. Phys.* **108,** 1645 (1998).

28. P. A. W. van der Heide, Q. D. Jiang, Y. S. Kim, and J. W. Rabalais, *Surf. Sci.* **473,** 59 (2001).

29. S. S. Kim, Y. Kim, H. I. Kim, S. H. Lee, T. R. Lee, S. S. Perry, and J. W. Rabalais, *J. Chem. Phys.* **109,** 9574 (1998).

30. F. Masson, C. S. Sass, O. Grizzi, and J. W. Rabalais, *Surf. Sci.* **221,** 299 (1989).

31. P. Bertrand, H. Bu, and J. W. Rabalais, *J. Phys. Chem.* **97,** 13788 (1993).

32. H. Bu, P. Bertrand, and J. W. Rabalais, *J. Chem. Phys.* **98,** 5855 (1993).

33. L. Houssiau, M. Graupe, R. Colorado, Jr., H. I. Kim, T. R. Lee, S. S. Perry, and J. W. Rabalais, *J. Chem. Phys.* **109,** 9134 (1998).

34. J. Ahn, M. M. Sung, J. W. Rabalais, D. D. Koleske, and A. E. Wickenden, *J. Chem. Phys.* **107,** 9577 (1997).

35. M. M. Sung, C. Kim, H. Bu, D. S. Karpuzov, and J. W. Rabalais, *Surf. Sci.* **322,** 116 (1995).

36. M. M. Sung, and J. W. Rabalais, *Surf. Sci.* **356,** 161 (1996).

37. M. M. Sung, and J. W. Rabalais, *Surf. Sci.* **365,** 136 (1996).

38. Y. Wang, H. Bu, T. E. Lytle, and J. W. Rabalais, *Surf. Sci.* **318,** 83 (1994).

39. Y. Wang, V. Bykov, and J. W. Rabalais, *Surf. Sci.* **319,** 329 (1994).

40. S. S. Todorov, H. Bu, K. J. Boyd, J. W. Rabalais, C. M. Gilmore, and J. A. Sprague, *Surf. Sci.* **429,** 63 (1999).

41. L. Houssiau, and J. W. Rabalais, *Nucl. Instrum. Meth. Phys. Res.* **B 157,** 274 (1999).

8

ION–SURFACE CHARGE EXCHANGE AND INELASTIC ENERGY LOSSES

When keV ion beams are scattered from surfaces, both neutrals and ions are observed in the scattered and recoiled flux. One of the main characteristics of scattering is the ion fraction Y, defined as the ratio of the number of ions to the total number of particles collected at a given solid angle

$$Y^{+,-} = \frac{I^{+,-}}{I^{+,-} + I^0},\tag{1}$$

where $I^{+,-}$ is a number of recoiled or scattered positive or negative ions and I^0 is the number of neutral particles. These ion fractions are known to be dependent upon a number of parameters, including the ion energy (or velocity), the scattering angle and trajectory, orientation of the target surface, the electronic structure of the solid surface, ion, and corresponding neutralized ion, and the adsorbate coverage of the surface. They vary from near 0 to $\sim 50\%$ for noble gas ions and are very high $> \sim 80\%$ for alkali ions. Ion fractions are experimentally measured integral quantities that are a function of the charge exchange probability between the scattered or recoiled particle and the solid.

Charge exchange processes during the collisions of ions and atoms with surfaces have been actively studied experimentally and theoretically for a long time.[1–4] The interest has been stimulated by the role that they play in phenomena from different research areas, such as heterogeneous catalysis, electronic emission, plasma wall interactions, film growth, and aerospace engineering. For example, ions that have lost energy in collisions with a thermonuclear reactor wall and were neutralized may return to plasma, since they are not affected by the magnetic field; this can lower the temperature of the plasma. The electron transfer between the particle and the solid is a very basic step in many of these processes. Therefore the study of charge exchange and related phenomena at a basic theoretical and experimental level is crucial for the

181

understanding and progress in these fields. The interest is maintained not only by the fundamental importance of the process, but also by it being a base for many analysis methods.

The interaction of the scattering particle with electrons from the solid has two aspects—inelastic energy losses and charge exchange. These phenomena are interrelated: the process of ionization of inner shells of the incoming particle or atom in the solid leads to loss of some of the kinetic energy of the projectile. The described range of the energies of scattered particles (several keV) is characterized by the domination of elastic losses over inelastic losses. Inelastic losses are composed mainly of losses due to interaction of the projectile with electrons in the solid; that is, the kinetic energy goes into ionization and excitation of the collision partners, kinetic energy of the released electrons, and energy carried away with radiation.

This chapter deals with basic processes responsible for the charge distribution of the scattered and recoiled particles and describes results of experimental work dedicated to charge exchange. It is concerned with inelastic scattering of atomic projectiles from surfaces in the energy range of few keV.

8.1. CHARGE EXCHANGE PROCESSES AND INTERACTIONS

The scattering of ions by a solid surface can be divided into three steps: (1) the incoming trajectory, (2) the collision with a surface atom, and (3) the outgoing trajectory. This division is rather conventional and serves to simplify the description of the process of charge exchange and reflects the dependence of the probability of various processes on the distance between the particle and the surface. For the incoming and outgoing parts of the trajectory, resonant and Auger processes first considered by Hagstrum[5] are dominant. In the close-collision encounters between atoms, the charge exchange probabilities are determined by a quasi-molecular (Fano-Lichten) mechanism.[6]

8.1.1. Resonant and Auger Processes

Figure 8.1 shows schematic potential energy diagrams of the processes occurring between the particle and the surface. The electrons of the solid are characterized by the Fermi level E_f, and the ion has an energy level E_i. The main difference between the processes illustrated is in whether the energy of the electron is changed as a result of the transition. One-electron processes (i.e., resonant ionization [RI]) and resonant capture or neutralization (RN) (reverse processes), having high probability at distances 5–7 Å can occur whenever they are energetically allowed. In these processes the energy of the electron participating in the transition does not change. Two-electron Auger processes (i.e., Auger-capture or Auger neutralization [AN]) and Auger de-excitation (AD), have characteristic distances of 1–3 Å and involve the simultaneous transition of two electrons.

In RN (Figure 8.1a), an electron tunnels from the occupied valence band of the solid into an unoccupied energy level of the atom. In the reverse process—RI—the

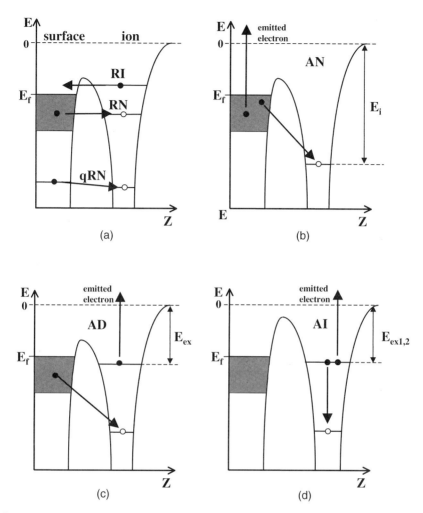

FIGURE 8.1. Electron energy diagrams of the charge transfer processes for a particle near a metal surface. The gray area shows the Fermi sea and E_f is the Fermi energy. The full dots represent occupied states, and the open circles represent unoccupied states. The arrows indicate directions of electronic transitions. RI, resonant ionization; RN, resonant neutralization; qRN, quasi-RN; AN, Auger neutralization; AD, Auger de-excitation; AI, autoionizaton.

electron tunnels from an occupied energy level of the ion or atom to the unoccupied conduction band in the solid. Core–level or quasi-RN (qRN) has a relatively small probability of occurrence and is possible only when an energy level of a surface atom has a core level with energy that closely matches that of the impinging ion.

If the energy difference between the level in the atom and valence band electrons is large enough ($|E_f| < |E_i|$), AN (Figure 8.1b), where two electrons of the valence band participate in the transition, is possible. An unoccupied level of the ion is filled with an electron from the valence band and the energy released goes to the excitation of the second electron, which can leave the solid. The maximum kinetic energy of the emitted secondary electrons is $(E_e)_{max} = |E_i| - 2|E_f|$. The AD (Figure 8.1c) process can cause an excited atom to leave the surface in the ground state; a surface valence electron falls into the empty atomic core level, and an electron is emitted as an Auger electron). The maximum energy of the emitted electrons is given by $(E_e)_{max} = |E_i| - |E_{ex}| - |E_f|$ (E_{ex} – energy of the electron in the excited level of the atom). De-excitation can also occur by photon emission; however, since the typical radioactive lifetimes are of the order of 10^{-8} s, whereas typical time scales for the decay of an excited state by an Auger transition are in the range of 10^{-13}–10^{-16}s, the fluorescence yield is very small. Only for highly charged heavy ions does the contribution of radioactive de-excitation increase to a more significant fraction. During the de-excitation of the particle, collective excitation of conduction band electrons (bulk and surface plasmons)[7–9] is possible. Autoionization (AI) or intra-atomic AD (Figure 8.1d) can occur during the decay of doubly or multiply excited states. AI is similar to the AD process, with the exception that all electrons taking part in the process belong to the projectile atom. The energy of the emitted electrons is $(E_e) = |E_i| - |E_{ex1}| - |E_{ex2}|$, where E_{ex1} and E_{ex2} are the binding energies of the atomic electrons.

The processes described above are competitive,[10] and their relative roles are determined by the energy position of the vacant levels of the ion relative to the valence band of the surface. For the case where the vacant level of the particle is close to the Fermi level of the target, such as alkali ions, the dominant mechanism for neutralization of incoming ions is the resonant tunneling process. For noble gas ions with large ionization potentials, neutralization occurs via AN or Auger transitions preceded by resonant tunneling processes between the valence band of the solid and the exited levels of the ion (AD).

Considering the processes of charge exchange, it is necessary to take into account the coulomb field of the ion near the surface leading to reconstruction of electrons in the solid. The resultant field is determined by the charge of the incoming ion (q) and the induced image charge (-q) placed inside the solid. The mirror plane of the charge does not coincide with the top atomic layer due to the tail of the electronic wavefunction that extends from the solid into the vacuum. The electron density outside of the surface drops exponentially, and the image plane is located about 1 Å in front of the first atomic layer. The ion-surface distance z is thus determined as a distance between the ion and the image plane. The influence of the image potential is manifested in two important phenomena: (1) acceleration of the ion towards the surface and (2) the shifting and broadening of the electron levels of the incoming or outgoing particle.

1. *Ion acceleration.* The interaction of the ion with its image is described by the attractive potential that accelerates ions towards the surface. The image force acts until complete ion neutralization is achieved, with the average kinetic energy gain being in the range $2.5q < \Delta E_{kin} < 4q$ [eV].[11] This effect is significant for ions whose vertical component of the velocity is very low (low energy or grazing collisions) or highly charged ions and puts a lower limit on this collisional velocity.

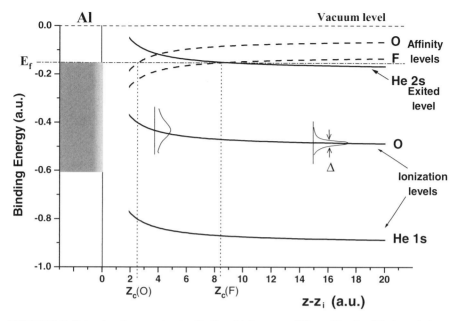

FIGURE 8.2. Behavior of energy levels of helium (He), oxygen (O), and fluorine (F) atoms in front of an aluminum (Al) surface. The affinity levels of O and F bend downwards as the distance z between the atom and surface decreases, while the ionization levels of He and O bend upwards. The low-lying 1s energy level of He does not cross the Al Fermi level, whereas the excited 2s level crosses the Fermi energy (E_f) at a long distance z_c.

2. *Level shift.* The image potential at a distance of several a.u. from the surface can be approximated as $e^2/4(z-z_i)$, where z_i is the position of the image plane. This potential causes displacement of the atomic level of the ion impinging on the surface as shown in Figure 8.2. For the energy of this level one can write

$$E_i(z) = -(E_i)_{inf} + (2q - 1)/4(z - z_i) \text{ [a.u.]},\qquad(2)$$

where $(E_i)_{inf}$ is energy level of the atom at infinity. The image charge perturbation thus leads to a shift of the projectile states. An important result of this level shift is that at a certain distance from the surface, the energy level of the particle is the same as the energy of the filled or unoccupied levels and resonant processes of electron capture or loss become possible. An ion energy level E_i, which can be the ionization level, excited level, or the electron affinity level is shifted up (ionization and excited level) or down (electron affinity level). The interaction between the incoming ion and the surface also leads to broadening of the energy level of the particle, with the level width $\Delta(z)$ being reciprocal to the lifetime of the state $\tau(z)$. The increase of the level width with decreasing ion-surface separation leads to a decrease in the state lifetime and increase in the transition probability. The position of the energy level of the particle with respect to the valence band of the solid determines the direction of the transition, whereas its width determines the transition probability. The level population also depends on the density of states in the solid. The width of the broadened atomic state is proportional to coupling between the metallic and atomic wavefunctions, and is directly related to the electron exchange transition rate $\Gamma(z)$, or the transition probability per unit time along the incoming and outgoing trajectories

$$\Gamma(z) = 1/\tau(z) = \Delta(z)/\hbar. \tag{3}$$

The atomic and metallic orbitals have a decaying exponential spatial dependency. As a result, the transition rate has an exponential dependence on the ion–surface distance z

$$\Gamma(z) = \Gamma(0)\exp(-az), \tag{4}$$

where a [distance^{-1}] determines the ion–surface interaction range (the decay rate length of the coupling matrix element) and $\Gamma(0)$ denotes the maximum (bulk) transition rate and has a typical value in the range from 0.01–0.1 a.u. The transition rate characterizes the efficiency of the process and provides direct information about its probability.

For a charged particle approaching or receding from the surface, the probability of neutralization in a time interval dt is $\Gamma(z)\,dt = (\Gamma/v_\perp)\,dz$, where $v_\perp$ is the component of the ion velocity perpendicular to the surface. The differential equation describing the probability that the particle with normal incoming velocity $v_\perp = v_{in}$ to the surface will reach z in its original charge state (survival probability) is

$$dP_{in}/dz = -(\Gamma_{in}/v_{in})P_{in}. \tag{5}$$

This equation has the solution

$$P_{in}(z, v_{in}) = \exp\left[\int_\infty^z (\Gamma_{in}(0)/v_{in})\exp(-az')dz'\right], \tag{6}$$

which upon integration yields

$$P_{in} = \exp\left\{-\left(\Gamma_{in}(0)/a\,|v_{in}|\right)\exp(-az)\right\}, \tag{7}$$

with the probability for the ion to become neutral

$$P^0 = 1 - P_{in}. \tag{8}$$

For the outgoing part of the trajectory, the probability that the particle with outgoing velocity perpendicular to the surface $v_\perp = v_{out}$ starting at z will reach infinity in its original state is

$$dP_{out}/dz = -(\Gamma_{out}/v_{out})P_{out}. \tag{9}$$

The solution of Eq. 9 is similar to that of Eq. 5, which, upon integration over the limits from z to ∞, yields

$$P_{out} = \exp\left\{-\left[\Gamma_{out}(0)/a v_{out}\right]\exp(-az)\right\}. \tag{10}$$

For an incoming or outgoing trajectory, the probability of an electron transition reaches its maximum at some distance $z = z_0$ from the surface, where $d^2P/dz^2 = 0$ and

$$z_0 = (1/a)\ln\left[\Gamma(0)/a v_\perp\right]. \tag{11}$$

The decay rate length of the coupling matrix element a is different for different processes and determines characteristic distances at which they are effective. For example, for helium (He) at a metal surface, the maximum probability for resonant neutralization is attained at a distance of ~5–6 Å and for AD at a distance ~2 Å.[5] In case of a two-way charge exchange, when both capture and loss of electrons are possible, the final charge state of the scattered atom is approximately determined by the level population at the distance $z_{fr} = z_0$, known as *freezing distance*.[12–16] The total probability of survival after the interaction is given by the product of Eqs. 7 and 10:

$$P = P_{in}P_{out} = \exp\left\{-\left[\Gamma_{in}(0)/a v_{in}\right]\exp(-az)\right\}\exp\left\{-\left[\Gamma_{out}(0)/a v_{out}\right]\exp(-az)\right\}. \tag{12}$$

In most cases, one can assume that the transition rates are equal on the incoming and outgoing parts of trajectory ($\Gamma_{in} = \Gamma_{out}$). Since the charge interaction takes place near $z = z_0$, for distances of closest approach less than z_0, z in Eq. 10 can be put equal to zero and the charge survival probability is represented by a simple formula

$$P = \exp\left[-v_0\left(1/v_{in} + 1/v_{out}\right)\right], \tag{13}$$

where v_0 (~10^{-7}cm/s)[12,13] is a characteristic velocity that depends on the projectile-target system and the electron density distribution. From Eq. 13 it is observed that a decrease in the velocity of the particle or increase of the interaction time will greatly increase the probability of neutralization. The resulting dependencies are approximate, but in many cases they describe the experimental data rather well.[14]

8.1.2. Close Encounter (Quasi-Molecular Mechanism)

The processes of charge exchange considered above are the processes of interaction of electrons from the outer shells of the incoming particle with electrons from the valence band and conduction band of the solid and depend on the properties of the electron subsystem of the solid as a whole. Another important process of charge exchange can occur in the close encounter (i.e., in binary collisions) and strongly depends on the actual distance between the colliding atoms and not the distance to the image plane. The interaction of a bombarding particle with a target atom leads to formation of a quasi-molecule. This process has been well studied in gas phase collisions.[17–20]

The term *quasimolecule* is used to describe a system of two atoms approaching and receding from each other in the process of the encounter. The lifetime of the quasi-molecule is equal to the time during which atoms are very near each other; for energies in the 0.1–10-keV range, it is about $10^{-13} - 10^{-15}$ s. During this time the axis of the quasi-molecule is turned by an angle $(\pi - \theta)$, where θ is a scattering angle. The minimum interatomic distance is much smaller than the size of normal molecules, and can even be smaller than the size of the inner electronic shells. This allows the use of a united atom limit when describing a quasi-molecule. Plots of the dependence of the distance of closest approach on the kinetic energy of the projectile are shown in Figure 8.3 along with the positions corresponding to the sum of the radii of maximum radial charge density for various combinations of electronic shells.[25] The plots show that significant penetration of some of the core atomic shells is achieved.

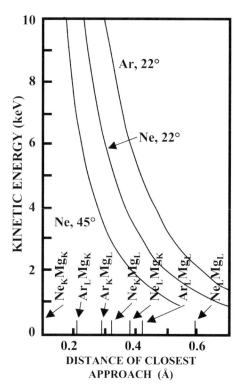

FIGURE 8.3. Calculated distances of closest approach vs. primary ion kinetic energy for neon–manganese (Ne/Mg) collisions at 22° and 45° and argon-manganese (Ar/Mg) collisions at 22°. The sums of the radii of maximum radial charge density for various combinations of the L and K shells are indicated on the figure (From Rabalais *et al.*, 1985, with permission).

The notion of quasi-molecule is based on the adiabatic approximation stating that electrons move much faster than nuclei and form orbitals corresponding to the fixed nuclei. The strong excitation of the electronic shells during atomic collisions results from the Pauli exclusion principle.[21] During close atomic encounters, the atomic levels of the united atom cannot host the increased amount of electrons. This leads to promotion and demotion of electronic states and exchange of vacancies and electrons between the crossing states, evoking changes such as charge exchange, excitation, and ionization in the electronic shells of the colliding particles. The velocity of the atoms does not play a significant role. It only serves for reaching the necessary short interatomic distances.

Specific electron promotions in the close encounter can be predicted by constructing separated-united atom diagrams[19] and using diabatic correlations[21,22] to determine electron configurations in the quasi-molecule. Such diagrams, as in Figure 8.4, show that as the separated atoms merge to form a molecule, filled molecular orbitals (MO) resulting from inner shells cross virtual MOs that correlate to higher principal quantum number n atomic orbitals (AO) of the separated atoms. As atoms approach and recede from each other, electronic transitions can occur at these crossings with the result that electrons can be trapped in AOs of higher n. The energy for the promotion of electrons is obtained from the kinetic energy of the nuclei. For Ne–Mg collisions the diagrams show that excitation energy can be channeled into Ne through its 2p level, which correlates with the highly promoted $4f\sigma$ MO. Electronic transitions[23] from the $4f\sigma$ MO to other σ MOs (e.g., 3s, 4p, 4s, etc.) can occur by radial coupling, and transitions to π MOs, in which the component of orbital angular momentum along the internuclear axis changes by one unit (e.g., 3p, 3d, 4f, etc.), proceed via rotational coupling. The resulting electronic configurations yield autoionizing and highly excited discrete states whose lifetimes (10^{-7}–10^{-9} sec) are longer than the collision times. For asymmetric atomic collisions, inner shell excitations go dominantly to the lighter lower atomic number atom.

The probability of charge-exchange (assuming neutralization probability equal to ionization probability) during level crossing in a close collision is expressed by the Landau-Zener formula[4]

$$P = P_{ion} = P_{neutr} = 2p(1 - p) \tag{14}$$

$$p = \exp(-v_L/v_r), \tag{15}$$

where p is the probability for electrons to survive in the same diabatic state after crossing, v_r is the relative velocity of the particles at the crossing distance, and v_L ($\sim$1–5 $\times$ 10^7 cm/s) is a characteristic velocity that depends on the crossing levels. Ionization during the close encounter and formation of autoionization states is of a threshold nature and leads to an increase in the number of scattered ions.[24–26]

The final charge state of a scattered atom is determined by contributions from both violent binary collisions of the impinging atom with a surface atom and electronic exchange processes with a surface as a whole.

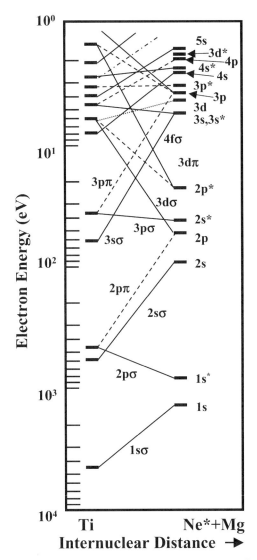

FIGURE 8.4. United atom (UA)-separated atom (SA) correlation diagram for electrons in the field of neon (Ne) and manganese (Mg) nuclei. Diabatic molecular orbitals (MOs) connect the levels of the infinitely separated atoms (right-hand side) with those of the united atom (left-hand side), maintaining the same number difference $(n-1)$. MOs with m = 0, 1, 2 (σ, π, δ) are denoted by solid, dashed, and dotted lines, respectively (From Rabalais *et al.*, 1985, with permission).

8.1.3. Inelastic Energy Losses

Although inelastic energy losses account for a small portion (5–10%) of total losses at low kinetic energies, they play an important role in electronic charge transfer. The interaction of a moving particle with an electronic system of a surface or single atom results in promotions of electrons to higher energy levels at the expense of the kinetic energy of the projectile. By measuring the discrete inelastic energy loss of a scattered particle, it is possible to associate the loss with specific electronic transitions involving changes in energy of the electrons in the projectile-solid system. In addition to direct measurements of scattered charge fractions and energies of ejected electrons, these characteristic losses provide a good source of information about the electronic transitions occurring during the collision.

As in the treatment of charge exchange, the inelastic energy loss can be divided into three parts: the incoming and outgoing parts, where the particle interacts with the surface as a whole, and the binary close encounter with a single surface atom. The close encounter is described using the Fano-Lichten theory,[6,19] while the loss on the approaching and exiting paths can be attributed to electronic friction. Correspondingly, the total inelastic loss of a projectile can be written as

$$\Delta E_{inel} = Q_1 + Q_2 + Q_3, \tag{16}$$

where Q_1 is an energy loss on the incoming part of the trajectory, Q_2 is a binary collision energy loss, and Q_3 is an energy loss on the outgoing part of the trajectory. Figure 8.5 schematically shows the distribution of the energy loss over the trajectory of the scattering particle. The most interesting component of ΔE_{inel} is the binary term Q_2 defined by electronic transitions in the violent collision. These affect the final charge distribution of the outgoing particles. It is easy to show[107] that the experimentally observed shift of the scattering peak from the binary collision position is

$$\Delta E = Q_1 K(\theta) + Q_2 + Q_3, \tag{17}$$

where $K(\theta)$ is the kinematic coefficient (Chapter 2, Eq. 4). The peak shift is less than the projectile inelastic scattering loss ΔE_{inel} (i.e., $\Delta E < \Delta E_{inel}$), due to the smaller *elastic* loss by a particle that has already lost a certain energy Q_1 before the binary collision. The inelasticity (the total energy loss partitioned between the projectile and target atoms) of the close collision can be found from the energy and momentum conservation laws

$$Q = 2\gamma \sqrt{(E_0 - Q_1)(E_{sc} + Q_3)} \cos\theta + (1 - \gamma)(E_0 - Q_1) - (1 + \gamma)(E_{sc} + Q_3), \tag{18}$$

where E_{sc} is the energy of the registered scattered particle and $\gamma = 1/A = M_1/M_2$ is the mass ratio of the projectile to target particle. The inelasticity Q is the energy that goes directly into the electronic excitations of the quasi-molecule. The inverse relationship between the peak shift and energy loss is

$$\Delta E = E_0 \left\{ \frac{A}{1 + A} \frac{Q}{E_0 - Q_1} + \frac{2\cos\theta\sqrt{A^2 - \sin^2\theta}}{(1 + A)^2} \right.$$
$$\left. \left(1 - \sqrt{1 - \frac{A(1 + A)}{(A^2 - \sin^2\theta)} \frac{Q}{E_0 - Q_1}} \right) \right\} + Q_3. \tag{19}$$

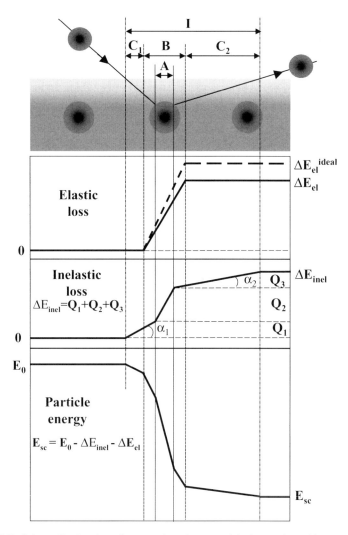

FIGURE 8.5. Schematic drawing of energy loss by a particle interacting with a surface. The interaction region (I) is divided into several parts: (A) the inelastic close encounter, (B) the elastic binary encounter, and (C) the continuous inelastic loss. The violent collision with the binary inelastic loss from MO promotions occurs in region A, whereas the elastic binary loss corresponds to a wider region B. C_1 and C_2 are the regions of continuous electronic friction from electrons of a surface. The rates of energy loss are α_1 and α_2, $\alpha_1 > \alpha_2$, reflecting the fact that $\alpha \sim E^n$, and the velocity of the particle is less on the outgoing part of the trajectory. The dashed line in the elastic energy loss plot shows an elastic loss of an ideal elastic binary collision.

In order to correctly estimate Q, the continuous losses Q_1 and Q_3 in Eq. 16 must be determined. They can be visualized as friction exerted by the electronic cloud of the solid along the incoming and outgoing paths. This friction is due to the interaction with electrons of the surface that have a continuous energy spectrum close to the vacuum level, for example, the conduction-band electrons. It is proportional to the velocity (or some power $\sim$1–2) of the velocity) and depends on the trajectory of the particle. If the total scattering angle is small ($<5°$), there may be no close encounter along the trajectory, and only the continuous terms remain in Eq. 16.

Several models accounting for a continuous energy loss have been developed. Firsov[27] considered the energy loss in a binary collision due to the flow of Thomas-Fermi electrons through the surface S normal to the line connecting the two colliding atoms. The inelastic force acting on each nucleus with coordinates R_a and R_b arises from the momentum transfer by electrons crossing the surface

$$F = \pm m \left(\dot{R}_a - \dot{R}_b \right) \int_S \frac{n v_e}{4} dS, \tag{20}$$

where n is the concentration of electrons, v_e is an average electron velocity, and m is the mass of an electron. Integration of Eq. 20 with respect to the coordinate along the path of the particle gives the continuous energy loss $Q_{1,3}$

$$Q_{1,3} = \frac{\hbar v}{4\pi^2 a_0} \int_{-\infty}^{\infty} \left(\int_S \varphi^2 dS \right) dx, \tag{21}$$

where φ is an electrostatic potential, $a_0 = 0.529 \ \mathring{A}$ is the first Bohr radius, and v is the velocity of the particle (assumed to be constant). Letting S be a plane in the middle of the line connecting the two atoms and using

$$\varphi = \frac{Z_a + Z_b}{r} \Phi(1.13 \, [Z_a + Z_b]^{1/3} \, r) \tag{22}$$

for the potential in that plane, the following approximate formula for the inelastic loss E (in electronvolts) is obtained

$$Q_{1,3} = \frac{(Z_a + Z_b)^{5/3} \, 4.3 \cdot 10^{-8} v}{\left[1 + 3.1 \, (Z_a + Z_b)^{1/3} \, 10^7 R_{min} \right]^5}, \tag{23}$$

where R_{min} is the distance of closest approach in centimeters and v is the velocity in centimeters per second. The Firsov energy loss is directly proportional to the velocity of the particle and is dependent on the distance of closest approach between the two atoms.

Lindhard et al.[28] obtained an estimate of the inelastic energy loss independent of the impact parameter. For the velocity of a projectile v that is comparable to or less than u_0, where $u_0 = e^2/\hbar = 21.88\text{Å/fs}$, it was assumed that the electronic stopping cross section $S_e = (1/n)dE/dx$ (n being the concentration of electrons) is nearly proportional to v if $v \leq Z_1^{2/3}u_0$. This also follows from the proportionality to v of the energy loss of a slow atom moving through an electron gas of constant density.[29] The variation of S_e with Z_1 and Z_2 is derived using a Thomas-Fermi model. To a first approximation

$$S_e = \xi_e \times 8\pi e^2 a_0 \frac{Z_1 Z_2}{\left(Z_1^{2/3} + Z_2^{2/3}\right)^{3/2}} \frac{v}{u_0}, \tag{24}$$

where ξ_e is of order of 1–2, but may vary with Z_1 as $\xi_e \approx Z_1^{1/6}$.

In order to achieve better quantitative agreement, Oen and Robinson[30] suggested a formula for the energy loss based on the electron density distribution in the hydrogen atom

$$Q_{1,3}(s, E) = \left(0.045bE^{1/2}/\pi a^2\right) \exp\left[-0.3 R_{\min}(s, E)/a\right], \tag{25}$$

where s is the impact parameter in the collision, E is the primary energy, a is the Molière screening length, $R_{\min}$ is the distance of closest approach, and b is a parameter. The magnitude of b is more important in determining the overall $Q_{1,3}$ than the shape given by the exponential factor. The theory of Lindhard[29] can be used to calculate the values of b.

The choice of the formula for continuous energy loss is largely based not on specific physical requirements, but on how good the formula reproduces the observed energy losses. In general, Eq. 25 provides the best fit to more experimental results.

8.2. EXAMPLES OF EXPERIMENTAL STUDIES—CHARGE EXCHANGE PHENOMENA

Information about charge exchange processes is obtained by measuring yields and energies of emitted electrons, photons, and scattered ions. For an absolute measurement of ion fractions of scattered particles, it is necessary to measure the number of neutral and charged scattered particles under the same experimental conditions. For the case of TOF measurements, the scattered particles are separated according to their charge using an electrostatic deflector or by accelerating or decelerating the charged component. In experiments with deflection plates, the total flux of the scattered particles (charged and neutrals) is measured first. Then voltage is applied to plates positioned between the sample and detector, which leads to the deflection of the charged fraction and registration of only the neutral particles. The signal from the ions is the difference between these two measurements. This simple and reliable method is applicable if the flux of the scattered particles has ions of only one polarity, but requires long data collection times for the case of small ion fractions (due to the differential statistics). In order to separate ions and neutral particles, an acceleration tube placed between the sample and detector can be used. The total flux of scattered particles goes through the acceleration tube. By applying a negative or positive voltage on this tube, ions can be separated according to their charge. Due to the resulting different velocities inside the acceleration tube, particles of the same species but having different charge will be separated in time. This allows collection of negative and positive ions as well as neutrals in the same measurement. In order to obtain quantitative information about the ion fractions from this method, it is necessary to take into account the focusing effect at the edges of the acceleration tube, since the acceleration tube focuses only the ions and not the neutrals. For low energies (<1 keV), the detector sensitivity for neutral and charged particles may be different and requires calibration.[31]

8.2.1. Negative Ion Formation via Resonance Processes

For negative ions, alkali atoms in their ground states, and rare gas atoms in exited states, the interaction with a surface is dominated by resonant charge transfer processes because the low binding energies of these electronic states make Auger transitions impossible. The formation of a negative ion is due to a resonant transition of an electron from the occupied valence band of a solid to the affinity level of the scattered particle, which is downshifted due to the effect of the image potential (Figure 8.2). The affinity levels cross the Fermi level at $z = z_c$, and at smaller distances, the affinity levels fall below the Fermi energy (E_f) and can be populated by resonant capture of electrons from the conduction band. At $z > z_c$, the affinity level is higher than E_f and the resonant charge transfer occurs in the opposite direction—a negative ion may decay by electron loss into unoccupied states of the solid. The population of the affinity level depends on its width and the time that the particle spends near the surface where resonant processes occur. The time spent near the surface is inversely proportional to the normal velocity of the particle $v_\perp$. Assuming that the capture and loss rates are equal and describing the charge exchange processes between the level and surface as in section 8.1.1, the probability of formation of negative ions can be written as

$$P^- = \{1 - \exp[u\Gamma(0)/av_\perp]\}\{\exp[-(u+1)\Gamma(0)/av_\perp]\}, \qquad (26)$$

where $u = \exp(-az_c) - 1$ and z_c is a crossing point of the affinity level of the particle with the Fermi level of the solid. In this equation, the first term describes the increase of the P^- fraction during the time spent inside the region $z < z_c$ when the affinity level is resonant with the conduction band and the second term describes the decay of the nascent negative ions by electron loss outside this region $z > z_c$.

The negative ion fractions of the scattered particles are independent of the charge state of the incident ion,[32,33] that is, the neutralization of a primary ion occurs on the incoming part of the trajectory at large distances from the surface. The scattered particle "forgets" its initial charge state, and its final charge state is defined by the interplay of electron loss and capture between the particle and surface on the outgoing part of the trajectory.

Resonant charge transfer is the most well-known process in studies of particle–solid interactions. The formation of negative ions during scattering from metal and insulator surfaces has been studied both experimentally[32–51] and theoretically[52–67] (for reviews, see also[68–72]). It was shown that the process of negative ion formation is determined by the surface electronic structure, the velocity of the particle, and by the multiple-level character of the interaction.

The effects of work function changes and local electron density on negative ion formation has been investigated in experiments with oxygen-covered[40–44] and chlorine-covered surfaces.[46,47] At grazing angles, the decrease in the work function strongly increases the negative ion formation. For large-angle scattering, the local effects play a dominant role. The resulting ion fraction is determined by the competition of work function and local effects.

The importance of the kinetic parallel velocity effect, especially in grazing angle scattering, has been demonstrated.[36,64–66] In the frame of reference associated with a moving atom, the energy levels of a surface are shifted from their values in the surface rest frame. Thus, the apparent occupation of the level suitable for the resonant charge transfer also changes. The distortion of the electronic system of the solid due to the parallel velocity effect is analogous to the effect of very high temperature in that it smears the Fermi edge between the occupied and unoccupied zones. The kinetic effect is important down to energies as low as 0.5 keV.[66]

The nature of the resonant charge transfer process is determined by the presence of several quasi-equivalent electrons in the incoming particle. In multielectron atoms, several states can correspond to the same configuration, for example ^3P, ^1D, and ^1S states corresponding to $2p^4$ in oxygen. In addition, in each state electrons are separated according to magnetic sublevels m. The different sublevels cross the Fermi level of a solid at different distances and, as a consequence, have different widths. One has to take into account the multielectron and multistate aspects of the interaction[52–56] in order to describe the ion-formation process.

In scattering of electronegative atoms such as fluorine (F) and chlorine (Cl) from metal surfaces, the negative ion fractions can be as high as 95%. These high ion fractions are explained by the relatively high energy of the affinity levels. On the outgoing part of the trajectory, the affinity level of the ion crosses the Fermi level at fairly large distances 8–9 a.u. (Figure 8.2). At these distances, the width of the affinity level Δ and rate of charge exchange Γ are small, and as a result, the electron loss probability for negative ions such as F^- and Cl^- moving away from the surface is low. The decrease of the ion fraction with increase of the particle energy[32] shows that only the capture process is responsible for the observed ionization probability. Contrary to the F^- case, the decay process is much more efficient for O^- formation, as the affinity levels cross the Fermi level at significantly smaller distances in this case.

A detailed study[39] of the formation of O^- ions in grazing scattering (with incidence angles $0.6–1.8°$) from an Al(111) surface over a wide range (1.5–700 keV) of collision energies showed a resonant structure for the O^- formation probability as a function of the projectile velocity (Figure 8.6). The experimental data were described using a multistate rate equation approach.[67] The finite level widths and parallel velocity

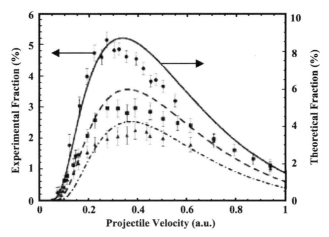

FIGURE 8.6. Negative-ion fractions as functions of the projectile velocity for the scattering of O atoms from an Al(111) surface with various angles of incidence α. The symbols represent the experimental data and the curves represent the theoretical results. Triangles and dashed-dotted line, $\alpha = 0.63°$; squares and dashed line $\alpha = 1°$; full dots and solid line, $\alpha = 1.8°$ (From Auth *et al.*, 1998, with permission).

broadening of the edge between the occupied and empty levels allows one to write the coupled equations of time evolution of the charge state containing the rates for capture and loss for all the ion and neutral states as

$$\frac{dP^i_{ion}}{dt} = -\left\{ \sum_j \Gamma^{loss}_{ij} \right\} P^i_{ion} + \sum_j \Gamma^{capt}_{ij} P^j_{neutral} \tag{27}$$

$$\frac{dP^j_{neutral}}{dt} = -\left\{ \sum_i \Gamma^{capture}_{ij} \right\} P^j_{neutral} + \sum_i \Gamma^{loss}_{ij} P^j_{ion},$$

where P^i_{ion} and $P^j_{neutral}$ are the populations of the negative-ion and neutral atom substates and Γ^{loss}_{ij} and $\Gamma^{capture}_{ij}$ are the electron loss and electron capture rates, respectively. Both experimental and theoretical dependencies show a typical resonance curve with a maximum at a certain value of the projectile velocity. The ion fraction is observed to increase with increasing angle of incidence. This effect is explained by a *freezing distance* approach when the increased perpendicular velocity causes a decrease in the freezing distance and a higher equilibrium ion population.

Tunneling rates and energy shifts are strongly influenced locally by impurities adsorbed on a surface. The lateral dependence of charge-transfer probabilities with atomic scale resolution has been demonstrated.[40] The negative-ion fractions of oxygen directly recoiled from a Ni{100}c(2 × 2)-O and a NiO{100} surface have been measured as a function of the crystal azimuthal angle and the recoiled atom exit trajectory from the surface (Figure 8.7). The determining factors in the O$^-$ yield are

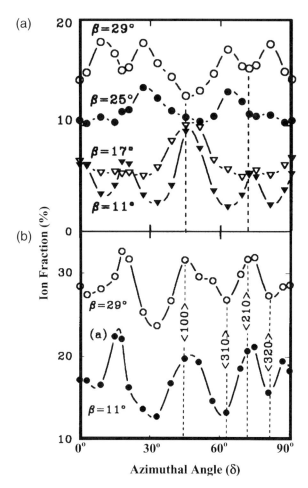

FIGURE 8.7. Experimental O^- ion fraction versus azimuthal angle δ for different exit angles β for (a) Ni{100}c(2 × 2)-O and (b) NiO{100} (From Hsu *et al.,* 1993, with permission).

oxygen–oxygen atomic spacing and spatial distribution of electronic density around oxygen. For high exit angles, O^- recoiled above rows with short interatomic spacings will neutralize more rapidly than above rows with long interatomic spacings due to the short distance between the recoiling trajectory and the chemisorbed O atom. This effect causes a reduction in ion fraction for O atoms moving along O-rich rows. At small exit angles, the recoiled O atoms initially pass very close to neighboring chemisorbed O atoms. During these close encounters, there is a possibility for transitions between the short-lived $m = 0$ state and the more long-lived $m = \pm 1$ states. The $m = \pm 1$ states thus formed have a high survival probability resulting in high ion fractions for grazing exit along O-rich rows. The data exhibit a pronounced angular dependence of the ion fractions. Thus, for Ni{100}c(2 × 2)-O at low exit angles, the maxima are along the $\langle 100 \rangle$, $\langle 210 \rangle$, and $\langle 110 \rangle$ azimuths (Figure 8.7) whereas the minima are along the $\langle 310 \rangle$ and $\langle 320 \rangle$ azimuths; these are reversed at high exit angles. For the NiO$\{100\}$ surface, the maxima and minima in the ion fractions occur at the same azimuthal angles for all exit angles. For this surface, the O–O distances are small for all trajectories, and there is no difference for large and small exit angles. These combined experimental and theoretical results for charge transfer provide a detailed microscopic map of the local electron tunneling rates between an individual atom and the surface.

8.2.2. Quasi-Resonant Charge-Exchange

In most cases the ions scattered from a surface neutralize via Auger and resonance processes by capturing electrons from the valence band of the target. In case of an atom on the surface having a core level energetically close to the empty level in the ion, a so-called quasi-resonant charge exchange can occur (Figure 8.1a). This effect has been observed[73–76] for scattering of He$^+$ from nine elements (Ga, Ge, As, In, Sb, Sn, Tl, Pb, Bi) that have d-electron core states with energy close to the He 1s level at 24.5 eV. Experiments showed oscillatory behavior of the scattered ion yield as a function of the ion velocity (Figure 8.8). The qRN neutralization can be described in a good approximation as a pure ion–atom collision process.[77–79] Experiments and theoretical analyses[80–82] show that the effect is dominated by a Landau–Zener-type charge-exchange effect known in atomic-collision physics as Stueckelberg oscillations.[83] Qualitatively, charge oscillations are explained by perturbation theory as interference between the asymmetric ψ_u and symmetric ψ_g molecular orbital states formed from atomic orbitals (e.g., He$^+$ + Pb and the He + Pb$^+$) eigenstates at close distances. The instantaneous frequency of beat oscillations is given by $(E_g - E_u)/h$, which is a function of the internuclear separation. The resulting charge state depends on the

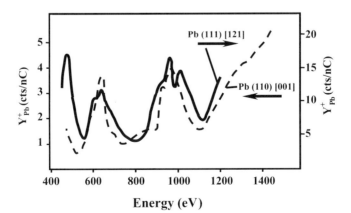

FIGURE 8.8. He$^+$ ion yields from He$^+$-Pb (110) (left scale) and He$^+$-Pb (111) (right scale). The beam was directed along the [001] and [121] directions, respectively (From Schippers *et al.*, 1991, with permission).

time that the particle has spent in the mixing region, which is inversely proportional to the velocity of the particle

$$P \sim \left[\exp\left(-v_c/v_\perp\right)\right] \sin^2 \left[\left(1/hv_\perp\right) \langle \Delta V(R)\rangle\right], \tag{28}$$

where $\langle \Delta V \rangle = \left(E_g - E_u\right) a$ and the exponential term is the contribution from the Auger and resonant neutralization processes.

The one-electron binary collision theory cannot be applied to explain the fine details of ion yield oscillations in the case of a crystalline target.[84] The experimentally observed dependence of oscillations on azimuth (target orientation) is explained by interference between the formation of an excited He state and He ground state. A general, more rigorous many-body method for calculating charge transfer probabilities, including intra-atomic correlation effects, has been proposed.[85] It was shown that qRN oscillations are very sensitive to the details of interaction between the excited atomic state and solid conduction band and provide information about the surface.

8.2.3. Slow, Highly Charged Ions

The potential energy of a slow, highly charged ion (SHCI) approaching a surface is much larger than its kinetic energy. This large potential energy influences the variety of processes that take place. Theoretical models and experimental results from multiply charged ion–surface interactions have been reviewed.[86–88]

As a highly charged ion approaches a surface, the *image charge* attraction leads to its acceleration towards the surface. The acquired additional kinetic energy depends on the ion charge and may reach several 10s of eV. The charge effectively lowers the potential barrier and changes the energy of projectile states. The resonant electron capture transitions from the surface to the outer unoccupied energy levels begin at a large distance from the surface. As an atom approaches the surface, the levels that are in resonance with occupied surface states begin to fill. This leads to formation of a so-called hollow multiply excited, essentially neutral atom with vacancies in the innermost shells and electrons in quasi-stationary high-energy Rydberg states. The resonance capture is accompanied by autoionization with electron emission resulting in constant shrinking of the electron cloud. The autoionization does not lead to change in the total charge because of the ongoing resonant capture.

Near the surface, electrons will screen the outer shell of the excited SHCI and a smaller hollow atom will be formed. The screening electrons from the solid will be in states with lower principal quantum number. Once below the surface, the hollow atom can relax via the following processes: AN, quasi-molecular transitions, and characteristic radiative de-excitation. The Auger processes (Auger cascades) and quasi-molecular transitions are responsible for electron emission during the interaction. Eventually, the major part of the potential energy of the highly charged ion transforms into the energy of the lattice, leading to sputtering of atoms and clusters from the surface. Information about the processes leading to formation and decay of the hollow atom can be obtained by measuring the charge states of ions scattered from the surface, electrons, and X-rays.

Recent investigations of the interaction of SHCI with single fullerene molecules[89] and fullerene thin films[90–91] are interesting due to the unique electronic structure of the fullerenes (i.e., high density of valence electrons and wide valence band). Studies of the interaction of SHCIs with fullerenes can close the gap between ion–atom and ion–surface collisions. The electronic transitions are fast and efficient with subsequent fragmentation of the fullerene. By studying the electronic stopping power oscillations as a function of the ionic charge, one can obtain information on the electron structure of the fullerene.[92]

A number of studies have been dedicated to large-angle scattering of highly charged ions.[93–97] For small-angle scattering, the majority of ions ($>90\%$) become neutralized due to the multiple nature of the collisions that the ions experience in the target. There is a small change in the ion yield for incident ions with and without an L-shell vacancy in this case. By using large scattering angles, one can register ions scattered from a single surface atom only. Contrary to the case of small-angle scattering, if ions carry an initial L-shell vacancy, the yield of multiply charged scattered ions increases by about three orders of magnitude as compared to ions with a completely filled L-shell.[97] The effect can be explained by incomplete filling of the L-shell during ion–surface interactions due to the large energy difference between the L-shell and the energy levels of the surface.

8.2.4. Complex Matrices

The surface composition rather than the bulk composition directly determines the ion fraction for scattered and recoiled particles. Complex matrices are particularly interesting in this respect since it is possible to achieve specific surface compositions through various sample treatment procedures. Among the important questions that need to be addressed about complex matrices is the composition attainable in the outermost atomic layer(s), along with the kinematics and electron exchange processes that occur when energetic ions collide with these surfaces. In order to address these problems, the ion neutralization probabilities on surfaces of different composition were studied.[50,98,99] The focus is to identify which properties of the surface will influence the neutralization of atoms near a semiconductor or insulator surface.

The full analysis of Auger and resonant transitions and electron promotions in close encounters was conducted for the case of scattered Ne^+ and directly recoiled Li^+ from Li and LiCu alloy surfaces over the range 1.5–10-keV Ne^+ energy.[98] The most significant observations as shown in Figure 8.9 are as follows: (1) both Y_{Li} and Y_{Ne} increase with increasing primary ion energy, and (2) the Y values are very sensitive to surface composition. Y_{Li} is low on the pure lithium (Li) surface and high on the surface containing both Li and (copper) Cu. Y_{Ne} exhibits the opposite behavior; it is significantly higher on the pure Li surface. In order to explain the observed behavior of the charge fraction, it is necessary to consider neutralization along both the incoming and outgoing trajectories as well as excitation in the close atomic encounter.

The higher Ne^+ ion fraction from the pure Li surface (Figure 8.9a) than from the LiCu surface is a result of (1) less neutralization along the incoming and outgoing

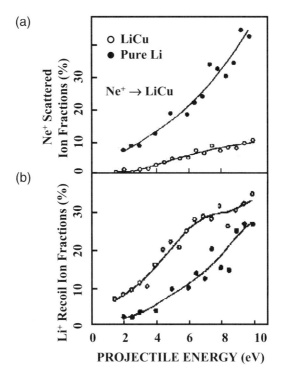

FIGURE 8.9. Scattered neon (Ne^+) (a) and recoiled lithium (Li^+) (b) ion fractions as a function of primary Ne^+ energy for a pure Li and LiCu surface (From Shi *et al.,* 1990, with permission).

trajectories and (2) strong promotion in the close encounter for Ne^+–Li collisions. The large increase in the Ne^+ ion fraction on the Li surface with increasing kinetic energy is similar to that observed in Ne–Mg inelastic collisions and indicates the formation of autoionizing states or direct formation of ionic states during (or shortly after) the close encounter.

The dominant mechanism for neutralization of Li^+ along the outgoing trajectory is RN from the Li valence band. RN transitions from the Cu valence band into the Li 2s ground state at 5.4 eV below the vacuum level are less probable for the same reasons as for the Ne^+ ions. During the close encounter, electron exchange with the promoted hole of the Ne^+ $3d\sigma$ orbital can occur, producing Ne^* and Li^+. The increase in the Li^+ ion fraction with increasing energy (closer distance of approach) is in agreement with this promotion in the close encounter. Since this close encounter is the same on both the pure Li and LiCu surfaces, the different Li^+ yields of Figure 8.9(b) are due to differences in the survival probabilities along the outgoing trajectories. The lower Li^+ fraction from the pure Li surface is a result of efficient RN with the Li valence band, which is diminished on the LiCu surface due to dilution of the Li. In both surfaces the survival probability increases with increasing outgoing velocity (i.e., increasing projectile velocity).

An experimental and theoretical investigation of the scattered Ne^+ and recoiled S^- on both the Cd- and S-terminated surfaces of $CdS\{0001\}$[50] has shown a strong correlation between these ion fractions and the lateral variation of the electrostatic potential along the outgoing atom trajectories. The scattered Ne^+ and recoiled S^- ion fractions in Figures 8.10 and 8.11 exhibit a periodic behavior as a function of azimuthal angle. The explanation of the periodic charge transfer patterns is obtained from the lateral dependence of the neutralization rates of a Ne^+ ion outside the surface. Neutralization of the Ne^+ ion can occur through resonant tunneling into the Ne 3s level and through Auger de-excitation and neutralization. In all cases, an electron is transferred from the crystal through the surface into the atom. An important factor in determining the neutralization rate is the electrostatic potential barrier of the surface. Neutralization during the close encounter is ruled out, as it is not likely to introduce an azimuthal dependence of the ion fractions. The electrostatic barriers created by impurity atoms can change the electron tunneling rates by orders of magnitude.[100] Large spatial variation of the electrostatic potential outside the $CdS\{0001\}$ surfaces implies the lateral variation of the electron transfer rates. The increased electrostatic potential in the vicinity of the cadmium (Cd) atoms translates into reduced potential

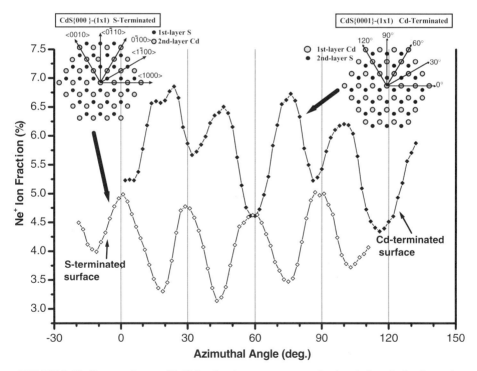

FIGURE 8.10. Scattered neon (Ne^+) ion fractions versus crystal azimuthal angle for the cadmium (Cd)- and sulfur (S)-terminated surfaces of CdS. Solid circles, Cd-terminated surface; open circles, S-terminated surface. Schematic diagrams of the two surface configurations are included (From Houssiau *et al.,* 1999, with permission).

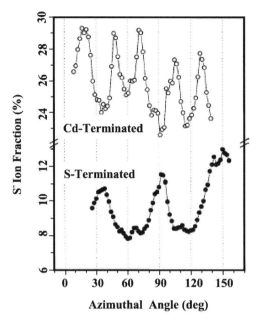

FIGURE 8.11. Recoiled sulfur (S^-) ion fractions versus crystal azimuthal angle for the cadmium (Cd) and sulfur (S)-terminated surfaces of CdS. Solid circles, S-terminated surface; open circles, Cd-terminated surface. (From Houssiau *et al.*, 1999, with permission).

energy barriers for electron tunneling between the surface and an atom in the vicinity of a surface Cd atom. The decreased electrostatic potential above the surface sulfur (S) atoms is equivalent to an increase in potential energy barrier for an electron tunneling between the surface and an atom above an S site. Therefore, electron transfer rates are smaller above S atoms than above Cd atoms in the top layer. The observed anisotropy in the ion fractions is a result of the variations in surface-to-atom electron transfer rates due to tunneling barriers introduced by the electrostatic potentials.

The situation with recoiled S^- is more complex because, for the Cd-terminated surface, S recoils originate from the second layer and are influenced by first-layer Cd atoms. The probability of S^- neutralization is enhanced along the dense Cd rows in this case. For the S-terminated surface, Y_S exhibits a 60° periodicity with the lowest values observed along the directions with the highest density of Cd atoms, following the periodicity of the closed-packed rows of the Cd layer. These results clearly demonstrate the sensitivity of processes of charge exchange to the surface structural corrugation as well as the elemental composition.

The effect of surface charging is of tremendous importance in the scattering analysis of insulator targets. An investigation of the scattered and recoiled ion fractions from 3-keV Ar^+ beams on $LiTaO_3(100)$ single crystals has shown a variation of the spectral features with the surface conductivity of the target material.[99] $LiTaO_3$ is an insulator at room temperature, and its conductivity increases by $\sim 10^3$ at temperatures in the range 100–200°C. The surface becomes positively charged by the ionic current of the primary beam. The scattered Ar^+ ion fraction is higher ($\sim 0.62\%$) for the surface analyzed at room temperature without electron beam compensation (i.e., in the charging regime), as compared to that for the surface analyzed at 160°C ($\sim 0.37\%$) and for the surface analyzed at room temperature with the electron beam on ($\sim 0.33\%$). A higher proportion of scattered Ar^+ ions compared to scattered neutrals from tantalum (Ta) atoms in the charging state of the surface compared to that of the charge-compensated surface is a reasonable consequence of the low neutralization probability for ions scattered from an electron-deficient, positively charged surface. The lower ion fraction from the uncharged surface is in accord with a higher neutralization probability from an uncharged conductive surface. The azimuthal anisotropy observed in scattering from the charged surface is a result of the influence of anisotropic electrical field gradients near the surface.

8.2.5. Rare Gas Ion Interactions with Silicon

The interaction of rare gas ions with silicon (Si) surfaces in the keV regime has been investigated using several different experimental techniques.[101–110] By analyzing the energy loss of the scattered Ne and the secondary electron spectra, evidence has been presented for the formation of doubly excited Ne^{**} due to electron promotion processes occurring at the close encounter between the impinging Ne and a substrate atom. The ion fractions are very similar for scattering of neutral Ne and Ne^+, indicating memory loss and that the charge transfer dynamics are determined by the evolution of the Ne states after the close encounter with a substrate atom. The evolution of the excited Ne during the outward trajectory of the Ne can be very complex. Various resonant and Auger processes can occur at close Ne–Si distances, leading to the formation of Ne^{**}, Ne^{+*}, Ne^{++}, Ne^*, and Ne^+.[107,108] A quantitative modeling of the complex dynamics involved requires the solution of coupled rate equations.[104,108] While the experimental data often does not allow identifying a particular neutralization channel, the results show how the neutralization probability of Ne^+ depends on such important parameters as the trajectory of the particle and local substrate charge density at the position of the ion.

The study of a correlation between the scattered particle neutralization rates and the local charge density encountered by the ion along its trajectory can be performed using the following simple assumptions: (1) The neutralization probability at ion position z is proportional to the local substrate charge density $\Gamma(z) = \gamma \rho(z)$; such an assumption is reasonable for both resonant and Auger neutralization.[111] (2) The ions that emerge from subsurface collisions will encounter regions of much larger valence electron density and for a longer duration than the ions scattering from surface atoms. These ions will therefore be neutralized with high probability.

The final ion fractions can be described by accounting for both the surface and subsurface collisions:

$$Y = P_{subsurf}\beta + P_{surf}(1 - \beta), \tag{29}$$

where β is the fraction of subsurface trajectories and P_{surf} and $P_{subsurf}$ are neutralization probabilities from the first and all other layers, respectively. The subsurface ion fraction, $P_{subsurf}\beta$ does not contribute, as all ions undergoing subsurface collisions are assumed to be neutralized. The resulting ion fraction can then be expressed as

$$Y = (1 - \beta) \exp\left[-\frac{\gamma}{v} \int_0^{+\infty} dz\rho(z)\right], \tag{30}$$

where $\rho(z)$ is a calculated charge density.

This approach allowed a description of the azimuthal dependence of the neutralization probability of Ne^+ ions scattered from a Si(100) surface as a function of azimuthal angle.[112] Azimuthal scans of the scattered Ne (neutrals and ions) and (neutrals only) are shown in Figure 8.12(a), and the ion fractions Y determined from these scans are shown in Figure 8.12(b). The intensity plots exhibit maxima along the 0°,

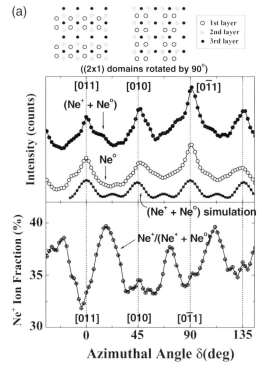

FIGURE 8.12. (a). (top) Azimuthal angle scans of neon (Ne^+) scattered from the Si(100)-(2 × 1) surface. The solid circles are ($Ne^+ + Ne^0$) neutrals plus ions, the open circles are (Ne^0) neutrals only, and the dotted line is a SARIC simulation of the total ($Ne^+ + Ne^0$) scattering intensity. The observed maxima exhibit a periodicity of 45°, corresponding to a surface with two domains. (bottom) Scattered Ne^+ ion fractions Y from the data of the upper graph. The figures above the graphs represent single (2 × 1) domains rotated by 90° (From Vaquila *et al.*, 2001, with permission).

(b)

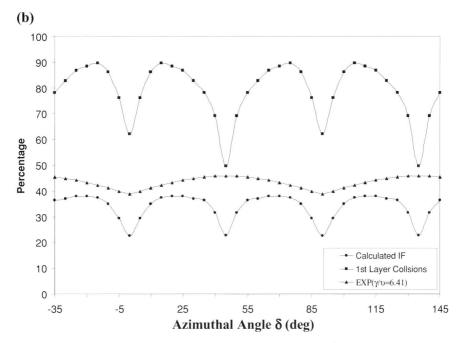

FIGURE 8.12.(b). *(Continued)* Calculated azimuthal variation of the Ne$^+$ ion fractions (IF) (domain averaged), the azimuthal variation of the fraction of particles scattering from the topmost surface layer $[1-\beta(\delta)]$, and the azimuthal variation of the exponential factor in Eq. (30) (From Vaquila *et al.,* 2001, with permission).

$45°$, $90°$, and $135°$ directions, whereas the ion fraction Y plot exhibits minima at the same positions.

Figure 8.12(b) shows various terms in Eq. 2 obtained from calculation—the exponential factor, fraction of surface trajectories $[1 - \beta(\delta)]$, and their product (i.e., the total ion fraction). The angular dependence of the scattered ion fractions is primarily caused by the angular dependence of the $[1 - \beta(\delta)]$ factor. The fraction of ions that are reflected from the first layer depends strongly on the azimuthal angle. The subsurface collisions therefore result in a very high probability of Ne$^+$ neutralization, and the contribution from the first layer determines the outgoing charge fraction.

The same model correlating local substrate charge density along the ion trajectory to the instantaneous neutralization probability of the scattered Ne$^+$ was used to explain the strong sensitivity of the ion neutralization probability to substrate doping.[113] The experimental Ne$^+$ ion fractions from p-doped Si(100)-(2 $\times$ 1) surfaces have been found to be larger than the ion fractions scattered from intrinsic and n-doped Si substrates by 22–29%, depending on the specific crystal azimuth as shown in Figure 8.13. In the intrinsic and n-doped Si, a depletion layer of majority carriers (electrons) near the surface is created that produces a positive space charge due to the presence of positively charged donor atoms at the surface. p-Doping results in the

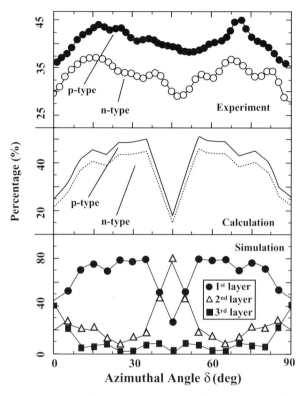

FIGURE 8.13. (top) Experimental ion fraction for Ne$^+$ scattered from an n-doped and p-doped Si(100)-(2 × 1) surface. (middle) Calculated ion fraction for Ne$^+$ scattered from an n-doped and p-doped Si(100)-(2 × 1) surface. (bottom) Calculated percentage of Ne$^+$ ions scattered from the first, second, and third Si(100)-(2 × 1) surface layers as a function of azimuthal angle (From Vaquila, *et al.*).

creation of a hole-depleted layer and a negative space charge region near the surface, leaving the surface states unpopulated.

Thus, collisions between Ne$^+$ and a p-doped Si substrate yield lower charge neutralization rates when compared to intrinsic and n-doped Si surfaces due to the presence of a repulsive surface potential and the low occupation of surface states. The comparatively low ion fraction observed in Ne$^+$ collisions on n-type and intrinsic Si surfaces are reflective of an increased charge density due to the occupation of surface states and a reduction of the charge-tunneling barrier arising from the positive surface space charge.

For extremely open surfaces, such as Si(100), an appreciable amount of ions may be generated in the second or deeper layers, notwithstanding the assumption of complete subsurface ion neutralization. In case of close-packed surfaces, the strong dependence of transition rates on the particle trajectory provides an effective mechanism for subsurface neutralization,[114–116] whereas in Si(100) the incomplete neutralization is one of the possible causes of the discrepancy between the model and experimental data.

When measuring the angular dependence of ion fractions, effects of blocking and shadowing between the target layers allow extraction of the contribution from each layer to the total ion fraction. For any specific experimental conditions, the ion fraction can be written as

$$Y = \frac{S^+}{S^+ + S^0} = \frac{\sum_i S_i^+}{\sum_i S_i} = \frac{\sum_i p_i S_i}{\sum_i S_i} = \sum_i p_i \left(\frac{S_i}{\sum_j S_j} \right) = \sum_i p_i \chi_i, \tag{31}$$

where p_i is the ion-survival probability for layer i. S_i and χ_i are correspondingly the number and the fraction of particles scattered from layer i. The sums in Eq. 31 are taken over all target layers. From the analysis of the simulated trajectories, the trajectory fraction χ_i^{sim} is

$$\chi_i^{sim} = \frac{S_i^{sim}}{\sum_j S_j^{sim}}. \tag{32}$$

Assuming a weak dependence of p_i on the azimuthal angle, a set of linear Eq. 31 can be solved with respect to the unknowns p_i,

$$p_i = \sum_\delta \chi_{i\delta}^{-1} Y_\delta, \tag{33}$$

where $\chi_{i\delta}^{-1}$ is the inverse of the matrix $\chi_{i\delta}$ of the fraction of particles scattered from layer i for azimuthal angle δ, and Y_δ are the measured ion fractions.

The use of SARIS (see Chapters 3 and 6) allows simultaneous acquisition of data from each of the surface layers. A three-dimensional ion fraction map of 5-keV He$^+$ scattered from a Si(100)-(2 × 1)-two domain surface was obtained with SARIS by capturing images of both scattered and recoiled ions and neutrals.[117] Due to the large solid angle subtended by the microchannel plate (MCP) detector, atoms and ions that are scattered and recoiled in both planar and nonplanar directions are detected simultaneously. This greatly reduces the data-collection time, allowing investigations of dynamic processes and phenomena that are sensitive to surface conditions, such as scattered ion yields and ion fractions. The extreme sensitivity of scattered ion fractions to the depth of the specific atomic layer from which the scattering occurs provides the ability to delineate scattering events and charge exchange probabilities from specific surface layers.

Figure 8.14(a and b) shows the simulated projection of 5-keV He$^+$ trajectories scattered from Si(100), together with the standard fcc(100) projection. Figure 8.14(d) shows that the ion fraction images are anisotropic and the maxima in Y correspond to the directions of the blocking cones where there are minima in the (neutrals + ions)

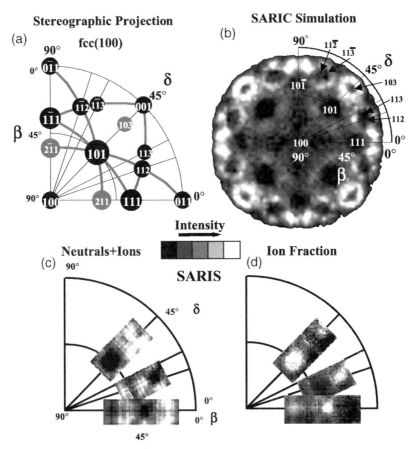

FIGURE 8.14. (a) Standard fcc (100) stereographic projection. (b) Simulated stereographic projection of classical ion trajectories for 5-keV He$^+$ ions scattering from a Si(100)-(2 × 1) two-domain surface using the SARIC program. The contour plot was formed by an azimuthal equidistant projection of the spatial intensity distribution onto the plane of the hemisphere base with angular coordinate equal to the crystal azimuthal angle δ and radial coordinate equal to the polar exit angle β. The number of detected trajectories increases from black to gray to white. (c) Experimental SARIS image of scattered neutrals plus ions in one quadrant of the simulated projection. The white areas correspond to focusing effects of the outgoing particles and the dark areas are due to blocking effects by surface atoms. (d) Experimental scattered ion fraction map determined as the intensity distribution Y = ions/(neutrals + ions). The white and black areas correspond, respectively, to high and low values of Y (From Vaquila *et al.,* 2002, with permission).

distribution of Figure 8.14(c). The scattered ion-survival probabilities are highest in the spatial regions corresponding to the center of the blocking cones.

More detailed analysis of the ion fraction data was performed for the azimuthal angles $\delta = 0°$ and $45°$ as shown in Figure 8.15. For $\delta = 0°$ (Figure 8.15a), two maxima are observed at $\beta \sim 35°$ and $\sim 55°$. These exit angles correspond to the $\langle 111 \rangle$ and $\langle 211 \rangle$ crystallographic directions. For $\delta = 45°$ (Figure 8.15b), a maximum at

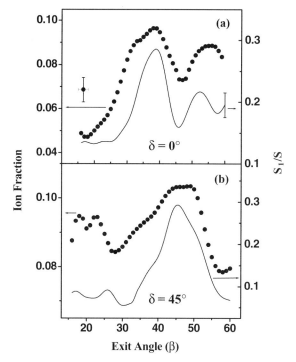

FIGURE 8.15. Black dots—Experimentally determined profiles of the helium (He$^+$)-scattered ion fractions for different azimuths. Solid lines, the fraction of scattered particles from the first layer in SARIC calculations for the azimuthal angles 0° and 45° (From Vaquila *et al.*, 2002, with permission).

$\beta = 45°$ with $Y \sim 10\%$ is observed corresponding to the $\langle 101 \rangle$ crystallographic direction. The solid curves in Figure 8.15 correspond to the calculated ratio of the number of particles scattered from the first layer to the total number of scattered particles from a thick target (calculated fraction). Using the experimental ion fractions Y and the number of scattering particles S from SARIC trajectory simulations allowed determination of the contribution of each layer to the total ion fraction.

The total survival probability for each layer is

$$p_i = P_i^{in}(1 - P_c)P_i^{out} + (1 - P_i^{in})P_c P_i^{out}, \qquad (34)$$

where $P^{in}(1 - P_c)P^{out}$ is the survival probability for ions that survived neutralization on the incoming trajectory P^{in}, close collision, and outgoing trajectory P^{out}, while $(1 - P^{in})P_c P^{out}$ is the survival probability for ions that were neutralized on the incoming trajectory, reionized during collision with a target atom P_c, and then survived neutralization on the outgoing trajectory P^{out}. It is assumed that the probability of charge exchange P_c during the close collision depends on the distance between the colliding particles and is independent of the layer number and that the probabilities for reionization of neutrals and neutralization of ions are equal.[118] Assuming constant electron density and taking into account the different times spent by a particle inside the surface when scattering from different layers, using Eqs. 33 and 34 with two adjustable parameters v_0 (Eq. 12) and P_c for the three exit angles $35°$, $45°$, and $55°$, the best agreement between experiment and theory was found for $v_o = (1.9 \pm 0.1) \times 10^7$ cm/sec and $P_c = 0.34 \pm 0.09$. The charge exchange probability determined in the close collision (i.e., $P_c = 0.34$) is significant. These results allow delineation of the amount of scattered ions from the various layers as follows: first layer $\sim 70\%$, second layer $\sim 25\%$, and third layer $\sim 5\%$. Thus, the survival probability p_i for particles scattered from the first layer is ~ 0.23, second layer ~ 0.08, and third layer ~ 0.015. Therefore, even for an open surface such as Si(100), the ion-fraction contribution from first layer scattering is dominant.

8.2.6. Measurements of Inelastic Energy Loss and Connection to the Charge Fraction

It is well known that core-electron excitations in a binary close encounter are responsible for high outgoing ion fractions. During a close encounter, one, two, or more electrons can be transferred into the higher lying atomic energy levels or the vacuum continuum. Promotions into the vacuum lead to single or multiple ionization, whereas transitions between levels lead to formation of excited atomic states. Doubly excited atoms may subsequently autoionize with one excited electron gaining energy from another and leaving the atom. Although all the above processes can occur, their probabilities differ significantly, so that there exists one or two dominant channels of ionization. Most of the studies of inelastic discrete losses have been performed for rare gases, especially Ne and He, scattered from Na, Al, Mg, Si, and Cu surfaces. These systems display a threshold behavior with considerable inelastic losses and high ion fractions observed for the distances of closest approach less than a certain value. It is now agreed that the high outgoing ion yields result from electronic excitations in the close encounters, although the discussion is ongoing, particularly for the case of Ne^+ projectiles, as to the exact nature of the excitation process. It should be noted that the molecular orbital approach is better suited for collisions in the gas phase, where the binary approximation can be used without reservations, while the presence of energy bands in solids affects the resultant molecular orbitals.

Two different treatments of electronic transitions in a violent collision are used to explain high outgoing ion fractions in scattering of rare gases from solids. One assumes direct ionization of the projectile in the violent collision as a result of the ejection of electrons into vacuum,[119–121] while the other treats the ionization as a two-step process.[122–127] The latter approach assumes the electronic promotions to higher lying atomic Rydberg states in the hard collision, with the overall preservation of the charge of the atom, followed by autoionization along the outgoing trajectory. Both approaches agree that particles arriving at the collision site have a high probability to be neutralized. This statement is confirmed by a large amount of experimental data.

In an experiment on grazing He^+ scattering from an Al(111) surface,[128] Auger electron capture has been shown to be a very efficient neutralization mechanism leading to a completely neutralized beam (ion fraction $\sim 10^{-4}$ below the threshold). Very small incidence angles $0.2° < \alpha < 1°$ ruled out the possibility of ionization due to MO crossings because of the small normal energy ~ 2 eV of the projectile. A threshold for the charge fraction was attributed to the kinematic shift of the conduction band Fermi-Dirac distribution due to the motion of the projectile.

In a series of works on He^+ and Ne^+ scattering from Al and Ne^+ and Ne^0 scattering from Si, Al, and Mg,[101,121] it was shown that Ne^+ ions, which survived neutralization (AN) contribute very little to the positive ion spectrum. This follows from the close similarity of the Ne^+ spectra for both Ne^+ and Ne^0 incidence. The intensity of outgoing Ne^+ does not depend on the charge state of the incoming Ne beam (Ne^+ or Ne^0), whereas the intensity of the outgoing Ne^{2+} strongly depends on the charge state of the primary beam.

The results of 2-keV Ne^+ and Ne^0 and 3-keV He^+ scattering on sodium (Na)-covered surface[122] have been compared with those obtained for other systems such as manganese (Mg), aluminum (Al), and Si. The incident angle of $6°$ was kept constant while the scattering angle was varied. The absence of a significant difference in the ion fractions between the ionic and neutral projectiles was also explained by a complete neutralization of the charged beam on the incoming part of the trajectory. The incident ions are shown to neutralize during the approach to the surface and then reionize in the inelastic process occurring in the violent single collision with an atom on the surface. Measurements of energies of electrons ejected as a result of the violent collision showed two peaks corresponding to $2p^4(^3P/^1D)3s^2$ in case of Ne and one peak due to He^{**} $2s^2$ de-excitation. Excitation mechanisms similar to those of gas-phase collisions have been detected. These works show that the Ne^{2+} ions are formed from the small portion of the original Ne^+ surviving neutralization on the incoming path. Autoionization decay of the neutralized Ne leads to formation of Ne^+.

In a 1–4-keV $Ne^+ \rightarrow Mg$ scattering experiment,[124] the first correlation between the scattered Ne charge fraction, inelastic energy loss, and characteristic electrons has confirmed the common origin of these phenomena. The charge fraction and inelastic peak shift were measured using a TOF technique, and the energies of the emitted electrons have been recorded by an electrostatic analyzer. For a given primary energy, a continuous range of minimum interatomic distances was accessed by varying the scattering angle. The charge fraction Y^+ was found to increase with decreasing minimal approach, with a sharp rise near $R_{min} \approx 0.55\,\mathring{A}$. In Figure 8.16, Y^+ is plotted against the distance of closest approach for different projectile energies with R_{min} calculated using the Molière potential. It is observed that the formation of the positive charge state is governed by a threshold mechanism which turned on after R_{min} decreased below a certain value. Below this activation threshold, the inelastic loss for the scattered ions was ~ 12 eV, whereas above the threshold, an additional Ne peak at ~ 30 eV was observed. Molecular orbital theory was used to explain this behavior in terms of electron promotions between the orbitals. For the case of the Ne–Mg system (Figure 8.4), Ne 2p electrons are promoted into the $4f\sigma$ MO. The electrons are promoted by gaining energy from the kinetic energy of colliding nuclei once the necessary interatomic distances are reached. De-excitation of doubly

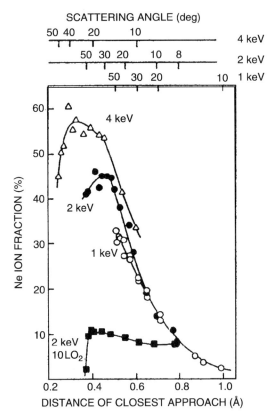

FIGURE 8.16. Scattered neon (Ne$^+$) ion fraction as a function of the distance of closest approach of the projectile and target atoms for 1-, 2-, and 4-keV energies on clean manganese (Mg) and for 2 keV on oxidized Mg (From Grizzi *et al.*, 1990, with permission).

excited Ne** into Ne$^+$ on the outgoing part of the trajectory far from the surface was cited as a reason for the abnormally high scattered ion yields. Upon chemisorption of oxygen, the threshold effects are simultaneously suppressed (i.e., the drastic decrease in the ion fraction is accompanied by a disappearance of autoionization electrons).

The mechanisms of threshold energy loss and electron excitation in a binary collision have been investigated[107] in measurements of inelastic losses of 500–1950-eV Ne$^+$ ions scattered from Si. In the scattering spectra, the Ne$^+$ single-scattering peak shifts from the binary position have been measured and their dependencies on the energy and scattering angle were fitted by Eq. 17, with Eq. 25 used to calculate the continuous losses Q_1 and Q_3 from Eq. 18. The three fitting parameters, friction constants b_1 and b_3 from Eq. 25 and the threshold inelastic loss Q_2, have been obtained for neutral, singly, and doubly charged Ne. The values of b were ordered according to $b^0 < b^+ < b^{2+}$, with $b_1^0 = b_3^0 = b^0 = 0.45$, $b_1^+ = b_3^+ = b^+ = 0.68$, and $b^{2+} = 0.74$. The total inelastic loss in the close collision is shown in Figure 8.17. It

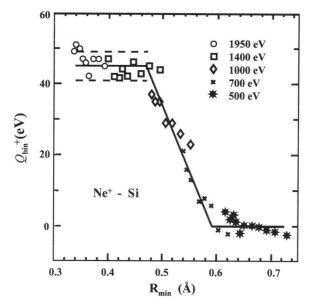

FIGURE 8.17. Inelasticity in the binary collision Q_2 for detected Ne^+ as a function of distance of closest approach obtained by subtracting the continuous inelastic energy loss. The solid line corresponds to the best fit, and the dashed lines mark the errors of Q_2. (From Xu *et al.*, 1998, with permission).

exhibits a sharp rise with a height of 45 ± 5 eV. This result favors the assumption that the energy loss goes into the double excitation of two 2p electrons (minimum excitation energy 47.35 eV).[129]

In a study of 1960-eV Ne^+ scattering from polycrystalline Al[125], inelastic losses of 45 ± 3 eV for Ne^+ and 86 ± 4 eV for Ne^{2+} have been observed. Eq. 25 was used to account for continuous loss on the incoming and outgoing parts of the trajectory. Fitting the inelastic loss obtained from the Ne → Al scattering peak displacement using Eq. 16, the constant b is 0.25 for both the incoming and outgoing trajectories, and the threshold loss in the violent collision is 45 ± 3 eV, also suggesting the formation of doubly excited Ne in this case.

Inelastic losses and their connection with the charge fraction[126] were investigated in a study of 2.4–10-keV Ne^+ scattered from a Cu_3Au sample. The inelastic peak displacements of Ne^+ single scattered from Cu were measured with respect to that of Ne^+ scattering from Au, which was assumed to be at the binary elastic scattering position. Using shadowing and blocking by target atoms, Ne^+ was preferentially scattered from only the top layer and then only the top two layers. The charge fraction of Ne^+ scattered on Au followed the conventional Auger or resonant neutralization processes (Eq. 10) and showed no threshold increase. The charge fraction from the Ne^+ → Cu scattering exhibits a threshold growth at approximately 4 keV (Figure 8.18a). At the same energies, the Ne^+ → Cu peak displacement showed a steep increase reaching as much as 130 eV for the higher energies near 10 keV

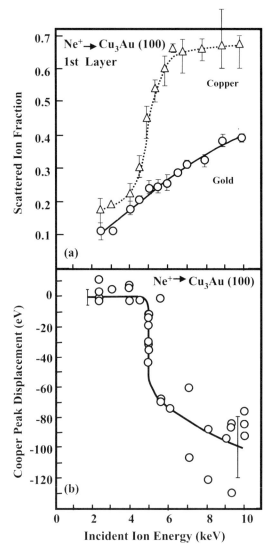

FIGURE 8.18. (a) Ion fraction of neon (Ne$^+$) scattered from first-layer copper (Cu) and gold (Au) atoms as a function of incident-ion energy. (b) Measured energy displacement of the low-energy ion scattering Cu scattering peak from the energy predicted from ideal elastic scattering. If inelastic electron effects were not present in Ne–Cu scattering, the energy displacement would be zero for all incident energies (From Buck *et al.*, 1993, with permission).

(Figure 8.18b). This displacement of 130 eV corresponds to ~171 eV energy loss in a single collision. The spectrum of electrons emitted from Ne as a result of a Cu collision had a familiar two-peak structure, with peaks located at 23 and 27 eV. The energy losses higher than ~45 eV required for Ne double excitation were explained by a sequence of complex processes preceding formation of Ne** and leading to electron promotions from lower energy levels with subsequent Auger de-excitation. Among possible processes are the formation of one or two 2s vacancies followed by an Auger de-excitation to a $2s^2 2p^4$, Ne^{2+} state, followed by capture of two electrons into the 3s states on the outgoing part of the trajectory. These rapid de-excitations occur near the surface and do not give sharp peaks in the electron spectra.

The inelastic losses and charge fractions of Ne^+ scattered from Cu and Au have been compared by scattering 1–9-keV Ne^+ and Na^+ from Cu in Cu(100), Cu_3Au crystals, and polycrystalline Cu surfaces.[123] Ne^+ ions scattered from the first layer, the first and the second layers, and in an arbitrary non-low index azimuthal direction were detected. The same dependencies of the inelastic peak displacement ΔE on the energy of the projectile were observed for all geometries and target compositions. The steep increase of Q and ion yield was observed beginning at energies 3.5–4 keV. The threshold behavior was absent for Ne^+ scattered from Au. In the case of Cu, the total energy loss was assumed to be a sum of two independent terms: a continuous loss contribution and a threshold. A continuous loss was extracted from the "prethreshold" part of the plot of the experimental energy loss when the second mechanism is turned off. The continuous loss was fitted with the expression

$$Q_{1,3} = CLv^n, \tag{35}$$

where C is a constant, v is the particle velocity, and L is the trajectory length inside the target. From the initial portion of the energy loss plot between 1 and 4 keV, it was found that $C = 3311.0\,\text{eV/Å}$ and $Q_{1,3} \sim E^{0.9}$. Subtracting the continuous background allowed a more precise determination of the threshold loss as

$$Q_2 = \Delta E_{inel} - Q_1 - Q_3. \tag{36}$$

The extracted value of the total inner loss is $Q = 112$ eV. While the total ion fraction is influenced by processes occurring along all parts of the trajectory and is an integral characteristic of the history of the particle, the characteristic threshold loss allows singling out charge transfer processes occurring in the violent collision.

8.3. THEORETICAL METHODS

The conventional theory for describing charge transfer processes in atom–surface scattering has been based on the time-dependent Anderson model.[67,130] This model describes charge transfer through resonant tunneling of electrons between atoms and surfaces. In this approach, the instantaneous populations of the atomic levels are calculated by integrating the Hamiltonian of the surface–atom system along the trajectory of the atom as it moves outside the surface. Although a general solution to the problem is difficult,[85,131–133] in certain limits such as the limit of low velocity and weak atom–surface interaction,[132,134] the problem reduces to the integration of simple rate equations along the trajectory of the projectile. As an example of such simple rate equations, we consider a hydrogen atom interacting with a metal surface.

As the atom approaches the surface, several different atomic configurations may be populated by resonant tunneling. For simplicity we will only consider atomic configurations involving electrons in the H(1s) level. This level may be empty (positive ion case), singly occupied by either a spin-up or spin-down electron, or doubly occupied (negative ion state). The instantanous fractions of atoms in these configurations are denoted n_+, $n_\uparrow$, $n_\downarrow$, and n_-. For simplicity, assume a spin degenerate problem (equal tunneling rates for different spin and spin degeneracy) and introduce the neutral ion fraction $n_0 = n_\uparrow + n_\downarrow$. In a typical atom–surface scattering situation, all of these atomic configurations may play a part in the charge transfer. Even though the initial state may be a postive ion (proton), because of the atomic level shifts as the atom or ion approaches a surface, different atomic states such as the negative ion state may be formed temporarily, and these may influence the final charge transfer. For this reason, several of the configurations of the atom must be taken into account.

As the atom moves in the surface region, the instantaneous fractions of atoms in the different atomic configurations are obtained by solving a set of coupled equations. The equation for the fraction of positive ions has the form

$$\frac{dn_+}{dt} = \Gamma_0^+ [1 - f(\varepsilon_{1s})]n_0 - 2\Gamma_0^+ f(\varepsilon_{1s})n_+. \tag{37}$$

In this equation, $\Gamma_0{}^+$ is the tunneling rate of an electron from a neutral atom into the surface (ionization rate), ε_{1s} is the energy of the H(1s) state, and f is the Fermi Dirac distribution function. All of these quantities are time dependent because they depend on the instantaneous position of the atom with respect to the surface. The first term on the right-hand side describes the rate of formation of positive ions. Positive ions are formed from neutral atoms by resonant tunneling of an electron into the conduction band of the surface. The $[1-f]$ factor comes from the Pauli exclusion principle; the positive ion cannot be formed by resonant tunneling of an electron from the atom to the surface if the energy of the H(1s) orbital lies below the Fermi level. The second term describes the reverse process (i.e., neutralization). The Fermi Dirac factor is present because the process can only occur if an electron of energy ε_{1s} is present in the substrate. The factor of two arises from spin degeneracy; neutralization of the proton can occur either through a spin-up or a spin-down electron. In order to solve Eq. (37), the equation for the neutral fraction of atoms is needed. This equation has the form:

$$\frac{dn_0}{dt} = +2\Gamma_0^+ f(\varepsilon_{1s})n_+ - \Gamma_0^+ [1 - f(\varepsilon_{1s})]n_0 + 2\Gamma_0^- [1 - f(\varepsilon_A)]n_- - \Gamma_0^- f(\varepsilon_A)n_0$$

$$\tag{38}$$

This equation is more complicated due to the fact that neutral atoms can be formed either by resonant tunneling of an electron into positive ions or by resonant tunneling of electrons from negative ions into the substrate. The first term describes the increase in neutral fraction due to neutralization of the positive ion state. Again, the factor of two comes from the fact that neutralization of the positive ion can occur by either a spin-up or spin-down electron. The second term describes the decrease of the neutral fraction due to ionization of the neutral atoms. The third term describes how the instantaneous neutral fraction is increased by electron tunneling of an electron from a negative ion. The instantaneous rate for this process, Γ_0^-, is different from the rate for electron tunneling between a neutral atom and the surface Γ_0^+. The $[1-f]$ factor comes from the Pauli principle; electron tunneling from the negative ion state can only occur if the affinity level ε_A lies above the Fermi level of the surface. The factor of two comes from spin degeneracy. The last term describes how the neutral fraction of the atoms is decreased by the resonant capture of an electron into the negative ion state. In order to solve Eq. (38), it is necessary to know the instantaneous fractions of negative ions outside the surface. This quantity is obtained by solving the following equation:

$$\frac{dn_-}{dt} = \Gamma_0^- f(\varepsilon_A)n_0 - 2\Gamma_0^-[1 - f(\varepsilon_A)]n_- \tag{39}$$

The first term on the right-hand side describes the increase of the instantaneous negative ion fraction due to resonant capture of an electron from the surface. The second term describes the reverse process. Again, the factor of two comes from the spin degeneracy. These three equations form a closed set of equations that can be solved numerically on a simple computer. The simple equations presented above only provide an approximate solution to the problem. The full numerical solution to the problem is much more complicated and can exhibit several interesting many-body effects that have been discussed elsewhere.[135-138]

The crucial input in the dynamical description discussed above is the knowledge of how the energies ε_{1s} and ε_A and the tunneling rates Γ_0^+ and Γ_0^- depend on the position of the atom outside the surface. In addition, some assumption of the atomic trajectory must be made to convert the atomic motion into a time-dependent problem.

The calculation of the energy shifts and broadening (electron tunneling rates) of atomic levels near metal surfaces represents a difficult problem in which it is crucial to properly describe the electronic structure of the adsorbate–substrate system. Since the excited adsorbate states in general will be resonances, conventional bound state techniques cannot be used directly for the calculations of the resonance energies. Due to the overlap between the atomic states and the surface states, perturbative techniques are not suitable except at very large atom–surface separations.

Several accurate nonperturbative methods have been developed for the calculation of the energies and widths of atomic and molecular resonances near clean and impurity-covered surfaces. The most widely used are the complex scaling method of Nordlander and Tully,[56,57,100,139-144] the coupled angular momentum method of Gauyacq and Teillet-Billy,[58,65,145-151] and the close-coupling method of Kurpick and Thumm.[38,60,152-158] While the methods are different in nature, when applied to similar systems, the results have been in agreement. A significant limitation of these theoretical methods is that the surface potential is constructed very crudely using empirical methods rather than calculated. For this reason, the applications of these methods have been restricted to surfaces described using simple approximate models.

The extension of the conventional methods for the calculation of energy shifts and broadening of atomic resonances to realistic corrugated surfaces represents a formidable task. The major obstacle preventing the application of these approaches to realistic corrugated surfaces is the difficulty in constructing the electron potential in the surface region. The electron potential that is responsible for the tunneling barrier between the atom and the surface depends on the interaction between the atom and the surfaces. This potential must be calculated self-consistently rather than estimated by adding substrate and adsorbate potentials and simple image potentials describing the interaction between the adsorbate and the surface.

Recently, first principles quantum chemical approaches have been proposed for the calculation of energy shifts and broadenings of atomic or molecular levels near realistic surfaces.[159–161] In these approaches, the energy shift and broadening of the atomic levels are obtained by investigating the density of states of the atom–substrate system projected on the atomic levels. These type of methods are in principle capable of an *ab initio* calculation of the energy shift and broadening of both atomic and molecular levels outside arbitrary surfaces including insulators, semiconductors, metals and disordered surfaces. The methods show considerable promise, but due to their complexity, applications to realistic systems have so far been scarce.

REFERENCES

1. J. W. Rabalais, Ed. "Low Energy Ion-Solid Interactions," New York, Wiley Interscience, 1994.

2. E. S. Parilis, I. M. Kishinevsky, N. Yu. Turaev, B. E. Baklitzky, F. F. Umarov, V. K. Verleger, S. L. Nizhnaya, and I. S. Bitensky, "Atomic Collisions on Solid Surfaces," North Holland, 1993.

3. M. Nastasi, J. W. Mayer, and J. K. Hirvonen, "Ion-Solid Interactions," Cambridge U. Press, 1996.

4. E. E. Nikitin and S. Ya. Umanskii, "Theory of Slow Atomic Collisions," Springer-Verlag, Berlin, 1984.

5. H. D. Hagstrum, P. Petrie, and E. E. Chaban, *Phys. Rev. B* **38**, 10264 (1988); N. H. Tolk, J. C. Tully, W. Heiland, and C. White, (Eds.), "Inelastic Ion Surface Collisions," Academic Press, New York, 1977; H. D. Hagstrum, *Phys. Rev.* **96**, 336 (1954).

6. U. Fano and W. Lichten, *Phys. Rev. Lett.* **14**, 627 (1965).

7. R. A. Baragiola and C. A. Dukes, *Phys. Rev. Lett.* **76**, 2547 (1996).

8. R. Hentschke, K. J. Snowdon, P. Hertel, and W. Heiland, *Surf. Sci.* **173**, 565 (1986).

9. N. Stolterfoht, J. H. Bremer, V. Hoffmann, M. Rosler, R. Baragiola, and I. De Gortari, *Nucl. Inst. Meth B* **182**, 89 (2001).

10. V. A. Esaulov, *J. Phys. Condens. Matter* **6**, L699 (1994).

11. C. Lemell, H. P. Winter, F. Aumayr, J. Burgdorfer, and F. Meyer, *Phys. Rev. A* **53**, 880 (1996).

12. W. Heiland and E. Taglauer, *Nucl. Inst. Meth.* **132**, 535 (1976).

13. A. B. Popov, B. N. Makarenko, and A. P. Shergin, *Tech. Phys. Lett.* **20**, 89 (1994).

14. M. L.Yu and B. N. Eldridge, *Phys. Rev. B* **42,** 1000 (1990).

15. G. A. Kimmel and B. H. Cooper, *Phys. Rev. B* **48,** 12164 (1993).

16. E. G. Overbosch, B. Rasser, A. D. Tenner, and J. Los, *Surf. Sci.* **92,** 310 (1980).

17. V. V. Afrosimov, Yu. S. Gordeev, M. N. Panov, and N. V. Fedorenko, *Soviet Phys.-Tech. Phys.* **9,** 1248, 1256, 1265 (1965); M. Ya. Amusia, Phys. Lett. **14,** 36 (1965).

18. E. Everhart and Q. C. Kessel, *Phys. Rev. Lett.* **14,** 247 (1965).

19. M. Barat and W. Lichten, *Phys. Rev. A* **6,** 211 (1972).

20. V. V. Afrosimov, G. G. Meskhi, N. N. Tsarev, and A. P. Shergin, *Sov. Phys.-JETP* **57,** 263 (1983).

21. W. Brandt and R. Laubert, *Phys. Rev. Lett.* **24,** 1037 (1970).

22. F. T. Smith, *Phys. Rev.* **179,** 111 (1969).

23. W. Lichten, *J. Phys. Chem.* **84,** 2102 (1980).

24. A. L. Boers, *Nucl. Instrum. Met. Phys. Res. B* **4,** 98 (1984).

25. J. W. Rabalais, J. N. Chen, R. Kumar, and M. Narayana, *J. Chem. Phys.* **83,** 6489 (1985).

26. J. W. Rabalais and J. N. Chen, *J. Chem. Phys.* **85,** 3615 (1986).

27. O. B. Firsov, *Sov. Phys. JETP* **9,** 1076 (1959).

28. J. Lindhard and M. Scharff, *Phys. Rev.* **124,** 128 (1961).

29. J. Lindhard, Kgl. Danske Videnskab. *Selskab, Mat.-fys. Medd.* **28,** No. 8 (1954), J. Lindhard and M. Scharff, *ibid.* **27** No. 15 (1953).

30. O. S. Oen and M. T. Robinson, *Nucl. Instr. Meth. Phys. Res.* **132,** 647 (1976).

31. E. M. Kuipers and A. L. Boers, *Nucl. Instr. Meth. Phys. Res. B* **29** 567 (1987).

32. J. Limburg, R. Morgenstern, B. N. Makarenko, and A. P. Shergin, *Izv. Akad. Nauk. Fiz.* (Proceedings of the Academy of Sciences) **62,** 797 (1998).

33. J. Folkers, S. Schippers, D. M. Zehner, and F. W. Meyer, *Phys. Rev. Lett.* **74,** 2204 (1995).

34. A. C. Lavery, C. E. Sosolik, and B. H. Cooper, *Nucl. Instr. Meth. B* **157,** 42 (1999).

35. M. Richard-Viard, C. Benazeth, P. Benoit-Cattin, P. Cafarelli, S. Abidi, and J. P. Ziesel, *Nucl. Instr. Meth. Phys. Res. B* **575,** 164 (2000).

36. F. Wyputta, R. Zimny, and H. Winter, *Nucl. Instr. Meth. Phys. Res. B* **58,** 379 (1991).

37. M. Maazouz, L. Guillemot, and V. A. Esaulov, *Phys. Rev. B* **56,** 9267 (1997).

38. M. Maazouz, S. Ustaze, L. Guillemot, and V. A. Esaulov, *Surf. Sci.* **409,** 189 (1998).

39. C. Auth, H. Winter, A. G. Borisov, B. Bahrim, D. Teillet-Billy, and J. P. Gauyacq, *Phys. Rev. B* **57,** 12579 (1998).

40. C. C. Hsu, A. Bousetta, J. W. Rabalais, and P. Nordlander, *Phys. Rev. B* **47,** 2369 (1993).

41. V. A. Esaulov, *Surf. Sci.* **415,** 95 (1998).

42. S. Ustaze, L. Guillemot, R.Verucchi, S. Lacombe, and V. A. Esaulov, *Nucl. Instr. Meth. Phys. Res. B* **135,** 319 (1998).

43. J. C. Tucek, S. G. Walton, and R. L. Champion, *Surf. Sci.* **410,** 258 (1998).

44. S. Ustaze, S. Lacombe, L. Guillemot, V. A. Esaulov, and M. Canepa, *Surf. Sci.* **414,** L938 (1998).

45. M. Maazouz, A. G. Borisov, V. A. Esaulov, J. P. Gauyacq, L. Guillemot, S. Lacombe, and D. Teillet–Billy, *Phys. Rev. B* **55,** 13869 (1997).

46. E. M. Staicu-Casagrande, S. Lacombe, L. Guillemot, V. A. Esaulov, L. Pasquali, S. Nannarone, and M. Canepa, *Surf. Sci.* **480,** L411 (2001).

47. M. Casagrande, S. Lacombe, L. Guillemot, and V. A. Esaulov, *Surf. Sci.* **445,** L29 (2000).

48. D. J. O'Connor, M. Maazouz, L. Guillemot, and V. A. Esaulov, *Surf. Sci.* **398,** 49 (1998).

49. L. Guillemot, S. Lacombe, and V. A. Esaulov, *Nucl. Instr. Meth. Phys. Res. B* **164,** 601 (2000).

50. L. Houssiau, J. W. Rabalais, J. Wolfgang, and P. Nordlander, *J. Chem. Phys.* **110,** 8139 (1999).

51. M. Maazouz, A. Borisov, V. A. Esaulov, J. P. Gauyacq, L. Guillemot, S. Lacombe, and D. Teillet-Billy, *Surf. Sci.* **364,** L568 (1996).

52. B. Bahrim, D. Teillet-Billy, and J. P. Gauyacq, *Surf. Sci.* **316,** 189 (1994).

53. A. K. Kazansky, A. G. Borisov, and J. P. Gauyacq, *Nucl. Instr. Meth. Phys. Res. B* **157,** 21 (1999).

54. R. Taranko and E. Taranko, *Surf. Sci.* **441,** 167 (1999).

55. S. Ustaze, L. Guillemot, V. A. Esaulov, P. Nordlander, and D. C. Langreth, *Surf. Sci.* **415,** L1027 (1998).

56. P. Nordlander and N. D. Lang, *Phys. Rev. B* **46,** 2584 (1992).

57. P. Nordlander and J. C. Tully, *Phys. Rev. Lett.* **61,** 990 (1988).

58. D. Teillet-Billy and J. P. Gauyacq, *Surf. Sci.* **239,** 343 (1990).

59. F. Martin and M. F. Politis, *Surf. Sci.* **356,** 247 (1996).

60. P. Kurpick, U. Thumm, and U. Wille, *Phys. Rev. A* **56,** 543 (1997).

61. H. Winter, Comments *At. Mol. Phys.* 26, 287 (1991).

62. A. G. Borisov, J. P. Gauyacq, and S. V. Shabanov, *Surf. Sci.* **487,** 243 (2001).

63. B. Bahrim, D. Teillet-Billy, and J. P. Gauyacq, *Surf. Sci.* **431,** 193 (1999).

64. J. N. M. van Wunnick, R. Brako, K. Makoshi, and D. M. Newns, *Surf. Sci.* **126,** 618 (1983).

65. A. G. Borisov, D. Teillet-Billy, and J. P. Gauyacq, *Phys. Rev. Lett.* **68,** 2842 (1992).

66. N. Lorente, A. G. Borisov, D. Teillet-Billy, and J. P. Gauyacq, *Surf. Sci.* **429,** 46 (1999).

67. D. C. Langreth and P. Nordlander, *Phys. Res. B* **43,** 2541 (1991).

68. R. Brako, D. M. Newns, *Rep. Prog. Phys.* **52,** 655 (1989).

69. H. Winter, Prog. *Surf. Sci.* **63,** 177 (2000).

70. H. Niehaus, W. Heiland, and E. Taglauer, *Surf. Sci. Rep.* **17,** 21 (1993).

71. A. G. Borisov and V. A. Esaulov, *J. Phys. Condens. Matter* **12,** R177 (2000).

72. J. Los, J. J. C. Geerlings, *Phys. Rep.* **190,** 133 (1990).

73. R. L. Erickson and D. P. Smith, *Phys. Rev. Lett.* **3**4, 297 (1975).

74. A. D. F. Kahn, D. J. O. O'Connor, and R. J. MacDonald, *Surf. Sci. Lett.* **262,** L83 (1992).

75. S. Schippers, C. Oelschig, W. Heiland, L. Folkerts, R. Morgenstern, P. Eeken, I. F. Urazgildin, and A. Niehaus, *Surf. Sci.* **257,** 289 (1991).

76. A. Zartner, E. Taglauer, and W. Heiland, *Phys. Rev. Lett.* **40,** 1259 (1978).

77. G. J. Lockwood and E. Everhart, *Phys. Rev.* **125,** 567 (1962).

78. E. Everhart, *Phys. Rev.* **132,** 2083 (1963).

79. H. F. Helbis and E. Everhart, *Phys. Rev.* **136,** A675 (1964).

80. W. Lichten, *Phys. Rev.* **139,** A27 (1965).

81. J. C. Tully, *Phys. Rev. B* **16,** 4324 (1977).

82. W. Bloss and D Hone, *Surf Sci.* **72,** 277 (1978).

83. E. C. G. Stueckelberg, *Helv. Phys. Acta* **5,** 370 (1932).

84. W. Heiland, *Proc. Nato-ASI,* Alicante, Spain, 1990, Plenum, New York, 1991.

85. H. Shao, D. C. Langreth, P. Nordlander, *Phys. Rev. B* **49,** 13948 (1994).

86. A. Arnau, F. Aumayr, P. M. Echenique, M. Grether, W. Heiland, J. Limburg, R. Morgenstern, P. Roncin, S. Shippers, R. Schuchterfoht, P. Varga, and T. J. M. Zouros, *Surf. Sci. Rep.* **27,** 113 (1997).

87. D. H. G. Schneider, and M. A. Briere, *Phys. Scripta* **53,** 228 (1996).

88. T. Schenkel, A. V. Hamza, A. V. Barnes, and D. H. Schneider, *Progr. Surf. Sci.* **61,** 23 (1999).

89. T. Schlatholter, O. Hadjar, R. Hoekstra, and R. Morgenstern, *Appl. Phys. A* **72,** 281 (2001).

90. H. Khemliche, C. Laulhe, S. Hoekstra, R. Hoekstra, and R. Morgenstern, *Phys. Scripta,* *T* **80,** 66 (1999).

91. C. Laulhe, R. Hoekstra, S. Hoekstra, H. Khemliche, R. Morgenstern, A. Narmann, and T. Schlatholter, *Nucl. Instr. Meth. Phys. Res. B* **157,** 304 (1999).

92. O. Hadjar, P. Foldi, R. Hoekstra, R. Morgenstern, and T Schlatholter, *Phys. Rev. Lett.* **84,** 4076 (2000).

93. F. W. Meyer and V. A. Morozov, *Nucl. Instr. Meth. Phys. Res. B* **157,** 297 (1999).

94. F. W. Meyer and V. A. Morozov, *Phys. Rev. Lett.* **86,** 736 (2001).

95. R. Schuch, W. Huang, H. Lebius, Z. Pesic, N. Stolterfoht, and G. Vikor, *Nucl. Instr. Meth. Phys. Res. B* **157,** 309 (1999).

96. J. W. Rabalais, J. A. Schultz, R. Kumar, and P. T. Murray, *J. Chem. Phys.* **78,** 5250 (1983).

97. W. Huang, H. Lebius, R. Schuch, M. Grether, A. Spieler, and N. Stolterfoht, *Nucl. Instr. Meth. Phys. Res. B* **135,** 336 (1998).

98. M. Shi, O. Grizzi, and J. W. Rabalais, *Surf. Sci.* **235,** 67 (1990).

99. Y. Kim, S. S. Kim, E. Ada, Y. L. Yang, A. J. Jacobson, and J. W. Rabalais, *J. Chem. Phys.* **111,** 2720 (1999).

100. P. Nordlander and N. D. Lang, *Phys. Rev. B* **44,** 13681 (1991).

101. R. Souda, K. Yamamoto, W. Hayami, T. Aizawa, and Y. Ishizawa, *Surf. Sci.* **363,** 139 (1996).

102. L. Guillemot, S. Lacombe, M. Maazouz, E. Sanchez, and V. A. Esaulov, *Surf. Sci.* **356,** 92 (1996).

103. S. V. Pepper and P. R. Aron, *Surf. Sci.* **169,** 14 (1986).

104. P. Eeken, J. M. Fluit, and A. Niehaus, I. Urazgildin, *Surf. Sci.* **273,** 160 (1992).

105. G. Zampieri, F. Meier, and R. Baragiola, *Phys. Rev. A* **29,** 116 (1984).

106. B. Hird, R. A. Armstrong, and P. Gauthier, *Phys. Rev. A* **49,** 1107 (1994).

107. G. Xu, G. Manico, F. Ascione, A. Bonanno, A. Oliva, and R.A. Baragiola, *Phys. Rev. A* **57,** 1096 (1998).

108. V. A. Esaulov, *J. Phys. Condens. Matter* **6,** L699 (1994).

109. B. Arezki, Y Boudouma, P. Benoit-Cattin, A. C. Chami, C. Benazeth, K. Khalal, and M. Boudjema, *J. Phys. Condens. Matter* **10,** 741 (1998).

110. S. Mouhammad, P. Benoit-Cattin, C. Benazeth, P. Cafarelli, M. Richard-Viard, and J. P. Ziesel, *Nucl. Instr. Meth. Phys. Res. B,* **125,** 297 (1997).

111. G. Verbist, J. T. Devresse, and H. H. Brongersma, *Surf. Sci.* **233,** 323 (1990).

112. I. Vaquila, K. M. Lui, J. W. Rabalais, J. Wolfgang, and P. Nordlander, *Surf. Sci.* **470,** 255 (2001).

113. I. Vaquila, J. W. Rabalais, J. Wolfgang, and P. Nordlander, *Surf. Sci.* **489,** L561 (2001).

114. E. Taglauer, W. Englert, and W. Heiland, *Phys. Rev. Lett.* **45,** 740 (1980).

115. A. Narman, H. Derks, W. Heiland, R. Monreal, E. Goldberg, and F. Flores, *Surf. Sci.* **217,** 255 (1989).

116. R. Beikler and E. Taglauer, *Nucl. Inst. Meth. B* **182,** 180 (2001).

117. I. Vaquila, I. L. Bolotin, T. Ito, B. N. Makarenko, and J. W. Rabalais, *Surf. Sci.* **496,** 187 (2002).

118. J. M. Beaken and P. Bertrand, *Nucl. Instr. Meth. Phys. Res. B* **67,** 340 (1992).

119. Y. Muda and D. M. Newns, *Phys. Rev. B* **37,** 7048 (1988).

120. R. Souda, K. Yamamoto, W. Hayami, T. Aizawa, and Y. Ishizawa, *Surf. Sci.* **363,** 139 (1996).

121. R. Souda, K. Yamamoto, W. Hayami, T. Aizawa, and Y, Ishizawa, *Phys. Rev. Lett.* **75,** 3552 (1995).

122. S. Lacombe, L. Guillemot, M. Maazouz, N. Mandarino, E. Sanchez, and V. A. Esaulov, *Surf. Sci.* **410,** 70 (1998).

123. T. Li and R. J. MacDonald, *Phys. Rev. B* **51,** 17876 (1995).

124. O. Grizzi, M. Shi, H. Bu, J. W. Rabalais, and R. A. Baragiola, *Phys. Rev. B* **41,** 4789 (1990).

125. F. Ascione, G. Manico, P. Alfano, A. Bonnano, N. Mandarino, A. Oliva, and F. Xu, *Nucl. Instr. Meth. Phys. Res. B* **135,** 401 (1998).

126. T. M. Buck, W. E. Wallace, R. A. Baragiola, G. H. Wheatley, J. B. Rothman, R. J. Gorte, and J. G. Tittensor, *Phys. Rev. B* **48,** 774 (1993).

127. G. Zampieri, F. Meier, and R. Baragiola, *Phys. Rev. A* **29,** 116 (1984).

128. H. Winter, G. Siekmann, H. W. Ortjohann, J. C. Poizat, and J. Remillieux, *Nucl. Instr. Meth. Phys. Res. B* **135,** 372 (1998).

129. J. Ø. Olsen, T. Andersen, M. Barat, Ch. Courbin-Gaussorgues, V. Sidis, J. Pommier, J. Agusti, N. Andersen, and A. Russek, *Phys. Rev. A* **19,** 1457 (1979).

130. R. Brako and D. M. Newns, *Surf. Sci.* **108,** 253 (1981).

131. H. Shao, D. C. Langreth, and P. Nordlander, *Phys. Rev. B* **49,** 13929 (1994).

132. H. Shao, D. C. Langreth, and P. Nordlander, in "Low Energy Ion-Surface Interactions," J. W. Rabalais (Ed.), John Wiley & Sons, Ltd., New York, 1994, pp. 117–186.

133. J. B. Marston, D. R. Andersson, E. R. Behringer, and B H. Cooper, *Phys. Rev. B* **48,** 7809 (1993).

134. H. Shao, P. Nordlander, and D. Langreth, *Nucl. Instr. and Meth. B* **100,** 260 (1995).

135. H. Shao, P. Nordlander, and D. C. Langreth, *Phys. Rev. B* **52,** 2988 (1995).

136. H. Shao, P. Nordlander, and D. C. Langreth, *Phys. Rev. Lett.* **77,** 948 (1996).

137. J. Merino and J. B. Marston, *Phys. Rev. B* **58,** 698 (1998).

138. M. Plihal, D. C. Langreth, and P. Nordlander, *Phys. Rev. B* **59,** 13322 (1999).

139. P. Nordlander and J. C. Tully, *Phys. Rev. B* **42,** 5564 (1990).

140. P. Nordlander, in "Laser Spectroscopy and Photochemistry on Metal Surfaces," H. L. Dai and W. Ho (Eds.), World Scientific Publishing, Singapore, 1995, pp. 347–368.

141. P. Nordlander, *Phys. Rev. B* **53,** 4125 (1996).

142. P. Nordlander and F. B. Dunning, *Phys. Rev. B* **53,** 8083 (1996).

143. M. Akbulut, T. E. Madey, and P. Nordlander, *J. Chem. Phys.* **106,** 2801 (1997).

144. J. Braun and P. Nordlander, *Surf. Sci. Lett.* **448,** L193 (2000).

145. A. G. Borisov, D. Teillet-Billy, and J. P. Gauyacq, *Surf. Sci.* **284,** 337 (1993).

146. B. Bahrim, D. Teillet-Billy, and J. P. Gauyacq, *Phys. Rev. B* **50,** 7860 (1994).

147. A. G. Borisov, G. Makmetov, D. Teillet-Billy, and J. P. Gauyacq, *Surf. Sci. Lett.* **350,** L205 (1996).

148. G. E. Makhmetov, A. G. Borisov, D. Teillet-Billy, and J. P. Gauyacq, *Surf. Sci.* **366,** L769 (1996).

149. D. Teillet-Billy, J. P. Gauyacq, and P. Nordlander, *Surf. Sci. Lett.* **371,** L235 (1997).

150. J. P. Gauyacq and A. G. Borisov, *J. Phys. Condens. Matter* **10,** 6585 (1998).

151. J. A. M. C. Silva, J. Wolfgang, A. G. Borisov, J. P. Gauyacq, P. Nordlander, and D. Teillet-Billy, *Nucl. Instrum. Meth. B* **157,** 55 (1999).

152. P. Kuerpick and U. Thumm, *Phys. Rev. A* **54,** 1487 (1996).

153. P. Kuerpick, U. Thumm, and U. Wille, *Nucl. Instrum. Meth. B* **125,** 273 (1997).

154. P. Kuerpick and U. Thumm, *Phys. Rev. A* **58,** 2174 (1998).

155. P. Kuerpick, U. Thumm, and U. Wille, *Phys. Rev. A* **57,** 1920 (1998).

156. J. Ducree, H. J. Anrda, and U. Thumm, *Phys. Rev. A* **60,** 3029 (1999).

157. U. Thumm, P. Kuerpick, and U. Wille, *Phys. Rev. B* **61,** 22722 (2000).

158. B. Bahrim and U. Thumm, *Surf. Sci.* **451,** 1 (2000).

159. J. Merino, N. Lorente, P. Pou, and F. Flores, *Phys. Rev. B* **54,** 10959 (1996).

160. T. Klamroth and P. Saalfrank, *Surf. Sci.* **410,** 21 (1998).

161. M. Taylor and P. Nordlander, *Phys. Rev. B* **64,** 115422 (2001).

HYPERTHERMAL REACTIVE ION SCATTERING FOR MOLECULAR ANALYSIS ON SURFACES

9.1. INTRODUCTION

A major concern in surface analysis is the identification of molecules on surfaces. While the existing surface techniques are powerful for studying atoms and simple molecules on surfaces, it is difficult to identify and characterize complex molecules. The need for molecular characterization is expected to increase as the trend of surface science research and application gradually moves from the study of simple molecules to more complex structures. This is obvious from the enhanced interest in complex molecular surfaces such as self-assembled monolayers and functionalized molecular films in relation to the prospect of molecular nanotechnology.

An interesting offspring of the ion scattering technique, called reactive ion scattering (RIS), has been recently developed.[1-39] RIS using cesium (Cs^+) projectile ions[21-39] has demonstrated the potential for surface molecular analysis. In general, RIS refers to all ion scattering techniques that exploit the reactive processes taking place between projectile ions and surfaces. Whereas conventional ion scattering, which is based on elastic scattering, performs elemental and structural analysis of surfaces, the use of active ions opens a variety of reactive channels, such as dissociation of molecular ions, transfer of an atom or molecule between a projectile ion and a surface, and ion–surface charge exchange. RIS adds rich chemistry to the physics of ion scattering. The chemical reactivity of a projectile ion can be highlighted most when its kinetic energy is relatively low. For this reason, RIS employs an ion beam at hyperthermal energies. Hyperthermal energy is loosely defined as the energy range in which the scattering process undergoes a transition from thermal to binary collision (BC) scattering; this is typically $\sim$1–100 eV. Chemical bond energies also fall into this region. The scope of RIS is quite diverse and is still expanding. This chapter focuses on the RIS method using hyperthermal Cs^+ projectiles, hereafter called Cs^+ RIS,[21-39] due to its ability for surface molecular analysis. The subject of molecular analysis is complementary to elemental and structural determination by conventional ion scattering.

Many interesting RIS phenomena other than Cs^+ RIS have been discovered in recent hyperthermal ion beam studies. Some of these topics have been reviewed.[2–5] Due to space limitations, they will only be mentioned briefly herein. Reactive ion–surface collisions can lead to dissociation of molecular projectiles in a process known as surface-induced dissociation.[1–13] The molecular dissociation occurs with a high yield even for hyperthermal ions due to efficient transfer of projectile translational energy to internal excitation during collision. Factors that affect the molecular dissociation yield include projectile translational[3–12] and vibrational energies,[10] nature of the target material,[3–5,8] and the molecular shape of the projectile.[13] The surface-induced dissociation method has been used for structural determination of gaseous projectile ions.[3–5] Another interesting type of reactive collision is the abstraction of atoms or groups from the surface by the incident projectile. Abstraction reactions have been observed in RIS experiments with organic projectile ions scattering from self-assembled monolayers and hydrocarbon surfaces[3–5] as well as in the scattering of O^+ at an oxidized Si(100) surface[14] and NO^+ at an O/Al(111) surface.[15] Ion–surface reactive collisions can involve charge exchange between the projectile ion and surface. This process is usually undesirable in RIS experiments, as it can lead to the loss of scattered ions through neutralization. Charge reversal phenomena have also been observed in RIS,[5,11,16,17] revealing interesting electron loss and capture processes between the projectile ions and surfaces or adsorbed species. When the incident ion energy is extremely low, it is possible to deposit gaseous molecular ions intact on a surface. Such soft-landing of molecular ions has been demonstrated on metal[8] and ice[20] surfaces as well as on soft molecular surfaces, such as self-assembled monolayers.[19] Due to the early stage of its development, RIS studies have been mainly focused on discovering new types of processes and finding their mechanistic interpretations. Important applications to surface science and related disciplines may be expected.

This chapter is organized as follows. Section 2 addresses nonreactive ion–surface collisions at hyperthermal energies and describe its two aspects: (2.1) surface scattering phenomena of hyperthermal ions with emphasis on heavy projectiles, and (2.2) sputtering yields in hyperthermal ion–surface collisions. Section 3.1 introduces the basic concepts of the Cs^+ RIS process, and section 3.2 discusses its mechanism, dynamics, and quantitation. Section 4 presents several case studies to which Cs^+ RIS has been applied, including identification of surface molecules (section 4.1), study of reaction kinetics and intermediates (section 4.2), and molecules on ice (section 4.3). Section 5 summarizes the features of the Cs^+RIS technique.

9.2. CHARACTERISTIC FEATURES OF HYPERTHERMAL ION–SURFACE COLLISIONS

9.2.1. Scattering of Hyperthermal Projectiles from Surfaces

When hyperthermal ions collide with a surface, the scattering process is different from that of keV collisions described in Chapter 2. Since the ion kinetic energy is

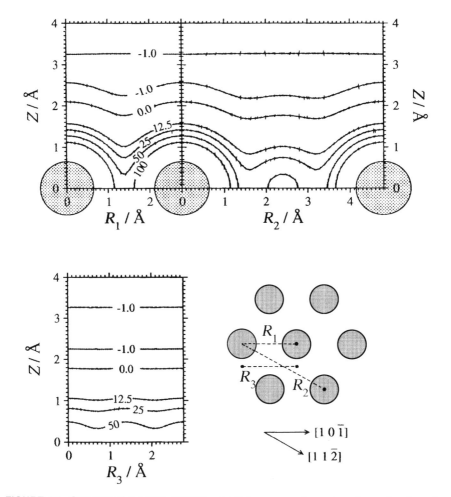

FIGURE 9.1. Contours of the Cs⁺–Pt(111) potential energy surface along three directions for the top view of the fcc(111) surface. Z is the distance perpendicular to the surface. The contour values are in eV (From Lahaye *et al.*, 2001, with permission).

relatively small compared with the ion–atom repulsive potential, hyperthermal ions do not approach near the core of individual surface atoms, but are reflected at a larger distance from the surface. The ions see the surface as a relatively smooth structure with relatively insignificant atomic corrugation. This effect becomes pronounced as the size of a projectile increases and the distance of closest approach to a surface increases. Figure 9.1 shows the potential energy surface that a Cs⁺ projectile experiences on a Pt(111) surface as contours along the crystal orientation axes.[38] The negative values of the potential contours are due to the image charge attraction between the ion and the metallic surface. The repulsive potentials due to Pt atoms overlap between neighboring atoms such that they cannot be considered as isolated, independent collision centers. Consequently, multiple scattering effects dominate in hyperthermal collisions. At hyperthermal energies, the projectiles have very low velocities, especially when the projectile is heavy. For example, Cs^+ with an energy of 10 eV has a velocity of 3.8×10^3 m/sec, which is of the same order as the velocity of sound in typical solids. This implies that during the collision the surface atoms connected through the bonding network can respond collectively to the projectile impact.[40] Thus, many-body effects play important roles for hyperthermal collisions.

The validity of the BC model in hyperthermal ion–surface collisions was one of the main issues in early scattering studies. The scattering behavior of hyperthermal beams has been explored with various types of atomic projectiles, including alkali metal ions,[40–50] noble gas ions,[40,50–55] and neutral atoms.[56–58] Although the kinetic energy accessible to a neutral beam is limited to less than a few eV in supersonic nozzle expansion, neutral beam experiments provide a useful reference for ion scattering results obtained in a comparable energy region. In an ion scattering experiment, the scattered ion energy distribution can be affected by neutralization of the ions during the collision. For this reason, some ion scattering researchers favored alkali metal ions due to their low neutralization probability; in many cases the projectiles were lighter than target atoms.[41–46,49,50] In the early work by Hulpke,[42] lithium (Li^+) ions were scattered from W(110) and Si(111) surfaces for beam energies of 2–20 eV, and the energies and angular distributions of the scattered ions were measured. The results showed that the scattering behavior can be explained in terms of BCs with single surface atoms if the influence of the attractive interaction between the ion and the surface is taken into account. Later, it was suggested[43] that the spectra of Li^+ ions might be distorted by ion neutralization effects. Aided by computer trajectory simulations, alkali ion scattering studies have since been used to investigate the potentials of ion–surface interactions.[43–46,49,50] Scattering of noble gas ions was examined at hyperthermal energy first by Tongson and Cooper.[51] They observed that the BC approximation is valid for beam energies down to 40 eV for helium (He^+) and 20 eV for neon (Ne^+) scattering from copper (Cu). The BC behavior in hyperthermal scattering was also observed in a recent study of Ne^+ collisions with gold (Au) and platinum (Pt) surfaces.[55]

There are, however, reports indicating that the BC approximation is invalid in hyperthermal scattering. Experiments on He^+ and Ne^+ scattering at a Si(100) surface in the range 20–300 eV[54] showed that the scattered ions have a broad energy distribution that is shifted to higher energy than that of a single BC collision. The energy shift increased with decreasing beam energy and, at beam energies of 20–30 eV, the scattered energy distribution peaked near the double-scattering position. The different observations for noble gas ion scattering from metals[51,55] and Si[54] suggest the possibility that the measured energy distributions were distorted by ion-neutralization effects. Noble gas ions can survive only single collisions with metal atoms since their neutralization probability is high on such low work-function surfaces. Multiply scattered ions are preferentially neutralized and do not contribute to the energy spectra. On a silicon (Si) surface, on the other hand, ions can survive even multiple collisions and scatter with higher energy. Classical trajectory simulations for the scattering of noble gas *atoms* (no neutralization of projectile) from a nickel Ni(100) surface[53] showed that the atoms indeed undergo prominent multiple scattering at hyperthermal energies, resulting in a scattered energy distribution that is significantly higher than that of a single BC collision. The multiple scattering effect increased as the projectile energy was decreased.

Heavy projectiles exhibit an interesting deviation from binary scattering behavior in the hyperthermal regime. According to the BC model, a projectile heavier than a target atom cannot backscatter at angles above a maximum critical angle θ_{max}, as shown by Eqs. 4 and 13 in Chapter 2. However, this is observed experimentally in hyperthermal ion–surface scattering. Eq. 13 (Chapter 2) shows that binary scattering of a heavy projectile does not occur at arbitrary angles, but is confined to $\theta_{max} \leq \sin^{-1} M_2/M_1$. For example, $\theta_{max} = 44°$ and $12°$ for argon (Ar) and xenon (Xe), respectively, scattering from silicon (Si). Therefore, according to the BC approximation (BCA), these projectiles should not be detected at higher scattering angles (e.g., 90°). In hyperthermal scattering experiments, however, these scattered projectiles are detected at such large angles.[40] This clearly demonstrates that the BCA is not valid in such a case and that the scattering events involve multiple collisions with surface atoms.

Experimental studies of heavy ions colliding with surfaces at hyperthermal energy are relatively rare. Evstifeev and coworkers[47,48] examined Cs^+ scattering from tungsten (W), molybdenum (Mo), cobalt (Co), and uranium (U) surfaces. They observed that the energies of ions scattered from these surfaces were unusually higher than expected from sequential BC with the surface atoms. Yang et al.[40] studied the scattering of 3–300-eV Cs^+, Ar^+, and xenon (Xe^+) ions from Si in a 45°/45° scattering geometry. They observed that the E_1/E_0 ratio varies drastically with E_0 as shown in Figure 9.2 for the Cs^+ scattering result. Ar^+ and Xe^+ may not be appropriate cases to

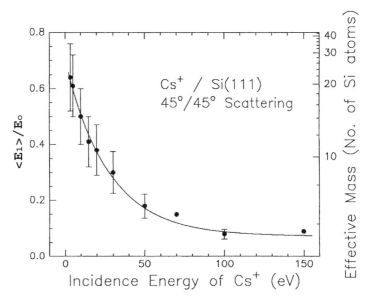

FIGURE 9.2. Ratio of the scattered to incident Cs^+ energy ($<E_1>/E_0$) as a function of E_0. The effective mass scale on the right-hand side was deduced from the $<E_1>/E_0$ ratio and the binary collision model (From Yang *et al.*, 1996, with permission).

deduce quantitative interpretation because their E_1 distributions may be affected by the ion neutralization effect. Figure 9.2 shows that the E_1/E_0 ratio for Cs^+ is initially very high (0.64 at $E_1 = 3$ eV) and drops sharply with increasing energy to 0.1 for $E_1 >$ 100 eV. Such a large variation in E_1/E_0 suggests that extensive multiple collisions take place at low incident energies. Another interesting feature of Figure 9.2 is that even when E_0 increases substantially above 20 eV, *the absolute E_1 value remains fairly constant and low.* As a result, the E_1 distribution of Cs^+ has a substantial population at low energies comparable to those of sputtered particles.

The extreme variation of E_1/E_0 with E_0 is unique to the case of a projectile that is heavier than the surface atoms and is attributed to the many-body nature of the collisions. According to the molecular dynamics classical trajectory simulations of Xe scattering from Ni(100),[53] the Xe atoms scatter by multiple interactions with top-layer Ni atoms. This multiple interaction increases the effective mass of the surface, thus giving rise to a high E_1/E_0 ratio. When E_0 becomes substantially high, Xe starts to penetrate into the surface layer by pushing Ni atoms away. Upon surface penetration, Xe atoms generate collision cascades in the near surface region, lose a large portion of their initial kinetic energy, and then scatter with lower energy. Thus, above the penetration threshold energy, the E_1 distribution is rapidly downshifted, and this phenomenon is accompanied by the sputtering of surface atoms. An interesting question may be the number of surface atoms that make direct contact with a projectile during such multiple collisions. Despite the strong multiple collision effects, molecular dynamics (MD) simulations of Cs scattering from Si[40] showed that only a few Si atoms experienced direct collisions. The neighboring lattice atoms, however, collectively respond to the impact through the bonding network, thereby dramatically increasing the effective mass of the surface. This is a dynamical effect caused by coupling between the motions of the projectile and the surface atoms during the long interaction time between a slow-moving heavy projectile and a surface.

A large projectile–target mass ratio gives rise to prominent angular dependency in the scattered particle intensity as well as in the E_1/E_0 distribution. Heavy projectiles scatter with a narrow angular distribution and towards a supraspecular direction.[38,53,56] The E_1/E_0 value of an angle-resolved scattered flux increases with increasing scattering angle.[38] However, even for heavy projectiles such as Cs^+, the scattering behavior can be BC-like if the surface atoms are heavier than the projectile. According to a recent theoretical study,[38] Cs^+ scattering from Pt(111) can be adequately described in terms of a BC model that is augmented with double collisions and an image charge correction.

9.2.2. Low-Energy Sputtering

Ion–surface collisions induce the removal of surface atoms in a process known as sputtering. Sputtering occurs as incoming projectiles transfer energy to target atoms such that they can overcome the surface binding energy. According to the theory of Sigmund,[59] the sputtering yield, or the number of sputtered target atoms per projectile, increases in proportion to the energy deposited into the surface layer. Reviews are available in the literature for sputtering phenomena in keV and higher energy regimes.[60,61]

The sputtering yield decreases sharply as the projectile energy is reduced, and below a certain energy, target atoms do not receive enough energy to overcome the surface barrier. This leads to the concept of a threshold sputtering energy (E_{th}). The sputtering threshold is defined theoretically as the projectile energy at which the sputtering yield becomes zero and, thus, is ultimately related to the surface binding energy. A simple sputtering theory based on the binary collision model[62] suggests a relationship between the sputtering threshold and the surface binding energy E_1 given by Eq. 1,

$$E_{th} = E_1/\gamma \quad \text{with} \quad \gamma = 4 M_1 M_2/(M_1 + M_2)^2, \tag{1}$$

where γ is the energy transfer ratio determined by the projectile (M_1) and target masses (M_2). Other variations of this expression have been derived from more elaborate theories.[63] However, experimentally one can only measure the energy at which the number of sputtered particles is reduced below the detection limit. An experimental threshold energy therefore is always higher than a theoretical one, and the extrapolation of sputtering yield data towards zero yield is required.

Efforts to measure E_{th} date back to the 1920s when Kingdon and Langmuir studied the removal of a thorium (Th) layer coated on a W wire in a low-pressure plasma.[64] It has been discovered since then that E_{th} for many metals lies in the hyperthermal regime. Wehner and co-workers[65–68] were first to provide some of the most reliable data for E_{th}. They measured the sputtering yield (i.e., the number of atoms removed from the surface divided by the number of atoms impinging on the surface) from the weight loss of the target material in a low-pressure discharge tube as well as from the spectral emission intensity of the sputtered atoms. The weight loss provided the absolute sputtering yield, and the spectral emission measurements increased the sensitivity down to 10^{-5} atoms per incident ion. In their experimental configuration, the ion-bombardment energy was defined by the voltage applied to the target with respect to the plasma, and the ion-incidence direction was near the surface normal. For metals, the sputtering yields were found generally to be lower than 1.0 for $E_0 \leq 300$ eV and dropped rapidly as E_0 approached E_{th}. The E_{th} values measured by Wehner and co-workers are summarized in Table 9.1. They deduced a relationship that $E_{th} \sim 4\Delta H_{sub}$, where ΔH_{sub} is the heat of sublimation. This relationship, however, agrees only with certain metals, which apparently indicates that the sputtering process cannot be described simply by thermodynamics. Trajectory calculations have been employed for detailed examination of the sputtering process,[53,59,69–71] revealing features such as collision dynamics leading to sputtering, the energy transfer efficiency from projectile to target atoms, and the mass and size effects of projectile and target atoms.

The low-energy sputtering of chemisorbed species has also been studied by several researchers.[72–76] Winters and Sigmund[72,73] examined the sputtering yields for N atoms chemisorbed on W by bombardment with noble gas ions at 20–500 eV. Sputtering

TABLE 9.1. Sputtering threshold energies (eV) measured by Wehner and co-workers (Reproduced with permission from ref. 68)

Target Ion	Ne	Ar	Kr	Xe	Hg	Target Ion	Ne	Ar	Kr	Xe	Hg
Be	12	15	15	15		Mo	24	24	28	27	32
Al	13	13	15	18	18	Rh	25	24	25	25	
Ti	22	20	17	18	25	Pd	20	20	20	15	20
V	21	23	25	28	25	Ag	12	15	15	17	
Cr	22	22	18	20	23	Ta	25	26	30	30	30
Fe	22	20	25	23	25	W	35	33	30	30	30
Co	20	25	22	22		Re	35	35	25	30	35
Ni	23	21	25	20		Pt	27	25	22	22	25
Cu	17	17	16	15	20	Au	20	20	20	18	
Ge	23	25	22	18	25	Th	20	24	25	25	
Zr	23	22	18	25	30	U	20	23	25	22	27
Nb	27	25	26	32							

of O atoms chemisorbed on Ni(111) in the p(2 × 2) phase has been examined by Rabalais *et al.*[74] Figure 9.3 shows the result of Winters and Sigmund[73] for sputtering of N atoms. The sputtering yield curve for Xe^+ increases continuously from 0.05–0.85 over the energy span of 25–500 eV, but for He^+, the yield saturates at 0.05 above 100 eV. Theoretical analysis of the results[73] led to a suggestion that the

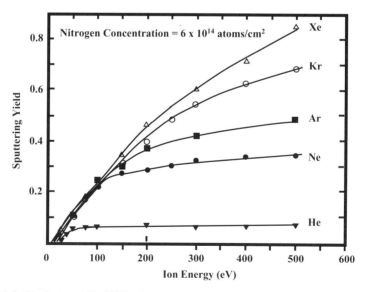

FIGURE 9.3. Sputtering yields (S) for nitrogen atoms chemisorbed on tungsten by bombardment with noble gas ion projectiles. λ is the secondary electron emission coefficient (From Winters *et al.*, 1974, with permission).

adsorbed N atoms are ejected as a consequence of direct knock-on collisions with impinging or reflected projectile ions. These direct knock-on events prevail at energies near the sputtering threshold, whereas sputtering due to complex collision cascades becomes important at much higher energies. The sputtering yield curve for oxygen on Ni(111)[74] exhibits a similar energy dependency and magnitude as those for N on W. The O sputtering yield was converted to an absolute sputtering cross section of $1-5 \times 10^{-16}$ cm^{-2} for Ne$^+$ in the 20–180-eV range. The sputtering yield of molecular CO on Ni(111) for noble gas ion energies of 10–500 eV and different incidence angles and projectile masses was measured by Diebold et $al.$[75,76] The sputtering yield for CO was in the range of 0.5–10 at a full CO coverage, corresponding to a sputtering cross section of $1-10 \times 10^{-15}$ cm^{-2}. This value is roughly 10 times higher than that for sputtering atomic nitrogen or oxygen. CO was sputtered more efficiently by ion impact at inclined angles, although the detailed angular dependency varied with the projectile.

Primarily neutral species are ejected in hyperthermal sputtering, since at this energy the yield for secondary particle ionization is very small or negligible for metals, semiconductors, and adsorbed molecules. In certain cases, however, such as when preformed ions exist on a surface, the ions can be ejected even by low-energy impact. This type of a hyperthermal ion ejection mechanism is different from that of the secondary ion emission occurring by keV ion bombardment that involves impact-induced ionization of neutral species. Kim et $al.$[77] examined the sputtering of potassium (K$^+$) ions from a K-covered Ni surface by noble ion bombardment and mass spectrometric detection. Potassium adsorbs ionically on a Ni surface because its ionization energy is lower than the Ni work function and, thus it is sputtered as K$^+$. The threshold for K$^+$ sputtering was observed at a much lower energy than for Ni$^+$ sputtering from the Ni substrate. This is due to the ionic adsorption of K$^+$ and/or the lower surface binding energy of K$^+$ than that of Ni$^+$. By scaling the sputtering threshold energies with the surface binding energies, it was found that the ion-sputtering process could be explained within the framework of a momentum-transfer collision model, similar to the case of neutral atom sputtering.

Kang and co-workers[31,33] investigated the low-energy sputtering of molecular ions such as hydronium (H$_3$O$^+$) and ammonium (NH$_4^+$) prepared on ice surfaces at low temperature. These ions are stabilized on an ice surface through solvation by surrounding water molecules. Figure 9.4 shows the sputtered ion intensities as a function of Cs$^+$ impact energy. NH$_4^+$ and H$_3$O$^+$ start to be emitted at energies of 17 and 19 eV, respectively, from an ice surface containing both preformed NH$_4^+$ and H$_3$O$^+$ ions. On pure H$_2$O and NH$_3$ surfaces, on which no H$_3$O$^+$ or NH$_4^+$ exists, H$_3$O$^+$ and NH$_4^+$ thresholds appear only above 60 eV. Evidently, a large energy gap exists for ejecting the same ions from two different surfaces, the one containing preformed ions and the other having only neutral species. In the former, the NH$_4^+$ and H$_3$O$^+$ ions are ejected from the preexisting species, whereas in the latter, ions are ejected by collision-induced secondary ionization processes. Interestingly, these preformed molecular ions are ejected without fragmentation, suggesting a possibility for identifying molecular ions on surfaces. This application of the low-energy sputtering method will be described further in section 4.3.

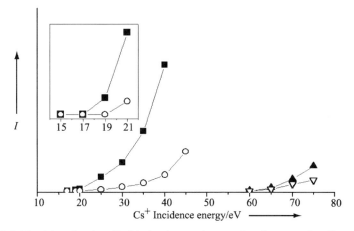

FIGURE 9.4. The intensity of emitted hydronium and ammonium ions as a function of collision energy from two different types of surfaces. For a surface prepared by co-exposure of 0.3 L HCl and 0.15 L NH$_3$ on a D$_2$O-ice layer at 100 K, ammonium (■) and hydronium (○) ions are emitted at energies above 17 and 19 eV, respectively. The inset magnifies the threshold region. On frozen films of pure D$_2$O or NH$_3$, the thresholds for hydronium (▲) and ammonium (▽) ions are much higher (60 eV). The beam incidence and detector angles were both 67.5° with respect to the surface normal (From Park *et al.*, 2001, with permission).

9.3. REACTIVE ION SCATTERING OF HYPERTHERMAL Cs⁺ BEAMS

9.3.1. Basic Description of Cs⁺ RIS

Among various RIS processes, Cs⁺ RIS makes a significant case due to its ability to analyze surface molecules. A hyperthermal Cs⁺ ion can pick up a molecule from a surface to scatter in the form of CsX⁺. This phenomenon was first observed on a Si surface with adsorbed water.[21] Since then, Cs⁺ RIS has been examined on a variety of surfaces and adsorbed molecules,[21–37] demonstrating that the Cs⁺ RIS phenomenon is quite universal and therefore can be applied to surface molecular analysis.

The Cs⁺-RIS process is illustrated schematically[22] in Figure 9.5 by showing the trajectory of a hyperthermal Cs⁺ ion colliding with a surface adsorbed with a molecule. A simple analogy of the process, which might be too simplified in rigorous scientific standards, is the action of a fisherman snatching a fish from water with a harpoon. A Cs⁺ ion ("harpoon") hits a surface and causes desorption of a molecule ("fish"). The desorbed molecule X is picked up by Cs⁺ on its scattering trajectory, forming a CsX⁺ ion complex ("speared fish"). CsX⁺ is detected by a mass spectrometer, thus leading to mass identification of the desorbed molecule ("weighing a fish"). More detailed consideration of the RIS process will be given in section 3.2, explaining its mechanism in relation to the scattering and sputtering characteristics of hyperthermal heavy projectiles given in sections 2.2 and 2.3.

A typical result of Cs⁺ RIS analysis of surface molecules is shown in the mass spectra of Fig. 9.6. In this study,[24,25] OH, CO, and benzene molecules were co-adsorbed

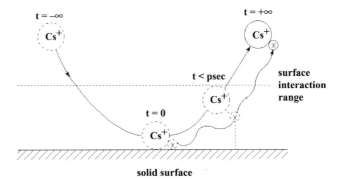

FIGURE 9.5. Pictorial illustration of the Cs^+ reactive scattering trajectory leading to CsX^+ formation. Collision of hyperthermal Cs^+ causes desorption of neutral X from the surface. The desorbed X forms a transient CsX^+ complex via ion–molecule attraction forces. Stabilization of the CsX^+ complex is achieved through interaction between the complex and the surface (simply depicted by the dotted line) while the complex remains near the surface (t < 1 psec) (From Yang *et al.*, 1996, with permission).

on Ni(111), and the surface was impacted with a Cs^+ beam at 30 eV. The spectra are characterized by two kinds of peaks: (1) Cs^+ peak elastically (nonreactively) scattered from a surface appearing at $m/z = 133$ *amu/charge* and (2) peaks above $m/z = 133$ due to Cs^+ RIS (i.e., the pickup of surface molecules by Cs^+). The Cs^+ RIS peaks are assigned as $m/z = 150$ $(CsOH^+)$, $m/z = 161$ $(CsCO^+)$, and $m/z = 211$ $(CsC_6H_6{}^+)$. These peaks indicate that all molecular species on the surface are detected in molecular states without fragmentation. There can be an ambiguity in the mass spectral interpretation; for example, the peak at $m/z = 161$ might alternatively be assigned to $CsC_2H_4^+$ that can be produced from C_6H_6 dissociation to C_2H_4 on Ni. This possibility was checked by a control experiment using C_6D_6 adsorption. The experiment showed that the peak at $m/z = 161$ is not shifted to $m/z = 165$ $(CsC_2D_4^+)$ by C_6D_6 adsorption, confirming that it is indeed $CsCO^+$.

Another noticeable feature in Figure 9.6 is the absence of a $CsNi^+$ signal. This indicates negligible sputtering of Ni surface atoms at this collision energy. Sputtered Ni^+ ions were also not observed in the lower mass region of the spectra. The peak intensity ratio of CsX^+/Cs^+ is also noteworthy. It represents the efficiency of molecular pickup by Cs^+, or the Cs^+ RIS yield (Y_{ris}). The ratio is of the order of 10^{-4} for these molecules, which corresponds to a typical Y_{ris} value for chemisorbed molecules. This value may seem small, but it is ~100 times higher than the ionization efficiency of an electron impact ionizer in a mass spectrometer $(\sim 10^{-6})$. That is, Cs^+ RIS efficiently converts neutral molecules (X) into ions (CsX^+) such that they can be identified by a mass spectrometer. With such an efficiency, one can detect surface molecules present at a coverage of a few percent of a monolayer by acquiring a Cs^+ RIS spectrum for a duration of ~1 min using a Cs^+ beam current of ~1 nA/cm^2.[25]

The experimental requirements for a Cs^+ RIS study are a low-energy Cs^+ ion gun, a mass spectrometer with a pulse-counting ion detector, and a ultrahigh vacuum (UHV) scattering chamber. Figure 9.7 shows a schematic drawing for an instrument equipped

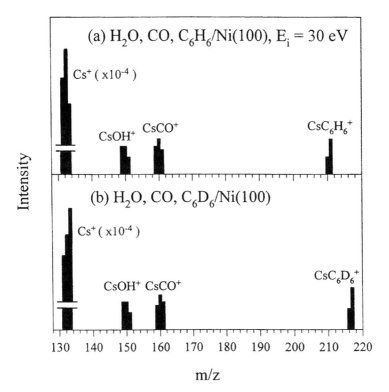

FIGURE 9.6. Mass spectra of the ions emitted upon 30-eV Cs$^+$ impact of a Ni(111) surface chemisorbed with OH, CO, and benzene. Mass resolution was lowered for increased sensitivity. (a) C$_6$H$_6$ was initially adsorbed to a saturation coverage at 300 K, then reacted with the residual gases CO and H$_2$O. (b) The result of a control experiment with C$_6$D$_6$ (From Kang et al., 1998, with permission).

with a quadrupole mass spectrometer (QMS) and a rotatable Cs$^+$ ion gun. Only a brief description of Cs$^+$ RIS instrumentation is given here; more details are given elsewhere.[23–35] The low-energy Cs$^+$ gun provides ion energies of 5–100 eV, which is the energy range most frequently used in the experiment, with an ion current density at a target of ~1 nA/cm^2. An intense Cs$^+$ beam increases the sensitivity for surface analysis, but it also causes contamination of the surface due to Cs deposition. The QMS detector is operated with the ionizer off because the RIS products are all ions. This has a clear advantage in that RIS signals are measured against zero background noise, if the stray ions inside the chamber are removed during the experiment (such as ions from an ionization gauge). The scattering angle is variable[35] via independent rotary motions of the Cs$^+$ gun and the sample surface. Such an angle-resolved instrument has advantages and versatility due to more degrees of freedom in the experimental configuration. For instance, it is useful to study the scattering dynamics as well as to examine the RIS phenomena from an unexplored type of surface. However, if one intends to perform only surface analysis, a fixed-angle instrument can be satisfactory. In this case, a specular scattering geometry is preferred.[25]

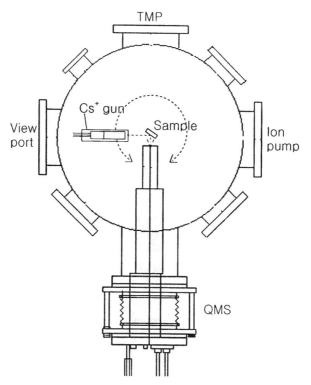

FIGURE 9.7. Typical components for an angle-resolved reactive ion scattering chamber (top view). The Cs$^+$ source rotates on a scattering plane defined by the positions of the source, sample center, and quadrupole mass spectrometer (QMS) (From Han *et al.*, 2001, with permission).

A Cs$^+$ RIS experiment can be performed in several different modes depending on the information desired. A mass spectrum of Cs$^+$ RIS products can be obtained by scanning a QMS, which identifies molecules on a surface. In order to study the kinetics of surface reactions, a QMS is tuned to several preselected masses corresponding to the reactants and products, and their intensity variation is measured as a function of reaction time or surface temperature. This type of experiment monitors real-time kinetics as well as temperature-dependent reaction behavior. The time resolution in such measurements is ~1 sec with an instrument employing a continuous Cs$^+$ beam.[25] Another variable in the experiment is the impact energy of Cs$^+$. Examination of RIS products as a function of Cs$^+$ energy provides information about the structure of surface molecules being examined.[36]

9.3.2. Mechanism, Dynamics, and Quantitation of the Cs$^+$-RIS Process

A rigorous treatment of the Cs$^+$ RIS process has been presented by Kang *et al.*[23,25] They suggested that RIS can conceptually be divided into two steps: the sputtering of adsorbed species by hyperthermal ions (reaction 1) and the association reaction between the sputtered molecule and the scattered Cs^+ (reaction 2).

$$Cs^+(g) + X\text{-}surface \to Cs^+(g) + X(g) + surface \qquad \text{(reaction 1)}$$

$$Cs^+(g) + X(g) \to [CsX^+(g)]^* \to CsX^+(g) \qquad \text{(reaction 2)}$$

Here, $[CsX^+(g)]^*$ denotes transient formation of a complex species by Cs^+ and X association. For reaction (1), the terminology *collisional desorption* or *low-energy sputtering* is used in place of sputtering. The word *sputtering* has been used for surface collisions at keV and higher energies, and a preconception exists that the phenomenon involves extensive damage to surfaces and molecular structures. The molecules are desorbed with much less fragmentation and surface damage by using hyperthermal beams. Desorption of adspecies by hyperthermal collisions (reaction 1) was described in section 2.1. The process occurs, for example, with a cross section (σ_d) of magnitude of $1\text{--}10 \times 10^{-15}$ cm^{-2} for CO on Ni(111) at ion impact energies of 10–500 eV.[75,76] This is a typical value of σ_d for weakly chemisorbed species and corresponds to a collisional desorption yield of 0.5–10 at full coverage of a CO monolayer on Ni(111). The magnitude of σ_d, of course, varies with the strength of molecule–surface bonding and the ion impact energy. σ_d for strongly chemisorbed species such as N and O atoms on W and Ni is $1\text{--}5 \times 10^{-16}$ cm^{-2} for Ne$^+$ at 20–180 eV,[73,74] which is lower by one order of magnitude than CO on Ni. Surface metal atoms can also be desorbed by hyperthermal collisions, but their threshold energies are somewhat higher and their desorption yields are lower than chemisorbed species.[65–68]

The desorbed X is attracted to the scattered Cs^+, forming a transient CsX^+ complex $[(CsX^+)^*]$ by means of a binary association reaction (reaction 2). Since Cs^+ is chemically inert, the attractive force between Cs^+ and X is electrostatic, such as ion-dipole and ion-induced dipole interactions.[78] The association reaction is believed to take place in the gas phase very near the surface. The reaction has the highest probability when Cs^+ and X are near each other in the phase space, i.e., when they are spatially near each other and move with similar velocities and directions. The energy requirement for association is that their relative energy of motion is smaller than the ion-molecule binding energy. If a Cs^+–X pair is born under the condition that the initial positions of Cs^+ and X are within the Cs^+–X potential well, then the CsX^+ species will be bound together and will remain stable along its outgoing trajectory. In other cases, the nascent CsX^+ complex needs to be stabilized by transferring its excess energy to the surface. The energy transfer may take place effectively when the complex remains within a distance of 3 Å from the surface according to MD calculations.[69] This means that the complex remains in the near-surface region for about 2×10^{-13} sec when it has an energy of 10 eV and a trajectory of 45° from the surface.[40] This defines the time scale of the Cs^+ RIS process (i.e., the time required for Cs^+–X association and energy stabilization).

The RIS yield (Y_{ris}) is typically 10^{-4} for chemisorbed species, as shown in Figure 9.6. This value may be worth consideration in relation to the mechanism of the

RIS process depicted by reactions (1) and (2). According to this mechanism, Y_{ris} is the product of the yields for collisional desorption (Y_d) and Cs^+–X association (Y_{assoc}). Assume that Y_d is of the order of 0.1, a typical value for chemisorbed molecules existing in partial monolayer coverages. Y_{assoc} will then have a value of about 10^{-3}. Now consider the association reaction taking place in the gas phase between an ion and a nonpolar molecule. The Langevin cross section for the reaction, $\sigma = \pi e(2\alpha/E_{rel})^{1/2}$, is estimated to have an order of 10^{-15} cm^2 when the relative energy of motion (E_{rel}) is 1 eV and the molecular polarizability (α) is 1×10^{-24}cm^3. From the experimental kinetic energy distribution that will be shown later and the maximum energy requirement for the relative Cs^+–X motion, one can estimate that about 10% of the scattered Cs^+ ions are able to form a stable complex by Cs^+–X interaction. In order to have an ion–molecule association probability of 10^{-3} between Cs^+ and gas molecules from this Langevin cross-section value, a molecular gas density of 10^{22} molecule/cm^3 is required. Its reciprocal value, 10^{-22} cm^3/molecule (100 Å^3/molecule), is a reasonable volume of space in which a desorbing molecule can spread out during the time scale of reactive scattering, or the volume of the interaction region depicted in Figure 9.5. This order-of-magnitude calculation shows that the observed RIS yields can reasonably be accounted for in terms of the proposed two-step mechanism.

In contrast to the observations on chemisorbed surfaces, enormously high RIS yields were observed on physisorbed molecular layers. On the surface of a frozen water layer, for example, Y_{ris} reaches 0.3, as shown in the RIS mass spectra[27] of Figure 9.8. These RIS experiments produced multihydrated cluster ions, $Cs(H_2O)_n^+$ with $n > 1$, with their intensities gradually decreasing with the increasing number of water molecules. Such large cluster ions did not appear from chemisorbed surfaces. These striking differences suggest that the nature of the Cs^+ RIS process changes greatly on a frozen water layer.

In order to explain the extremely large Y_{ris} on ice surfaces, it was suggested[27] that the clustering reactions occur through multiple ion–molecule association steps;

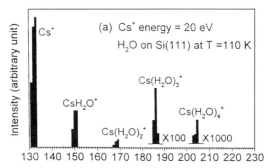

FIGURE 9.8. Mass spectra of Cs^+ reactive ion scattering (RIS) products observed on a frozen water layer upon 20-eV Cs^+ collision. The intensities of larger clusters are magnified by the factors indicated (From Shin *et al.*, 1999, with permission).

in other words, $Cs(H_2O)_n^+$ clusters grow through successive Cs^+-H_2O association reactions with water molecules (reactions 3–1 to 3–n).

$$Cs^+(g) + H_2O\,(g) \rightarrow Cs(H_2O)^+(g) \qquad\qquad \text{(reaction 3–1)}$$

$$Cs(H_2O)_{n-1}^+\,(g) + H_2O\,(g) \rightarrow Cs(H_2O)_n^+(g) \qquad\qquad \text{(reaction 3–n)}$$

Analysis of the experimental cluster distributions suggests that $Y_{assoc}[Cs^+ (H_2O)_{n-1} - H_2O] = 0.16-0.43$, which is far greater than the Y_{assoc} value for chemisorbed systems. The Y_{assoc} value again suggests that $Cs(H_2O)_n^+$ clusters are created in a condensed phase, since in order to have $Y_{assoc} = 0.16-0.43$, the water molecular density in the reaction media needs to reach $\sim 10^{24}$ molecule/cm^3 (1 molecule/Å^3). Such a high molecular density indicates a condensed phase. Alternative interpretations may also be possible for the observation. For instance, multiply hydrated Cs^+ ions are initially ejected from the ice surface impacted with Cs^+. Since the ejected clusters may be internally excited upon their formation, denoted by $[Cs(H_2O)_n{}^+(g)]^*$ in reaction (4), they become fragmented into smaller clusters in the gas phase and thus are cooled down.

$$Cs^+(g) + H_2O\,(s) \rightarrow [Cs(H_2O)_n^+(g)]^*$$

$$\rightarrow Cs(H_2O)_m^+(g) + (n\text{-}m)\,H_2O\,(g),\ m < n \qquad\qquad \text{(reaction 4)}$$

The detailed mechanistic aspects of the clustering reactions of Cs^+ ions on frozen molecular surfaces are not very clear at the moment, and other explanations are possible.

Section 2.1 described the nature of heavy ion–surface collisions. The prominent feature is that heavy ions scatter with relatively low kinetic energies. The E_1 distribution of heavy ions has a substantial population at energies as low as those of sputtered particles. Kinetic energy distributions have been measured in a Cs^+ RIS experiment on a Si surface with chemisorbed water (Figure 9.9).[23] The Cs^+ ions scattered upon 100-eV collision with the surface exhibit an energy distribution peaking at ~ 10 eV. Evidently, the energy transfer from Cs^+ to the Si surface is efficient. $CsSi^+$, $CsSiO^+$, and Cs_2^+ ions are ejected with relatively lower kinetic energies, peaking at 3–5 eV. The peak energies for these cluster species are substantially lower than that for Cs^+, but there exists an overlapping region between the energy distributions of the scattered Cs^+ ions and the CsX^+ complexes. The low-energy Cs^+ ions belonging to this overlapped region, which is estimated to be roughly 10% in the energy spectrum of Figure 9.9, are able to meet the energy requirement for CsX^+ formation via reaction (2). It is to be noted that the low kinetic energy distribution is unique to Cs^+ scattering. Other alkali metal ions have been tested as projectiles in RIS experiments[23]; however, Cs^+ has been found to be most effective in forming the RIS products, apparently due to its low kinetic energy distribution.

An important aspect in surface analysis is whether a technique can quantitate the concentrations of surface species. In the following, a relationship between the RIS

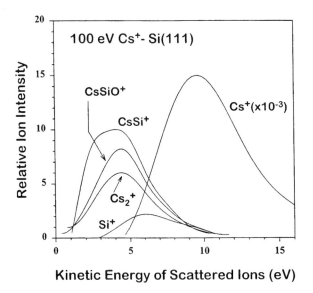

FIGURE 9.9. Kinetic energy distributions for scattered Cs^+, sputtered Si^+, and the Cs^+ RIS products $CsSi^+$, $CsSiO^+$, and Cs_2^+ produced from a Si surface partially covered with H_2O. The incident Cs^+ energy is 100 eV in a 45°/45° specular scattering geometry. The area under the curves represents relative intensity of the ions (From Yang et al., 1997, with permission).

yield and the surface coverage of molecules will be described.[25] The rate for a Cs^+–X association reaction, $\upsilon(CsX^+)$, is proportional to the gas-phase concentrations of scattered Cs^+ and desorbed X as

$$\upsilon(CsX^+) = k_{assoc}[Cs^+][X]. \tag{2}$$

Here, k_{assoc} represents the rate constant for the bimolecular association step (molecules^{-1}s^{-1}) and $\upsilon(CsX^+)$ is in units of ions/s. The velocity and angular distributions of Cs^+ and X are neglected in this expression. $\upsilon(CsX^+)$ is proportional to the RIS signal intensity, $I(CsX^+)$, or the flux of CsX^+ ions reaching a detector. In an experiment with a continuous Cs^+ beam, the gas-phase concentrations of $[Cs^+]$ and $[X]$ stay constant with time in the reaction region, provided that the molecule loss due to sputtering is negligible. Therefore, these concentrations can be replaced with the corresponding fluxes, $I(Cs^+)$ and $I(X)$, leading to Eq. 3,

$$I(CsX^+) = C\, k_{assoc}\, I(Cs^+)\, I(X), \tag{3}$$

where C is a proportionality factor. Since $Y_{rs}(X)$ is defined as the ratio $I(CsX^+)/I(Cs^+)$, it can be written as,

$$Y_{rs}(X) = [I(CsX^+/I(Cs^+)] = C k_{assoc} I(X). \tag{4}$$

I(X) corresponds to the collisional desorption rate of X_{ads} under continuous ion bombardment and is expressed as

$$I(X) = d[X(g)]/dt = \sigma_d F \theta_x, \tag{5}$$

where σ_d is the sputtering cross section for X (cm^2), F is the fluence of the Cs$^+$ beam (ions cm^{-2} s^{-1}), and θ_X is the surface coverage. Thus,

$$Y_{rs}(X) = C' k_{assoc} \sigma_d F \theta_X. \tag{6}$$

Eq. 6 shows that $Y_{rs}(X)$ is proportional to k_{assoc}, σ_d, θ_X, and F. F is fixed under given instrumental conditions and k_{assoc} and σ_d are the parameters specific to the molecules. If k_{assoc} and σ_d are determined through a calibration experiment, quantitative measurement of θ_X is possible for adsorbed species. When several different species coexist on a surface, their relative concentrations are not proportional to their RIS intensities, since the species have different values of k_{assoc} and σ_d. In this case, separate calibration experiments for each species will be able to achieve quantitative analysis of the co-adsorbed species.

9.4. APPLICATION OF THE Cs$^+$ RIS TECHNIQUE

9.4.1. Identification of Surface Molecules

An example of identification of molecules on surfaces by Cs$^+$ RIS has been given[24,25] in Figure 9.6. The result showed that OH, CO, and C_6H_6 molecules chemisorbed on a Ni(111) surface were detected as CsOH$^+$, CsCO$^+$, and $CsC_6H_6{}^+$, respectively, and the molecules were not fragmented by RIS detection. This example represents an application of RIS to chemisorbed molecules already well characterized by other spectroscopic methods.

An example of a controversial problem to which the RIS technique has been applied is the chemisorption state of O$_2$ molecules on a Si(111)-(7 × 7) surface. This is the so-called precursor state in the initial oxidation stage of the Si surface.[26] The subject has attracted much attention in surface science research.[79–86] A central dispute has been whether O$_2$ is adsorbed in dissociated[79,80] or molecular forms.[81–84] The reason for this controversy is apparently due to the lack of conclusive evidence that can resolve the difference in the two proposed states. Figure 9.10 presents the result of a Cs$^+$ RIS study of O$_2$ adsorbed on a Si(111)-(7 × 7) surface. The mass spectrum of Fig. 10(a), obtained for a Cs$^+$ impact energy of 5 eV, shows CsSiO$^+$ as the only RIS signal. An increase in the Cs$^+$ energy to 50 eV [Fig. 9.10(b)] produced an additional peak of CsSi$^+$, along with a substantial increase in the CsSiO$^+$ yield. The CsSiO$^+$ peak provides clear evidence for the presence of SiO on the surface, indicating O$_2$ dissociation. CsSiO$^+$ is produced even at low-impact energy (5 eV), suggesting that SiO is easily desorbed intact. The Si–O bond is strong and difficult to dissociate. This observation supports the diatomic molecular character of SiO and its weak bonding to the

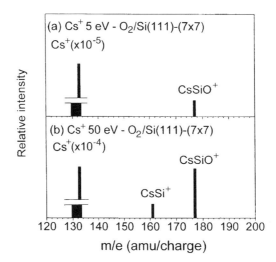

FIGURE 9.10. Mass spectra of the Cs^+ reactive ion scattering (RIS) products from a Si(111)-(7 × 7) surface exposed to 0.3 L O_2. (a) Cs^+ energy of 5 eV at sample temperature of 273 K. (b) 50-eV energy at 153 K. The intensity of the Cs^+ peak is reduced by a factor of 1×10^{-5} (a) and 1×10^{-4} (b). No ions were detected in the mass region below m/e = 120 amu/charge (not shown) (From Kim *et al.*, 1999, with permission).

Si surface. Another important feature of Figure 9.10 is the absence of the signals expected from molecular oxygen species on the surface, such as CsO_2^+, $CsSiO_2^+$, and CsO^+. Molecular oxygen species would have generated at least one of these products via collisional desorption of O_2 or its related fragments. Thus, the Cs^+ RIS experiment unambiguously shows that SiO is the precursor state. This conclusion has been further supported by theoretical calculation[85] and experiment.[86]

9.4.2. Surface Reactions and Intermediates

The sensitivity of the Cs^+ RIS technique makes it possible to monitor the concentration change of surface species that occurs on a time scale longer than ~1 sec. Thus, RIS can be a useful tool for kinetic studies of many surface reactions. The concentration change can be monitored simultaneously for several species on a surface by using the multiple-ion-detection mode, a feature available in many mass spectrometers. Adsorption and desorption kinetics of multiple adsorbates have been examined in real-time by RIS for CO, OH, and benzene adsorbed on Ni.[24,25] Cs^+ RIS can also be used for investigation of reaction intermediates. If an intermediate has a sufficiently long lifetime (>1 sec), which can often be made so by lowering the substrate temperature, it can be detected by Cs^+ RIS. The intermediates in many chemical transformations have a unique molecular structure and, for this reason, the molecular identification ability of Cs^+ RIS can be particularly useful. Surface reaction studies with Cs^+ RIS include Cs diffusion and clustering reactions on Si,[28] decomposition of organosilanes on Si,[29,30] dehydrogenation of ethylene on Pt,[36] and H/D exchange processes between ethylene and hydrogen on Pt.[37]

An example to be described in more detail here is the RIS study of the ethylene dehydrogenation reaction on Pt(111).[36] This is one of the model reactions of heterogeneous catalysis, and its mechanism is now well established.[87] Ethylene physisorbs as a molecule on Pt(111) at temperatures below 240 K and dehydrogenates to ethylidyne (Pt≡C–CH₃) in the temperature range 240–310 K. Above 450 K, ethylidyne decomposes to C_2H, CH, and then eventually to surface carbon. Ethylene and ethylidyne species on Pt(111) have been clearly identified by various experimental techniques, but the path of ethylene dehydrogenation to ethylidyne, taking place at 240–310 K, has not been well established. Several intermediates have been proposed for this path, including surface ethyl (Pt–CH₂–CH₃), vinyl (Pt–CH=CH₂), and ethylidene species (Pt=CH–CH₃).

Figure 9.11 shows evolution of the RIS spectra taken as the reaction proceeds with increasing temperature.[36] Initial adsorption of C_2H_4 at 220 K produces a $CsC_2H_4^+$ signal (spectrum a), confirming that ethylene exists in a molecular state at low temperature. An extra peak due to physisorption of background water (CsH_2O^+) is also observed. Upon increasing the temperature to 280 K (spectrum b), a $CsCH_3^+$ peak appears with a concomitant decrease of the $CsC_2H_4^+$ peak. At 320 K, only the $CsCH_3^+$ peak appears (spectrum c). This peak is assigned to ethylidyne produced

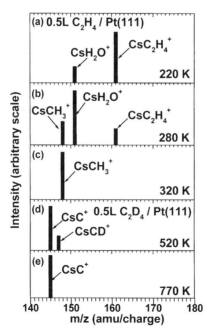

FIGURE 9.11. Reactive ion scattering (RIS) mass spectra taken at intermediate temperatures during ethylene decomposition on Pt(111). (a) An ethylene-adsorbed Pt surface was prepared by exposing to 0.5 L of C_2H_4 at 220 K. The surface temperature was then raised to 280 K (b), 320 K (c), 520 K (d), and 770 K (e). D-substituted ethylene (C_2D_4) was adsorbed on surfaces (d) and (e). Cs⁺ impact energy: 30 eV in (a)–(c), 50 eV in (d), and 100 eV in (e). A 45°/45° specular scattering geometry was used (From Hwang *et al.*, 2001, with permission).

by collisional rupture of the C–CH$_3$ bond and pickup of the methyl group. Ethylidyne is strongly bound to Pt through a metal–carbon triple bond, making it reasonable that the collision induced the C–CH$_3$ bond cleavage instead of intact desorption of the CCH$_3$ unit. A further increase in temperature to 520 K (spectrum d) gives rise to CsC$^+$ and CsCD$^+$ signals (C$_2$D$_4$ was adsorbed instead of C$_2$H$_4$ in this experiment). In spectrum (e), taken at 770 K, only the CsC$^+$ peak remains. Spectra (d) and (e) indicate that ethylidyne undergoes gradual decomposition at high temperature to C$_2$H, CH, and then eventually to C$_n$ species. These RIS spectra are consistent with the dehydrogenation path of ethylene on Pt(111) established in previous studies.

Spectrum (b) corresponds to the region in which the intermediate from ethylene to ethylidyne is formed; it shows CsCH$_3^+$ and CsC$_2$H$_4^+$ peaks. This observation suggests that the intermediate is ethylidene, but denies the possibility of vinyl and ethyl species. Ethylidene can produce CsC$_2$H$_4^+$ by molecular desorption and CsCH$_3^+$ by C–C fragmentation, the latter process being analogous to CsCH$_3^+$ production from ethylidyne. This interpretation was supported by a control experiment that detected CsC$_2$H$_5^+$ upon ethyl group adsorption on Pt(111) and studies of RIS products as functions of Cs$^+$ impact energy, reaction time, and temperature.[36] Figure 9.12 shows the result of a real-time kinetic study for CsCH$_3^+$ and CsC$_2$H$_4^+$ signals, obtained while

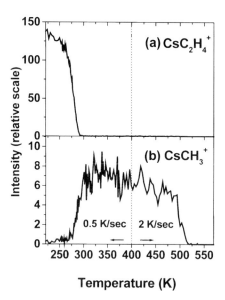

FIGURE 9.12. Variation of the signal intensities for CsC$_2$H$_4^+$ (a) and CsCH$_3^+$ (b) as a function of temperature from 220–570 K. The vertical scales indicate the relative intensities of the two signals. The surface was initially adsorbed with 0.5 L C$_2$H$_4$ at 220K. The curves were obtained by real-time monitoring of the signals during the temperature increase. The temperature scan rate was 0.5 K/sec below 400 K and 2 K/sec above 400 K. The change of a scan rate does not affect the RIS signal intensity. The Cs$^+$ beam energy was 30 eV with 45°/45° specular geometry (From Hwang *et al.*, 2001, with permission).

the reaction temperature was varied from 220–570 K. In Figure 9.12(a), the high $CsC_2H_4^+$ intensity below 260 K is due to molecular ethylene adsorption. Ethylene is transformed to ethylidene in the region 270–290 K, where the $CsC_2H_4^+$ intensity decreases rapidly. In Figure 9.12(b), a low-intensity $CsCH_3^+$ signal at 270–290 K orig-inates from ethylidene, and the intense signal above 290 K is due to stable ethylidyne species at these temperatures.

Figures 9.11 and 9.12 demonstrate the ability of Cs$^+$ RIS to identify various molec-ular species that appeared during the reaction of ethylene. The mechanistic reaction path was examined by kinetic study and detection of the ethylidene intermediate. The system also illustrates some guidelines regarding whether RIS can detect surface molecules by intact desorption or by fragmentation. When a molecule is adsorbed weakly on a surface (e.g., through a metal–carbon single bond or a weaker bond—C_2H_4, C_2H_5, and H_2O), it can be identified by intact desorption. On the other hand, a molecule adsorbed with a strong triple bond ($\equiv C-CH_3$) cannot be desorbed as a molecule. Instead, incoming Cs$^+$ breaks the weaker C–C bonds and picks up the CH_3 group. The fragmentation patterns of the molecules, however, are usually simple enough to find reasonable correlations between the functionality of the molecules and the RIS spectra.[29,30,36]

Mass spectrometric detection of Cs$^+$ RIS products makes the technique particularly useful for identifying isotopically substituted molecules. Figure 9.13 illustrates such

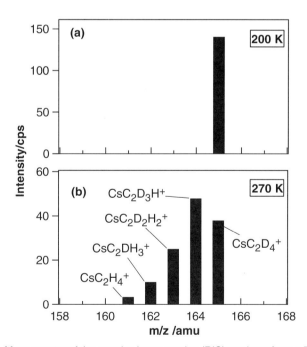

FIGURE 9.13. Mass spectra of the reactive ion scattering (RIS) products from a Pt(111) surface dosed with 5 L H_2 and 0.5 L C_2D_4 at 200 K (a) and 270 K (b) (From Kim *et al.*, 2002, with permission).

an example, by showing the result obtained for isotopic exchange reactions between C_2D_4 and H on Pt(111).[37] In this experiment, a Pt(111) surface was co-adsorbed with H and C_2D_4 at two different temperatures: (a) 200 K, at which isotope exchange is prohibited, and (b) 270 K, at which isotope exchange is facile. Spectrum (a) verifies molecular adsorption of ethylene without H/D exchange. Spectrum (b) shows the presence of various H/D exchange products (C_2D_3H, $C_2D_2H_2$, C_2DH_3, and C_2H_4) at 270 K. Note that the intensity distribution in spectrum (b) truly represents the relative amounts of the corresponding H/D substituted ethylenes on the surface. In the absence of the ionizer function of the mass spectrometer, the possibility of molecular cracking and isotopic scrambling in a mass spectrometer is eliminated. Also, these substituted ethylenes should exhibit almost the same RIS sensitivity due to their identical geometric structures. Considering that quantitative determination of multiple H/D exchange products is a difficult problem in surface analysis, Cs^+ RIS has unique ability for this application.

9.4.3. Application to Molecular Films: Reaction on Ice

The sensitivity of the Cs^+ RIS method is highest on soft, condensed molecular films. For example, Y_{ris} is of the order of one for picking up water molecules from an ice surface (section 3.2). This feature makes molecular films particularly attractive systems to study by Cs^+ RIS. The technique probes only neutral molecules on surfaces. Some molecular films, however, contain ionic species as well, since the films are electrically insulating and can stabilize the ions through solvation by surface molecules. Good examples of stable ionic species on a molecular surface are strong electrolytes on water-ice surfaces. Such preexisting ions can be detected by the method of low-energy sputtering described in section 2.2. The low-energy sputtering method employs ion impact at energies slightly above the sputtering threshold, and thus the ions can be ejected with minimal fragmentation to reveal their identity. The low-energy sputtering method is complementary to Cs^+ RIS, which selectively detects neutrals on surfaces. Combined usage of the two methods allows one to detect both neutrals and ions that exist on a surface. Note that Cs^+ RIS and low-energy sputtering experiments can be carried out simultaneously with a single Cs^+ beam, and their signals can be acquired in one mass spectrum.

Park et al.[31–34] employed Cs^+ RIS and low-energy sputtering techniques to examine the chemical states of simple electrolytes such as HCl, NH_3, and NaCl deposited on ice surfaces at low temperature. Figure 9.14(a) shows a mass spectrum obtained from a D_2O-ice surface dosed with HCl gas at 100 K.[33] The spectrum is characterized by two types of peaks apart from the reflected Cs^+ primaries that are shown on a reduced scale. (1) RIS products above $m/z = 133$ *amu/charge*, (i.e., the species due to pickup of surface molecules by Cs^+). CsD_2O^+ and $Cs(D_2O)_2^+$ are due to pickup of a water molecule and a dimer, respectively. $CsHCl^+$ is due to pickup of an HCl molecule. (2) Peaks due to low-energy sputtering, corresponding to the ejection of preexisting ions on the surface. The ions detected are HD_2O^+ and its hydration clusters ($HD_4O_2^+$). The presence of HD_2O^+ reveals ionic dissociation of HCl. The RIS products and the sputtered ions exhibit characteristic H/D isotope exchange. The results indicate that

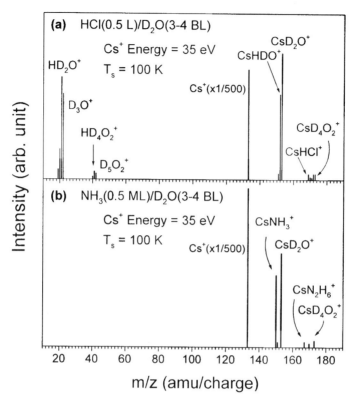

FIGURE 9.14. Reactive ion scattering (RIS) mass spectra obtained on D_2O-ice surfaces exposed to 0.5 L HCl (a) and NH_3 (b) at 100 K. The D_2O layer was 3-4 bilayers thick on a Ru(001) surface. The observed species are labeled. The Cs^+ peak intensity was reduced by 1/500 and the Cs^+ energy was 35 eV (From Park et al., 2001, with permission).

at 100 K, HCl exists in both the dissociated ionic state (HD_2O^+) and the molecular state ($CsHCl^+$). The ionization of HCl was examined by these methods as a function of the ice temperature.[31] It was found that the degree of ionization varies from the almost exclusive presence of molecular HCl at 50 K to complete ionization above 140 K.

Figure 9.14(b) presents a RIS mass spectrum from an NH_3-deposited ice surface. The only signals related to ammonia are $CsNH_3^+$ and $CsN_2H_6^+$. The NH_4^+ ion is not ejected by low-energy sputtering. The isotope exchange between NH_3 and D_2O is insignificant as well. This indicates that NH_3 exists as a neutral molecule at the surface but not as NH_4^+. Figures 9.14(a) and (b) demonstrate how a strong acid (HCl) and a weak base (NH_3) behave on ice at low temperature.

The proton transfer reaction between hydronium ion and ammonia, depicted by reaction (5), was investigated on ice surfaces by RIS and low-energy sputtering:[33,34]

$$H_3O^+ + NH_3 \leftrightarrow H_2O + NH_4^+. \qquad \text{(reaction 5)}$$

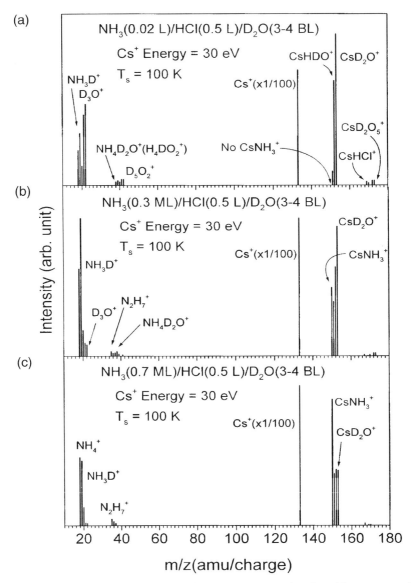

FIGURE 9.15. Reactive ion scattering (RIS) mass spectra as a function of the progress of the H_3O^+-NH_3 reaction on ice. The D_2O (3–4 bilayers) surface was exposed first to 0.5 L of HCl to generate hydronium ions and then to NH_3 at varying exposures: (a) 0.02 L, (b) 0.3 L, and (c) 0.7 L. The sample temperature was 100 K, and the Cs^+ energy was 30 eV (From Park *et al.*, 2001, with permission).

This simple prototypical reaction proceeds instantaneously in aqueous phase at 298 K with an equilibrium constant $K_{eq} = 1.7 \times 10^9$. It is even more facile in gas phase with $K_{eq} = 1 \times 10^{30}$. The hydronium ion-containing ice surfaces were prepared for the experiment by HCl exposure to ice, followed by addition of increasing amounts of NH$_3$ to examine the proton transfer between H$_3$O$^+$ and NH$_3$. The ratio of the proton donor to acceptor was changed by varying the NH$_3$ concentration. Figure 9.15 shows the results of the H$_3$O$^+$–NH$_3$ titration measurements performed at 100 K. Upon addition of a small amount of NH$_3$ (Figure 9.15a), the spectrum immediately showed a peak due to NH$_3$D$^+$, indicating proton transfer from D$_3$O$^+$ to NH$_3$ to form the ammonium ion. Absence of a CsNH$_3^+$ peak at this stage indicates that all of the added NH$_3$ molecules have reacted. Hydronium ions still have a large excess concentration, as can be seen from a strong D$_3$O$^+$ peak. The hydrated cluster ions are also observed along with their isotope exchange species. An increase in the NH$_3$ exposure (Figure 9.15b) resulted in an increase in the NH$_3$D$^+$ peak and a concomitant decrease in the D$_3$O$^+$ intensity. The appearance of the CsNH$_3^+$ peak indicates that a substantial portion of NH$_3$ remains unreacted on ice, despite the coexistence of D$_3$O$^+$. When NH$_3$ was added in much excess (Figure 9.15c), ammonium ions were the dominant ionic species at the surface, and the concentration of the remaining hydronium ion remained small.

A profound feature of the experiment of Figure 9.15 is that one can calculate the quotient (Q) of reaction (5) from the spectral intensities. Consider Eq. 8:

$$Q = [H_2O][NH_4^+]/[H_3O^+][NH_3]$$
$$= I(CsH_2O^+)I(NH_4^+)/I(H_3O^+)I(CsNH_3^+). \tag{8}$$

$I(X^+)$ represents the signal intensity of X^+. The relationship of Eq. 8 is semiquantitative, despite the different detection sensitivities for low-energy ions (low-energy sputtering) and neutrals (Cs$^+$ RIS). This is because the different sensitivity factors appear both in the numerator and denominator of Eq. 8 and cancel out automatically. The Q value for ice obtained in this manner is ca. 20. This value is much smaller than the equilibrium constant of the reaction in water (1.7×10^9) or in gas phase (1×10^{30}). Apparently, the reaction does not go to completion on ice, and all of the components of the reaction, H$_3$O$^+$, NH$_3$, H$_2$O, and NH$_4^+$, coexist on the surface. The result illustrates that the reaction on ice does not reach thermodynamic equilibrium but is instead kinetically trapped in metastable states. These findings may have relevance to the chemistry taking place on ice particles in the upper atmosphere or interstellar space. Frozen molecular surfaces may also provide a unique environment for studying low-temperature, two-dimensional reaction behavior.

9.5. SUMMARY

Although Cs$^+$ RIS is a relatively new technique in surface science research, it has already shown several promising applications. Its greatest advantage is the ability to probe surface molecules, which is often difficult to achieve with other surface spectroscopic methods. The unique features of the technique deduced from this chapter are as follows:

1. Cs^+ RIS detects neutral species desorbed from a surface. This is a great advantage over techniques that require postionization of desorbed molecules or secondary ion mass spectrometry.
2. Molecular fragmentation and substrate sputtering are greatly suppressed by employing hyperthermal primary beams.
3. Surface molecules are picked up by Cs^+ on an ultrafast time scale (<1 psec). This feature apparently improves the molecular identification ability of the technique, since the dissociated species formed by slow secondary surface reactions do not have time to form CsX^+ products. The ultrafast nature of the RIS process is yet to be further examined for other usages, such as the possibility of detecting the transient species of a reaction.
4. The technique is sensitive to the first monolayer of a surface, as hyperthermal ion impact ejects atoms and molecules exclusively from the top layer.
5. Isotopically substituted molecules can be quantitatively identified for their mass without complications due to isotopic scrambling.
6. Kinetics of surface reactions can be monitored in real time with a resolution of ~ 1 sec.
7. Both neutrals and ions existing on a surface can be probed by a combination of Cs^+ RIS and low-energy sputtering.

On the other hand, the technique currently has drawbacks and features that need to be improved, as listed below:

1. Quantitation is difficult for different co-adsorbed species because the RIS yields vary significantly with the nature of adspecies. Separate calibrations are necessary for each adspecies.
2. The technique cannot determine the molecular mass of strongly adsorbed molecules that are fragmented by Cs^+ collision rather than desorbed intact. It is possible, however, to identify the end functional groups of such molecules.
3. Surface contamination by the incoming Cs^+ beam must be avoided.
4. A conceivable improvement in the technique would be to use a pulsed ion beam in conjunction with TOF mass selection. This would reduce the Cs^+ contamination effect as well as increase the sensitivity.

REFERENCES

1. S. R. Kasi, M. A. Kilburn, H. Kang, J. W. Rabalais, L. Tavernini, and P. Hochmann, *J. Chem. Phys.* **88**, 5902 (1988).
2. S. R. Kasi, H. Kang, C. S. Sass, and J. W. Rabalais, *Surf. Sci. Rep.* **10**, 1 (1989).

3. R. G. Cooks, T. Ast, and Md. A. Mabud, *Int. J. Mass Spectrom. Ion Proc.* **100**, 209 (1990).

4. M. R. Morris, D. E. Riederer, Jr., B. E. Winger, R. G. Cooks, T. Ast, and C. E. D. Chidsey, *Int. J. Mass Spectrom. Ion Phys.* **122**, 181 (1992).

5. R. G. Cooks, T. Ast, T. Pradeep, and V. Wysocki, *Acc. Chem. Res.* **27**, 316 (1994).

6. H. Akazawa and Y. Murata, *J. Chem. Phys.* **92**, 5560 (1990).

7. Y. Murata, in "Unimolecular and Bimolecular Reaction Dynamics," C. Y. Ng, T. Baer, and I. Powis (Eds.), John Wiley & Sons, New York, 1994, Chapter 9.

8. J. A. Burroughs, S. B. Wainhaus, and L. Hanley, *J. Chem. Phys.* **103**, 6706 (1995).

9. S. B. Wainhaus, E. A. Gislason, and L. Hanley, *J. Am. Chem. Soc.* **119**, 4001 (1997).

10. J. S. Martin, J. N. Greeley, J. R. Morris, B. T. Feranchak, and D. C. Jacobs, *J. Chem. Phys.,* **100**, 6791 (1994).

11. W. R. Koppers, J. H. M. Beijersbergen, T. L. Weeding, P. G. Kistemaker, and A. W. Kleyn, *J. Chem. Phys.* **107**, 10736 (1997).

12. K. H. Park, B. C. Kim, and H. Kang, *J. Chem. Phys.* **97**, 2742 (1992).

13. H. Kang, H. W. Lee, W. R. Cho, and S. M. Lee, *Chem. Phys. Lett.* **292**, 213 (1998).

14. C. L. Quinteros, T. Tzvetkov, and D. C. Jacobs, *J. Chem. Phys.* **113**, 5119 (2000).

15. M. Maazouz, T. L. O. Barstis, P. L. Maazouz, and D. C. Jacobs, *Phys. Rev. Lett.* **84**, 1331 (2000).

16. P. Haochang, T. C. M. Horn, and A. W. Kleyn, *Phys. Rev. Lett.* **57**, 3035 (1986).

17. R. Souda, T. Suzuki, H. Kawanowa, and E. Asari, *J. Chem. Phys.* **110**, 2226 (1999).

18. H. Kang, S. R. Kasi, and J. W. Rabalais, *J. Chem. Phys.* **88**, 5882 (1988).

19. S. A. Miller, H. Luo, S. J. Pachuta, and R. G. Cooks, *Science* **275**, 1447 (1997).

20. J. P. Cowin, A. A. Tsekouras, M. J. Ledema, K. Wu, G. B. Ellison, *Nature* **398**, 405 (1999).

21. M. C. Yang, H. W. Lee, and H. Kang, *J. Chem. Phys.* **103**, 5149 (1995).

22. M. C. Yang, C. H. Hwang, J. K. Ku, and H. Kang, *Surf. Sci.* **366**, L719 (1996).

23. M. C. Yang, C. H. Hwang, and H. Kang, *J. Chem. Phys.* **107**, 2611 (1997).

24. H. Kang, K. D. Kim, and K. Y. Kim, *J. Am. Chem. Soc.* **119**, 12002 (1997).

25. H. Kang, M. C. Yang, K. D. Kim, and K. Y. Kim, *Int. J. Mass Spectrom. Ion Proc.* **174**, 143 (1998).

26. K.-Y. Kim, T.-H. Shin, S.-J. Han, and H. Kang, *Phys. Rev. Lett.* **82**, 1329 (1999).

27. T.-H. Shin, S.-J. Han, and H. Kang, *Nucl. Instrum. Meth. Phys. Res. B* **157**, 191 (1999).

28. S-J. Han, S. C. Park, J.-G. Lee, and H. Kang, *J. Chem. Phys.* **112**, 8660 (2000).

29. S.-C. Park, H. Kang, and S. B. Lee, *Surf. Sci.* **450**, 117 (2000).

30. H. G. Yoon, J.-H. Boo, W. L. Liu, S.-B. Lee, S.-C. Park, H. Kang, and Y. Kim, *J. Vac. Sci. Tech. A* **18**, 1464 (2000).

31. H. Kang, T.-H. Shin, S.-C. Park, I. K. Kim, and S.-J. Han, *J. Am. Chem. Soc.* **122**, 9842 (2000).

32. S.-C. Park, T. Pradeep, and H. Kang, *J. Chem. Phys.* **113**, 9373 (2000).

33. S.-C. Park, K.-W. Maeng, T. Pradeep, and H. Kang, *Angew. Chem. Int. Ed.* **40**, 1497 (2001).

34. S.-C. Park, K.-W. Maeng, T. Pradeep, and H. Kang, *Nucl. Instrum. Meth. Phys. Res. B* **182**, 193 (2001).

35. S-J. Han, C.-W. Lee, C.-H. Hwang, K.-H. Lee, M. C. Yang, and H. Kang, Bull. *Korean Chem. Soc.* **22**, 883 (2001).

36. C.-H. Hwang, C.-W. Lee, H. Kang, and C. M. Kim, *Surf. Sci.* **490**, 144 (2001).

37. C. M. Kim, C.-H. Hwang, C.-W. Lee, and H. Kang, *Angew. Chem. Int. Ed.* **41,** 146 (2002).

38. R. J. W. E. Lahaye and H. Kang, *Surf. Sci.* **490**, 144 (2001).

39. R. J. W. E. Lahaye and H. Kang, *Nucl. Instrum. Meth. Phys. Res. B* **182**, 207 (2001).

40. M. C. Yang, H. W. Lee, C. Kim, and H. Kang, *Surf. Sci.* **357**, 595 (1996).

41. V. I. Veksler, *Soviet Phys.-Solid State* **6**, 1767 (1965).

42. E. Hulpke, *Surf. Sci.* **52**, 615 (1975).

43. E. Hulpke and K. Mann, *Surf. Sci.* **133**, 171 (1983).

44. A. D. Tenner, K. T. Gillen, T. C. M. Horn, J. Los, and A. W. Kleyn, *Phys. Rev. Lett.* **52**, 2183 (1984).

45. A. D. Tenner, K. T. Gillen, T. C. M. Horn, J. Los, and A. W. Kleyn, *Surf. Sci.* **172**, 90 (1986).

46. A. W. Kleyn and T. C. M. Horn, *Phys. Rep.* **199**, 191 (1991).

47. N. N. Bazarbayev, V. V. Evstifeev, N. M. Krylov, and L. B. Kudryaschova, *Sov. J. Surf.* **9,** 170 (1988).

48. V. V. Evstifeev and I. V. Ivanov, *Surf. Sci.* **217**, L373 (1989).

49. R. L. McEachern, D. M. Goodstein, and B. H. Cooper, *Phys. Rev. B* **39**, 10503 (1989).

50. C. A. DiRubio, R. L. McEachern, J. G. McLean, and B. H. Cooper, *Phys. Rev. B* **54**, 8862 (1996).

51. L. L. Tongson and C. B. Cooper, *Surf. Sci.* **52**, 263 (1975).

52. H. Akazawa and Y. Murata, *Phys. Rev. Lett.* **61**, 1218 (1988).

53. C. Kim, H. Kang, and S. C. Park, *Nucl. Instrum. Meth. Phys. Res. B* **95**, 171 (1995).

54. H. W. Lee and H. Kang, *Bull. Korean Chem. Soc.* **16**, 101 (1995).

55. A. Tolstogouzov, S. Daolio, C. Pagura, *Nucl. Instrum. Meth. Phys. Res. B* **183**, 116 (2001).

56. E. Kolodney, A. Amirav, R. Elber, and R. B. Gerber, *Chem. Phys. Lett.* **113**, 303 (1985).

57. A. Amirav, M. J. Cardillo, P. L. Trevor, C. Lim, J. C. Tully, *J. Chem. Phys.* **87**, 1796 (1987).

58. C. T. Rettner, J. A. Barker, and D. S. Bethune, *Phys. Rev. Lett.* **67**, 2183 (1991).

59. P. Sigmund, *Phys. Rev.* **184**, 383 (1969).

60. H. F. Winters, "Advances in Chemistry Series," M. Kaminsky (Ed.) No. 158 (Amer. Chem. Soc., 1976).

61. R. Behrisch (Ed.), "Sputtering by Particle Bombardment I," Springer-Verlag, Berlin, 1981.

62. E. Hotston, *Nucl. Fusion* **15**, 544 (1975).

63. W. Eckstein, C. Garcia-Rosales, and J. Roth, *Nucl. Instrum. Meth. Phys. Res. B* **83**, 955 (1993).

64. K. H. Kingdon and I. Langmuir, *Phys. Rev.* **22**, 148 (1923).

65. G. K. Wehner, *Phys. Rev.* **108**, 35 (1957).

66. G. K. Wehner, *Phys. Rev.* **112**, 1120 (1958).

67. N. Laegreid and G. K. Wehner, *J. Appl. Phys.* **32**, 365 (1961).

68. R. V. Stuart and G. K. Wehner, *J. Appl. Phys.* **33**, 2345 (1962).

69. B. J. Garrison, N. Winograd, and D. E. Harrison, Jr., *J. Chem. Phys.* **69,** 1440 (1978).

70. S. B. M. Bosio and W. L. Hase, *J. Chem. Phys.* **107,** 9677 (1997).

71. H. Coufal, H. F. Winters, H. L. Bay, and W. Eckstein, *Phys. Rev. B* **44**, 4747 (1991).

72. H. F. Winters, *J. Vac. Sci. Technol.* **8**, 17 (1971).

73. H. F. Winters and P. Sigmund, *J. Appl. Phys.* **45**, 4760 (1974).

74. J. W. Rabalais, T. R. Schuler, and O. Grizzi, *Nucl. Instrum. Meth. Phys. Res. B* **28**, 185 (1987).

75. U. Diebold and P. Varga, *Vacuum,* **41**, 210 (1990).

76. U. Diebold, W. Moller, and P. Varga, *Surf. Sci.* **248**, 147 (1991).

77. C. Kim, J. R. Han, and H. Kang, *Surf. Sci.* **320**, L76 (1994).

78. P. Kebarle, *Ann. Rev. Phys. Chem.* **28**, 445 (1977).

79. R. Ludeke and A. Koma, *Phys. Rev. Lett.* **34**, 1170 (1975).

80. A. J. Schell-Sorokin and J. E. Demuth, *Surf. Sci.* **157**, 273 (1985).

81. H. Ibach, K. Horn, R. Dorn, and H. Luth, *Surf. Sci.* **38**, 433 (1973).

82. U. Hofer, P. Morgen, W. Wurth, and E. Umbach, *Phys. Rev. Lett.* **55**, 2979 (1985).

83. B. Schubert, Ph. Avouris, and R. Hoffmann, *J. Chem. Phys.* **98**, 7593 (1993).

84. G. Dujardin, G. Comtet, L. Hellner, T. Hirayama, M. Rose, L. Philippe, and M. J. Besnard-Ramage, *Phys. Rev. Lett.* **73**, 1727 (1994).

85. S. H. Lee and M. H. Kang, *Phys. Rev. B* **61**, 8250 (2000).

86. F. Matsui, H. W. Yeom, K. Amemiya, K. Tono, and T. Ohta, *Phys. Rev. Lett.* **85**, 630 (2000).

87. N. Sheppard, *Ann. Rev. Phys. Chem.* **39**, 589 (1988).

10

BIBLIOGRAPHY OF ION SCATTERING PUBLICATIONS

S. Aduru, and J. W. Rabalais, "Beam Energy and Glancing Angle Dependence of the Neutral Fraction of Scattered K and Directly Recoiled Si From Si(100)," *J. Vac. Sci. Technol. A* **6,** 805 (1988).

S. Aduru, and J. W. Rabalais, "Direct Recoil Spectra Without Interfering Scattering Spectra: K^+ on Si(100)," *Surface Sci.* **205,** 269 (1988).

S. Aduru, and J. W. Rabalais, "Initial Stage of Titanium Oxidation Studied by Direct Recoil Spectrometry," *Langmuir* **3,** 543 (1987).

J. Ahn, H. Bu, C. Kim, V. Bykov, M. M. Sung, and J. W. Rabalais, "Structural Study of Ni{100}-(2×2)-C Surface by Time-of-Flight Scattering and Recoiling Spectrometry (TOF-SARS)," *J. Phys. Chem.* **100,** 9088 (1996).

J. Ahn, and J. W. Rabalais, "Composition and Structure of the Cd- and S-Terminated CdS{0001}-(1×1) Surfaces," *J. Phys. Chem.* **102,** 223 (1998).

J. Ahn, and J. W. Rabalais, "Composition and Structure of the Al_2O_3{0001}-(1×1) Surface," *Surface Sci.* **388,** 121 (1997).

J. Ahn, M. M. Sung, J. W. Rabalais, D. D. Koleske, and H. E. Wickenden, "Surface Composition and Structure of GaN Epilayers on Sapphire," *J. Chem. Phys.* **107,** 9577 (1997).

A. J. Algra, E. V. Loenen, E. P. Suurmeije, and A. L. Boers, "The Ion Fractions of 2–10 keV Lithium, Sodium, and Potassium Scattered from A Copper(100) Surface," *Rad. Eff. Def. Sol.* **60,** 173 (1982).

A. J. Algra, S. B. Luitjens, H. Norggreve, E. P. Suurmeijer, and A. L. Boers, "The Ratio of the Single and Double Scattering Intensities in Ion Scattering Spectroscopy as a Quantitative Measure of Surface Structures," *Rad. Eff. Def. Sol.* **62,** 7 (1982).

A. J. Algra, S. B. Luitjens, E. P. Suurmeijer, and A. L. Boers, "Structure Analysis of Solid Surfaces by Multiple Scattering of Low Energy Ions," *Nucl. Instrum. Meth. Phys. Res. B* **203,** 515 (1982).

A. J. Algra, S. B. Luitjens, E. P. Suurmeijer, and A. L. Boers, "Determination of Interatomic Distances at Solid Surfaces with ISS: A Study of Stepped Cu(410)," *Proc. ECOSS III Cannes Suppl. Le Vide* **201,** 703 (1980).

A. J. Algra, S. B. Luitjens, E. P. Suurmeijer, and A. L. Boers, "The Structure of a Stepped Copper (410) Surface Determined by Ion Scattering Spectroscopy," *Nucl. Instrum. Meth. Phys. Res. B* **132,** 623 (1980).

A. J. Algra, P. P. Maaskant, S. B. Luitjens, E. P. Suurmeijer, and A. L. Boers, "The Two Scattering Energies in a Single Collision," *J. Phys. D.,* **13,** 2363 (1980).

A. J. Algra, E. P. Suurmeijer, and A. L. Boers, "The Position of Oxygen Adsorbed at the Steps of a Copper (410) Surface Studied with Low Energy Ion Scattering," *Surf. Sci.* **128,** 207 (1983).

P. F. Alkemade, W. C. Turkenburg, and W. F. Van der Weg, "The Energy Loss of Medium-Energy He^+ Ions Backscattered from a Cu(100) Surface," *Nucl. Instrum. Meth. Phys. Res. B* **28,** 161 (1987).

H. J. Andra, H. Winter, R. Frohling, N. Kirchner, H. J. Plohn, W. Wittmann, W. Graser, and C. Varelas, "Ion Beam Surface Interaction at Grazing Incidence," *Nucl. Instrum. Meth. Res. B* **170,** 527 (1980).

M. Aono, Y. Hou, C. Oshima, and Y. Ishizawa, "Low Energy Ion Scattering from the Si(001) Surface," *Phys. Rev. Lett.* **49,** 567 (1982).

M. Aono, Y. Hou, R. Souda, C. Oshima, S. Otani, and Y. Ishizawa, "Direct Analysis of the Structure, Concentration, and Chemical Activity of Surface Atomic Vacancies by Specialized Low-Energy Ion-Scattering Spectroscopy: TiC(001)," *Phys. Rev. Lett.* **50,** 1293 (1983).

M. Aono, Y. Hou, R. Souda, C. Oshima, S. Otani, Y. Ishizawa, K. Matsuda, and R. Shimizu, "Interaction Potential Between He^+ and Ti in the keV Range as Revealed by a Specialized Technique in Ion Scattering Spectrometry," *Jap. J. Appl. Phys.* **21,** L670 (1982).

M. Aono, and M. Katayama, "A Novel Method for Real-Time Structural Monitoring of Molecular Beam Epitaxy (MBE) Processes," *Proc. of the Japan Academy 65* **Ser. B** (1989).

M. Aono, M. Katayama, and E. Nomura, "Exploring Surface Structures by Coaxial Impact-Collision Ion Scattering Spectroscopy (CAICISS)," *Nucl. Instrum. Meth. Phys. Res. B* **64,** 29 (1992).

M. Aono, M. Katayama, E. Nomura, T. Chasse, D. Choi, and M. Kato, "Recent Developments in Low Energy Ion Scattering Spectroscopy," *Proc. of the Japan Academy 65* **Ser. B** (1989).

M. Aono, and R. Souda, "Quantitative Surface Atomic Structure Analysis by Low Energy Ion Scattering," *Jap. J. Appl. Phys.* **24,** 1249 (1985).

M. Aono, R. Souda, C. Oshima, and Y. Ishizawa, "Structure Analysis of the Si(111) 7×7 Surfaces by Low Energy Ion Scattering," *Phys. Rev. Lett.* **51,** 801 (1983).

P. Apell, "Multiply Charged Ion-Solid Interaction," *Nucl. Instrum. Meth. Phys. Res. B* **23,** 242 (1987).

U. A. Arifov, L. M. Kishinevskii, E. S. Mukhamadiev, and E. S. Parilis, "Auger Neutralization of Highly Charged Ions at the Surface of a Metal," *Soviet Phys. Tech.* **18,** 118 (1973).

A. Arnau, F. Aumayr, P. M. Echenique, M. Grether, W. Heiland, J. Limburg, R. Morgenstern, P. Roncin, S. Schippers, R. Schuch, N. Stolterfoht, P. Varga, T. J. Zouros, and H. P. Winter, "Interaction of Slow Multi-Charged Ions with Solid Surfaces," *Surf. Sci. Reports* **27,** 113 (1997).

A. Arnau, P. A. Zeijlmans van Emmichoven, J. I. Juaristi, and E. Zaremba, "Nonlinear Screening Effects in the Interaction of Slow Multicharged Ions with Metal Surfaces," *Nucl. Instrum. Meth. Phys. Res. B* **100,** 279 (1995).

E. Asari, W. Hayami, and R. Souda, "Transitional Structures of the $TiO_2(110)$ Surface from $p(1 \times 1)$ to $p(1 \times 2)$ Studied by Impact Collision Ion Scattering Spectroscopy," *Appl. Surf. Sci.* **167,** 169 (2000).

E. Asari, and R. Souda, "Atomic Structure of TiO_2 (110)-$p(1 \times 2)$ and $p(1 \times 3)$ Surfaces Studied by Impact Collision Ion-Scattering Spectroscopy," *Phys. Rev. B* **60,** 10719 (1999).

E. Asari, T. Suzuki, H. Kawanowa, J. Ahn, W. Hayami, T. Aizawa, and R. Souda, "$TiO_2(110)$-$p(1 \times 1)$ Surface Structure Analyzed by Impact-Collision Ion-Scattering Spectroscopy," *Phys. Rev. B* **61,** 5679 (2000).

M. Aschoff, G. Piaszenski, S. Speller, and W. Heiland, "The Structure of the $Au_3Pd(113)$-Surface Studied by Low Energy Ion Scattering," *Surf. Sci.* **402,** 770 (1998).

F. Ascione, G. Manicó, P. Alfano, A. Bonanno, N. Mandarino, A. Oliva, and F. Xu, "Scattering of 1960 eV Ne^+ from Al Surface," *Nucl. Instrum. Meth. Phys. Res. B* **135,** 401 (1998).

F. Ascione, G. Manicó, A. Bonanno, A. Oliva, and F. Xu, "On the Origin of Ne^{++} Production in Low-Energy Ne^+ Scattering from Al and Si Surfaces," *Surf. Sci.* **394,** L145 (1997).

M. J. Ashwin, and D. P. Woodruff, "Charge Exchange Processes in Li^+ and He^+ Ion Scattering from Alkali Adsorbates on Cu(110)," *Surf. Sci.* **244,** 247 (1991).

M. J. Ashwin, and D. P. Woodruff, "Low Energy Ion Scattering Study of the Cu(110)-(2×3)-N Structure," *Surf. Sci.* **237,** 108 (1990).

C. Bahrim, A. G. Borisov, D. Teillet-Billy, J. P. Gauyacq, F. Wiegershaus, St. Krischok, and V. Kempter, "Theoretical and Experimental Study of the $N^-(^1D)$ Formation in Nitrogen Collisions with a Metal Surface," *Surf. Sci.* **380,** 556 (1997).

C. Bahrim, P. Kürpick, U. Thumm, and U. Wille, "Electron Dynamics in Slow Atomic Interactions with Metal Surfaces and Thin Metallic Films," *Nucl. Instrum. Meth. Res. B* **164,** 614 (2000).

C. Bahrim, and U. Thumm, "$^3S^e$ and $^1S^e$ Scattering Lengths for e^- + Rb, Cs, and Fr Collisions," *J. Phys. B* **34,** L195 (2001).

C. Bahrim, and U. Thumm, "Charge Transfer Dynamics in Slow Atom-Surface Collisions: A New Close-Coupling Approach Including Continuum Discretization," *Surf. Sci. B* **451,** 1 (2000).

C. Bahrim, and U. Thumm, "The Low-Lying Negative-Ion States of Rb, Cs, and Fr," *Phys. Rev. A* **61,** 22722 (2000).

C. Bahrim, U. Thumm, and I. I. Fabrikant, "Negative Ion Resonances in Cross Sections for Slow Electron-Heavy Alkali Atom Scattering," *Phys. Rev. A* **63,** 42710 (2001).

L. L. Balashova, "Dependence of the Survival Fractions on Reflection Angle of Molecular Nitrogen Ions from the Copper Single-Crystal Surface," *Rad. Eff. Def. Sol.* **2,** 141 (1989).

L. L. Balashova, A. M. Borisov, A. I. Dodonov, E. S. Mashkova, and V. A. Molchanov, "Effect of Surface Semichannels in Fast Recoil Energy Distributions," *Nucl. Instrum. Meth. Res. B* **170,** 157 (1980).

L. L. Balashova, A. M. Borisov, E. S. Mashkova, and V. A. Molchanov, "Energy Features of the Blocking Effect for Fast Ionized Recoils," *Phys. Rev. A* **21,** 1185 (1980).

L. L. Balashova, A. M. Borisov, E. S. Mashkova, and V. A. Molchanov, "Blocking Effect for Fast Ionized Recoils," *Surf. Sci.* **77,** L643 (1978).

L. L. Balashova, E. S. Mashkova, and V. A. Molchanov, "Argon Ion Double Scattering from Polycrystalline Copper," *Surf. Sci.* **76,** L590 (1978).

R. Baragiola, "Comment on Collisional De-Excitation at Ion Bombarded Surfaces," *Phys. Rev. Letters* **7,** 408 (1996).

R. Baragiola, "Auger De-excitation of $Na(2p^{-1})$ Near Alkali-Halide Surfaces," *Nucl. Instrum. Meth. Phys. Res. B* **100,** 242 (1995).

R. Baragiola, "Ion Induced Kinetic and Auger Electron Emission from Solids," in *Low Energy Ion Surface Interactions,* J. W. Rabalais (Ed.), Wiley, New York, 1994.

R. Baragiola, E. V. Alonso, A. Oliva, A. Bonanno, and F. Xu, "Fast Electrons From Slow Atomic Collisions," *Phys. Rev. A* **45,** 5286 (1992).

R. Baragiola, and C. A. Dukes, "Plasmon-Assisted Electron Emission from Al and Mg Surfaces by Slow Ions," *Phys. Rev. Lett.* **76,** 2547 (1996).

R. Baragiola, L. Nair, and T. Madey, "Symmetric versus Asymmetric Collision in Ion-Induced Auger Emission from Silicon," *Nucl. Instrum. Meth. Phys. Res. B* **58,** 322 (1991).

B. Baretzky, W. Möller, and E. Taglauer, "Collision Dominated Preferential Sputtering of Tantalum Oxide," *Vacuum* **43,** 1207 (1992).

K. Bartschat, U. Thumm, and D. W. Norcross, "Characteristics of Light Emission after Low-Energy Electron-Impact-Excitation of Cesium Atoms," *J. Phys. B* **25,** L641 (1992).

R. Bastasz, T. E. Felter, and W. P. Ellis, "Low Energy He^+ Scattering from Deuterium Adsorbed on Pd(110)," *Phys. Rev. Lett.* **63,** 558 (1989).

M. Beckschulte, D. Mehl, and E. Taglauer, "The Adsorption of CO on Ni(100) Studied by Low Energy Ion Scattering," *Vacuum* **41,** 67 (1990).

M. Beckschulte, and E. Taglauer, "The Influence of Work Function Changes on the Charge Exchange in Low-Energy Ion Scattering," *Nucl. Instrum. Meth. Phys. Res. B* **78,** 29 (1993).

A. H. Begemann, and A. L. Boers, "Small Angle Multiple Reflection of Low Energy (6 keV) Noble Gas Ions from Single Crystal Surfaces as a Means to Study Surface Texture and Contamination," *Surf. Sci.* **30,** 134 (1972).

R. Beikler, and E. Taglauer, "Ion Scattering of Ordered Alloy Surfaces: CuAu(100) and NiAl," *Nucl. Instrum. Meth. Phys. Res. B* **161,** 390 (2000).

P. Bertrand, H. Bu, and J. W. Rabalais, "Orientation of Single Crystalline Paraffin Thin Films from Time-of-Flight Direct Recoil Spectrometry," *J. Phys. Chem.* **97,** 13788 (1993).

P. Bertrand, and Y. De Puydt, "Ion Scattering by Polymer Surfaces: Elastic and Inelastic Effects," *Nucl. Instrum. Meth. Phys. Res. B* **78,** 181 (1993).

P. Bertrand, P. Lambert, and Y. Travaly, "Polymer Metallization: Low Energy Ion Beam Surface Modification to Improve Adhesion," *Nucl. Instrum. Meth. Phys. Res. B* **131,** 71 (1997).

P. Bertrand, and J. W. Rabalais, "Ion Scattering and Recoiling for Elemental Analysis and Structure Determination," in *Low Energy Ion-Surface Interactions,* J. W. Rabalais (Ed.), Wiley, New York, 1994.

J. M. Beuken, and P. Bertrand, "An ISS Study of Thin Metallic Layer Evaporated on Gold Substrates: Experimentation vs. Simulation," *Nucl. Instrum. Meth. Phys. Res. B* **67,** 340 (1992).

R. Blum, D. Ahlbehrendt, and H. Niehus, "Preparation-Dependent Surface Composition and Structure of NiAl(001) SPA-LEED and NICISS Study," *Surf. Sci.* **366,** 107 (1996).

R. Blum, and H. Niehus, "Initial Growth of Al_2O_3 on NiAl(001)," *Applied Physics A* **66,** 529 (1998).

D. O. Boerma, "Materials Analysis Using Ion Beam Techniques," *Nucl. Instrum. Meth. Phys. Res. B* **50,** 77 (1990).

A. L. Boers, "Charge State of Low Energy Reflected Alkalis," *Nucl. Instrum. Meth. Phys. Res. B* **2,** 353 (1984).

A. L. Boers, "Charge State of Low Energy Reflected Particles," *Nucl. Instrum. Meth. Phys. Res. B* **4,** 98 (1984).

I. L. Bolotin, L. Houssiau, and J. W. Rabalais, "Real–Space Surface Crystallography: Experimental Stereographic Projections from Ion Scattering," *J. Chem. Phys.* **112,** 7181 (2000).

I. L. Bolotin, A. Kutana, B. Makarenko, and J. W. Rabalais, "Kinetics and Structure of O_2 Chemisorption on Ni(111)," *Surf. Sci.*, **472,** 205–222 (2001).

A. Bonanno, A. Amoddeo, and A. Oliva, "Kinetic Threshold in Ion-Induced Electron Emission from Polycrystalline W," *Nucl. Instrum. Meth. Phys. Res. B* **46,** 456 (1990).

A. Bonanno, N. Mandarino, A. Oliva, and F. Xu, "Atomic versus Bulk Al Auger Electron Emission Induced by Noble Gas Ion Bombardment," *Nucl. Instrum. Meth. Phys. Res. B* **71,** 161 (1992).

A. Bonanno, F. Xu, M. Camarca, R. Siciliano, and A. Oliva, "Directional Ejection of Fast Excited Si Atoms During 10 keV Ar^+ Ion Bombardment," *Nucl. Instrum. Meth. Phys. Res. B* **48,** 371 (1990).

A. M. Borisov, A. I. Dodonov, E. S. Mashkova, and V. A. Molchanov, "The Effect of Thermal Vibrations of Crystal Lattice on the Energy Distributions of Fast Ionized Recoils," *Surf. Sci.* **95,** L289 (1980).

A. M. Borisov, A. I. Dodonov, and V. A. Molchanov, "Temperature Effects in Fast Recoil Ejection," *Rad. Eff. Def. Sol.* **80,** 105 (1984).

A. Borisov, V. A. Esaulov, S. Lacombe, J. P. Gauyacq, L. Guillemot, D. Teillet-Billy, M. Maazouz, and R. Baragiola, "H^- Formation in the Scattering of Hydrogen Ions on an Al Surface," *Surf. Sci.* **364,** L568 (1996).

T. Bremer, W. Heiland, A. Klekamp, and E. Kratzig, "SIMS Investigations of Iron Profiles in $LiNbO_3$: Ti Waveguides," *Phys. Stat. Sol. A* **105,** K17 (1988).

H. Brenten, K. H. Knorr, D. Kruse, H. Müller, and V. Kempter, "Harpooning in Collisions of Li^+ Ions with Oxygenated Partially Cesiated W(110) Surfaces," *Nucl. Instrum. Meth. Phys. Res. B* **48,** 344 (1990).

H. Brenten, H. Müller, and V. Kempter, "Electrons from Intra- and Interatomic Auger Processes in Low-Energy Collisions of Singly and Doubly Charged Inert Gas Ions with W(110) Surfaces Partially Covered by Alkali Atoms and NaCl Molecules," *NATO–ASI Series B: Physics,* 306, Plenum Press, New York, 1993.

H. Brenten, H. Müller, and V. Kempter, "Electron Emission in Slow Collisions of Alkali Ions with Alkalated W(110) Surfaces," *Surf. Sci.* **271,** 103 (1992).

H. Brenten, H. Müller, and V. Kempter, "Electrons from Intra- and Interatomic Auger Processes in Low-Energy He^{++} Collisions with Partially Alkalated W(110)," *Surf. Sci.* **274,** 309 (1992).

H. Brenten, H. Müller, and V. Kempter, "Formation of Feshbach Resonances Associated with Doubly Excited He States in Slow Collisions with Low Work Function Surfaces," *Phys. Rev. Lett.* **70,** 25 (1993).

H. Brenten, H. Müller, and V. Kempter, "Intra- and Inter-Atomic Auger Processes in Collisions of Low-Energy He^{++}, He$^+$, He* (2^3S, 2^1S) and Li$^+$ with Li Covered W(110) Surfaces," *Phys. D* **22,** 563 (1992).

H. Brenten, H. Müller, and V. Kempter, "Electron Emission in Slow Collisions of K$^+$ Ions with Alkalated (K, Cs) W(110) Surfaces," *Z. Phys. D* **21,** 11 (1991).

H. Brenten, H. Müller, K. H. Knorr, D. Kruse, H. Schall, and V. Kempter, "The Role of Auger Processes in Slow Collisions of Li$^+$ Ions with Cesiated W(110) Surfaces," *Surf. Sci.* **234,** 309 (1991).

H. Brenten, H. Müller, D. Kruse, and V. Kempter, "Auger Electron Emission Induced by Slow Alkali Ion Collisions with Cesiated Surfaces," *Nucl. Instrum. Meth. Phys. Res. B* **58,** 328 (1991).

H. Brenten, H. Müller, A. Niehaus, and V. Kempter, "Autoionization and Autodetachment in Collisions of Slow Inert Gas Ions with Partially K Covered W(110) Surfaces," *Surf. Sci.* **278,** 183 (1992).

R. P. Bronckers, and A. G. J. De Wit, "Reconstruction of the Oxygen-Covered Cu(110) Surface Identified with Low Energy Ne$^+$ and H$_2$O$^+$ Ion Scattering," *Surf. Sci.* **112,** 133 (1981).

R. P. Bronckers, and A. G. J. De Wit, "Shadowing, Focusing and Charge Exchange Effects in the Angular Distributions of keV Ne and H$_2$O Scattered from Cu(110)," *Surf. Sci.* **104,** 384 (1981).

R. P. Bronckers, Th. M. Hupkens, A. G. De Wit, and W. C. Post, "A Versatile Target Manipulator for Use in Ultra-High Vacuum," *Nucl. Instrum. Meth. Res. B* **179,** 125 (1981).

H. H. Brongersma, and T. M. Buck, "Low Energy Ion Scattering (LEIS) for Composition and Structure Analysis of the Outer Surface," *Nucl. Instrum. Meth. Res. B* **149,** 569 (1978).

H. H. Brongersma, and T. M. Buck, "Neutralization Behavior in Scattering of Low Energy Ions from Solid Surfaces," *Nucl. Instrum. Meth. Res. B* **132,** 559 (1976).

H. H. Brongersma, and T. M. Buck, "Selected Topics in Low Energy Ion Scattering: Surface Segregation in Cu/Ni Alloys and Ion Neutralization," *Surf. Sci.* **53,** 649 (1975).

H. H. Brogersma, and J. B. Theeten, "The Structure of Oxygen Adsorbed on Ni(001) as Determined by Ion Scattering Spectroscopy," *Surf. Sci.* **54,** 519 (1976).

K. Brüning, J. Granow, M. Reiniger, W. Heiland, M. Vicanek, and T. Schlathölter, "Dissociation of Fast N$_2^-$ Molecules at a Pd(110) Surface," *Surf. Sci.* **402,** 215 (1998).

K. Brüning, W. Heiland, T. Schlathölter, I. A. Wojciechowski, M. B. Medvedeva, and V. Kh. Ferleger, "Dissociation of Fast N$_2$ Molecules Scattered from Different FCC(110) Surfaces," *J. Chem. Phys.* **113,** 2456 (2000).

H. Bu, P. Bertrand, and J. W. Rabalais, "The Structure of Benzene and Phenol Chemisorbed on Ni{110}," *J. Chem. Phys.* **98,** 5855 (1993).

H. Bu, O. Grizzi, M. Shi, and J. W. Rabalais, "Time-Of-Flight Scattering and Recoiling. II. The Structure of Oxygen on the W(211) Surface," *Phys. Rev. B* **40,** 10147 (1989).

H. Bu, and J. W. Rabalais, "Structure Analysis of O$_2$ and H$_2$O Chemisorption on a Si{100} Surface," *Surface Sci.* **301,** 285–294 (1994).

H. Bu, C. D. Roux, and J. W. Rabalais, "Hydrogen Adsorption Site on the Ni{110}-p(1×2)-H Surface from Time-of-Flight Scattering and Recoiling Spectrometry (TOF-SARS)," *Surf. Sci.*, **271,** 68 (1992).

H. Bu, C. D. Roux, and J. W. Rabalais, "Oxygen Induced Added-Row Reconstruction of the Ni{100} Surface," *J. Chem. Phys.* **97,** 1465 (1992).

H. Bu, M. Shi, K. Boyd, and J. W. Rabalais, "Scattering and Recoiling Analysis of Oxygen Adsorption Site on the Ir{110}-c(2 × 2)-O Surface," *J. Chem. Phys.* **95,** 2882 (1991).

H. Bu, M. Shi, F. Masson, and J. W. Rabalais, "Reconstruction of the Ir(110) Surface: A Mixed Faceted (1 × 3) and (1 × 1) Structure," *Surf. Sci. Lett.* **230,** L140 (1990).

H. Bu, M. Shi, and J. W. Rabalais, "O_2 Induced (1 × 3)->(1 × 1) Phase Change of an Ir{110} Surface from TOF-SARS," *The Structure of Surfaces III,* **24,** 456 (1991).

H. Bu, M. Shi, and J. W. Rabalais, "Sampling Depth of Time-of-Flight Ion Scattering: Primary Ion Type and Energy Dependence," *Nucl. Inst. Meth. B* **61,** 337 (1991).

H. Bu, M. Shi, and J. W. Rabalais, "Surface Periodicity of Ir(110) From Time-of-Flight Scattering and Recoiling Spectrometry (TOF-SARS)," *Surface Sci.* **244,** 96 (1991).

H. Bu, M. Shi, and J. W. Rabalais, "Coexisting (1 × 3) and (1 × 1) Structures in the O_2 Induced Phase Change of Ir(110)," *Surf. Sci.* **236,** 135 (1990).

H. Bu, S. V. Teplov, V. Zastavnjuk, M. Shi, and J. W. Rabalais, "Angular Anisotropy of Surface Multiple Scattering: Analysis of Carbon Contamination on an Ir{110} Surface," *Surf. Sci.* **275,** 332 (1992).

T. M. Buck, Y. S. Chen, and G. H. Wheatley, "Energy Spectra of 6–32 keV Neutral and Ionized Ar and He Scattered from Au Targets: Ionized Fractions as Functions of Energy," *Surf. Sci.* **47,** 244 (1975).

T. M. Buck, I. Stensgaard, G. H. Wheatley, and L. Marchut, "Scattering of Low Energy Ne^+ Ions on Ni (001) and Ni (001) Au (Segregated) Surfaces," *Nucl. Instrum. Meth. Res. B* **170,** 519 (1980).

T. M. Buck, W. E. Wallace, R. A. Baragiola, G. H. Wheatley, J. B. Rothman, R. J. Girte, and J. G. Tittensor, "Differences in the Neutralization of 2.4 to 10 keV Ne^+ Scattered from the Cu and Au Atoms of an Alloy Surface," *Phys. Rev. B* **48,** 774 (1993).

T. M. Buck, and G. H. Wheatley, "Order-Disorder and Segregation Behavior at the $Cu_3Au(001)$ Surface," *Phys. Rev. Lett.* **51,** 43 (1983).

T. M. Buck, G. H. Wheatley, G. L. Miller, D. A. Robinson, and Y. S. Chen, "Comparison of a Time-of-Flight System with an Electrostatic Analyzer in Low-Energy Ion Scattering," *Nucl. Instrum. Meth. Phys. Res. B* **149,** 591 (1978).

T. M. Buck, G. H. Wheatley, and L. K. Verheij, "Low Energy Neon Ion Scattering and Neutralization on First and Second Layers of a Ni(001) Surface," *Surf. Sci.* **90,** 635 (1979).

V. Bykov, L. Houssiau, and J. W. Rabalais, "Real-Space Surface Crystallography from Ion Scattering," *J. Phys. Chem. B* **104,** 6340 (2000).

V. Bykov, C. Kim, M. M. Sung, K. G. Boyd, S. S. Todorov, and J. W. Rabalais, "Scattering and Recoiling Imaging Code (SARIC)," *Nucl. Instrum. Meth. Phys. Res. B* **114,** 371 (1996).

M. A. Cazalilla, N. Lorente, R. Díez-Muiño, J. P. Gauyacq, D. Teillet-Billy, and P. M. Echenique, "Theory of Auger Neutralization and De-Excitation of Slow Ions at Metal Surfaces," *Phys. Rev. B* **58,** 13991 (1998).

B. T. Chait, and K. G. Standing, "A Time-of-Flight Mass Spectrometer for Measurement of Secondary Ion Mass Spectra," *Int. J. Mass Spectrom. Ion Phys.* **40,** 185 (1981).

C. S. Chang, U. Knipping, and I. S. Tsong, "Shadow Cones Formed by Target Atoms Calculated for 1 to 3 keV H^+, He^+, Li^+, Ne^+, and Na^+ Ion Bombardment," *Nucl. Instrum. Meth. Phys. Res. B* **28,** 493 (1987).

C. S. Chang, U. Knipping, and I. S. Tsong, "Shadow Cones Formed by Target Atoms Bombarded by 1 to 3 keV H$^+$, He$^+$, Li$^+$, Ne$^+$, and Na$^+$ Ions," *Nucl. Instrum. Meth. Phys. Res. B* **18,** 11 (1986).

C. S. Chang, T. L. Porter, and I. S. Tsong, "In-Plane Geometry of the Si(111)-($\sqrt{3} \times \sqrt{3}$) Ag Surface," *J. Vac. Sci. Tech. A* **7,** 1906 (1989).

S. Chaudhury, R. S. Williams, M. Katayama, and M. Aono, "Quantitative Analysis of the Azimuthal Dependence of Ion Scattering from Si(111)-($\sqrt{3} \times \sqrt{3}$)R30°-Ag," *Surf. Sci.* **294,** 93 (1993).

J. N. Chen, and J. W. Rabalais, "Effects of Chemical Environment on Direct Recoil Ion Fractions," *J. Amer. Chem. Soc.* **110,** 46 (1988).

J. N. Chen, and J. W. Rabalais, "Direct Recoil Ion Fractions Resulting From Ar$^+$ Collisions," *Nucl. Instrum. Meth. B* **13,** 597 (1986).

J. N. Chen, and J. W. Rabalais, "Preferential Sputtering of LiF Surfaces Monitored by Photoelectron Spectroscopy and Direct Recoil Spectrometry," *Surf. Sci.* **176,** L879 (1986).

J. N. Chen, and J. W. Rabalais, "Vacuum Ultraviolet Photon Emission Stimulated by H$_2^+$, N$_2^+$, and CO$^+$ Collisions With Surfaces," *Chem. Phys. Lett.* **124,** 409 (1986).

J. N. Chen, M. Shi, and J. W. Rabalais, "Primary Ion Dependence of LiF Direct Recoil Intensities and Ion-Fractions," *J. Chem. Phys.* **86,** 2403 (1987).

J. N. Chen, M. Shi, S. Tachi, and J. W. Rabalais, "Detection of Low Energy Neutrals by a Channel Electron Multiplier," *Nucl. Instrum. Meth. B* **16,** 91 (1986).

Y. S. Chen, G. L. Miller, D. A. Robinson, and G. H. Wheatley, "Energy Spectra of 6–32 keV Neutral and Ionized Ar and He Scattered from Au Targets: Ionized Fractions as Functions of Energy," *Surf. Sci.* **47,** 244 (1975).

M. Chester, and T. Gustafsson, "Geometric Structure of the Si(111)-$\sqrt{3} \times \sqrt{3}$ R30°-Au Surface," *Surf. Sci.* **256,** 135 (1991).

M. Copel, and T. Gustafsson, "Structure of Au(110) Determined with Medium Energy Ion Scattering," *Phys. Rev. Lett.* **57,** 723 (1986).

D. M. Cornelison, M. S. Worthington, and I. S. Tsong, "Si(111)-(4 × 1) Surface Reconstruction Studies by Impact Collision Ion Scattering Spectrometry," *Phys. Rev. B* **43,** 4051 (1991).

R. Daley, D. Farrelly, and R. S. Williams, "Rainbow Scattering at Shadowing and Blocking Critical Angles," *Surf. Sci.* **234,** 355 (1990).

R. Daley, J. H. Huang, and R. S. Williams, "Neutralization of Low Energy Helium Ions Scattered from Au Adatoms on Si(111)-$\sqrt{3} \times \sqrt{3}$-Au," *Surf. Sci.* **202,** L577 (1988).

R. Daley, and R. S. Williams, "Summary Abstract: Impact Collision Ion Scattering Spectroscopy of Ag(11) Using Li$^+$ and 4He$^+$," *J. Vac. Sci. Tech. A* **6,** 808 (1988).

A. G. De Wit, R. P. Bronckers, and J. M. Fluit, "Oxygen Adsorption on Cu(110): Determination of Atom Positions with Low Energy Ion Scattering," *Surf. Sci.* **82,** 177 (1979).

S. T. de Zwart, T. Fried, D. O. Boerma, R. Hoekstra, A. G. Drentje, and A. L. Boers, "Sputtering of Silicon by Multiply Charged Ions," *Surf. Sci.* **177,** L939 (1986).

H. Derks, H. Hemme, W. Heiland, and S. H. Overbury, "Low Energy Ion Scattering from the Au(110) Surface-Structural Results," *Nucl. Instrum. Meth. Phys. Res. B* **23,** 374 (1987).

H. Derks, W. Hetterich, E. van de Riet, H. Niehus, and W. Heiland, "Studies of the Structure of FCC(110) Surfaces Ir, Pt, Au and Ni by Low Energy Ion Scattering," *Nucl. Instrum. Meth. Phys. Res. B* **48,** 315 (1990).

H. Derks, J. Moller, and W. Heiland, "The Short Range Order Surface Structure of Au(110) in the Range of the (1 × 2) to (1 × 1) Phase Transition," *Surf. Sci.* **1988,** L685 (1987).

H. Derks, A. Närmann, and W. Heiland, "The Scattering of He Ions Off Ni(110) at Grazing Incidences: Surface Channeling," *Nucl. Instrum. Meth. Phys. Res. B* **44,** 125 (1989).

S. Dieckhoff, W. Maus-Friedrichs, and V. Kempter, "The Change of the Electronic Structure of Alkali Halide Films on W(110) Under Electron Bombardment," *Nucl. Instrum. Meth. Phys. Res. B* **65,** 488 (1992).

S. Dieckhoff, H. Müller, H. Brenten, W. Maus-Friedrichs, and V. Kempter, "The Electronic Structure of NaCl Adlayers on W(110) Studied by Soft Ionizing Radiation," *Surf. Sci.* **279,** 233 (1992).

M. Dirska, J. Manske, G. Lubinski, M. Schleberger, R. Hoekstra, and A. Närmann, "Ion Scattering of Magnetic Surfaces," *AIP Conf. Proc.* **392,** 1381 (1997).

A. I. Dodonoy, E. S. Mashkova, and V. A. Molchanov, "Ejection of Fast Recoil Atoms from Solids Under Ion Bombardment: Medium-Energy Ion Scattering by Solid Surfaces: Part III," *Rad. Eff. Def. Sol.* **110,** 227 (1989).

S. E. Donnelly, R. G. Elliman, D. J. O'Conor, and R. J. MacDonald, "Low Energy Ion Scattering as a Probe of Neon Radiation Damage in Nickel," *Nucl. Instrum. Meth. Phys. Res. B* **15,** 130 (1986).

J. Ducrée, H. Andrä, and U. Thumm, "Improved Simulation of Highly Charged Ion-Surface Collisions," *Phys. Sci. T* **80,** 220 (1999).

J. Ducrée, H. Andrä, and U. Thumm, "Neutralization of Hyperthermal Multiply Charged Ions at Surfaces: Comparison Between the Extended Dynamical Over-Barrier Model and Experiment," *Phys. Rev. A* **60,** 3029 (1999).

J. Ducrée, F. Casali, and U. Thumm, "Extended Classical Over-Barrier Model for Collisions of Highly Charged Ions with Conducting and Insulating Ionic Crystal Surfaces," *Phys. Rev. A* **57,** 338 (1998).

W. P. Ellis, and R. Bastasz, "Structural Determination of Oxygen Chemisorption-Site Geometry on W(211) by Low Energy He$^+$ ISS," in *The Structure of Surfaces II,* J. F. van der Veen and M. A. Van Hove (Eds.), Springer, Berlin, 1988.

W. P. Ellis, and R. R. Rye, "Low-Energy ^{3}He$^+$ Ion Scattering Spectroscopy Studies of D$_2$ Adsorption on W(211)," *Surf. Sci.* **161,** 278 (1985).

W. Englert, W. Heiland, E. Taglauer, and D. Menzel, "The Sorption of CO and O on Ni(111) Studied by LEED and ISS," *Surf. Sci.* **83,** 243 (1979).

I. N. Evdokimov, J. A. Van den Berg, and D. G. Armour, "Proper Surface Channeling of Low Energy Argon Ions Incident on a Nickel(110) Crystal," *Rad. Eff. Def. Sol.* **41,** 33 (1979).

Th. Fauster, "Surface Geometry Determination by Large-Angle Ion Scattering," *Vacuum* **38,** 129 (1988).

Th. Fauster, H. Durr, and D. Hartwig, "Determination of the Geometry of Sulfur on Nickel Surfaces by Low Energy Ion Scattering," *Surf. Sci.* **178,** 657 (1986).

P. Fenter, and T. Gustafsson, "Structural Analysis of the Pt(110)-(1 × 2) Surface Using Medium-Energy Ion Scattering," *Phys. Rev. B* **38,** 10197 (1988).

Th. Fauster, and M. H. Metzner, "Low Energy Ion Scattering from a Ni(001) Surface," *Surf. Sci.* **166,** 29 (1986).

L. Folkerts, S. Schippers, D. M. Zehner, and F. W. Meyer, "Fine Scales for Charge Equilibration of O^{9+}(3 < q < 8) Ions During Surface Channeling Interactions with Au(110)," *Phys. Rev. Lett.* **74,** 2204 (1995).

H. Franke, K. Schmidt, and W. Heiland, "The Interaction of Swift C^+, CH^+ and C^{+2} Ions with a Ni(110) Surface at Grazing Incidence," *Surf. Sci.* **269,** 219 (1992).

H. Franke, K. Schmidt, and W. Heiland, "The Scattering of CH Ions from a Ni(110) Surface: Dissociation and Charge Exchange," *Surf. Sci.* **272,** 189 (1992).

T. Fujino, T. Fuse, J. T. Ryu, K. Inudzuka, Y. Yamazaki, M. Katayama, and K. Oura, "Adsorption of Atomic Hydrogen on Ag-Covered 6H-SiC(0001) Surface," *Jpn. J. Appl. Phys.* **39,** 4340 (2000).

T. Fujino, T. Fuse, J. T. Ryu, K. Inudzuka, Y. Yamazaki, M. Katayama, and K. Oura, "Structural Analysis of the 6H-SiC(0001)$\sqrt{3} \times \sqrt{3}$ Reconstructed Surface," *Jpn. J. Appl. Phys.* **39,** 6410 (2000).

T. Fujino, T. Fuse, E. Tazou, T. Nakano, K. Inudzuka, K. Goto, Y. Yamazaki, M. Katayama, and K. Oura, "In-Situ Monitoring of Hydrogen-Surfactant Effect During Ge Growth on Si(001) Using Coaxial Impact-Collision Ion Scattering Spectroscopy and Time-of-Flight Elastic Detection Analysis," *Nucl. Instrum. Meth. Phys. Res. B* **161,** 419 (2000).

T. Fujino, M. Katayama, Y. Yamazaki, S. Inoue, J. T. Ryu, and K. Oura, "Ion Scattering and Recoiling Spectroscopy for Real Time Monitoring of Surface Processes in a Gas Phase Atmosphere," *Surf. Rev. Lett.* **7,** 657 (2000).

T. Fuse, K. Kawamoto, M. Katayama, and K. Oura, "In-Situ Observation of Ge δ-layer in Si(001) Using Quasi Medium Energy Ion Scattering Spectroscopy," *Mat. Sci. in Semiconductor Proc.* **2,** 159 (1999).

T. Fuse, K. Kawamoto, S. Kujime, T. Shiizaki, M. Katayama, and K. Oura, "Quasi-Medium Energy Ion Scattering Spectroscopy Observation of Surface Segregation of Ge δ-doped Layer During Si Molecular Beam Epitaxy," *Surf. Sci.* **393,** L93 (1997).

T. Fuse, K. Kawamoto, T. Shiizaki, M. Katayama, and K. Oura, "Quasi-Medium Energy Ion Scattering Spectroscopy Study of Ge δ-layer on Si(001)," *Appl. Surf. Sci.* **121,** 218 (1997).

T. Fuse, K. Kawamoto, T. Shiizaki, E. Tazou, M. Katayama, and K. Oura, "Quasi-Medium Energy Ion Scattering Spectroscopy Observation of a Ge δ-doped Layer Fabricated by Hydrogen Mediated Epitaxy," *Jpn. J. Appl. Phys.* **37,** 2625 (1998).

T. Fuse, J. T. Ryu, T. Fujino, K. Inudzuka, M. Katayama, and K. Oura, "Adsorption of H on the Ge/Si(001) Surface as Studied by Time-of-Flight Elastic Recoil Detection Analysis and Coaxial Impact Collision Ion Scattering Spectroscopy," *Jpn. J. Appl. Phys.* **38,** 1359 (1999).

J. W. Gadzuk, "Theory of Atom-Metal Interactions: I: Alkali Atom Adsorption," *Surf. Sci.* **6,** 133 (1967).

T. J. Gannon, G. Law, P. R. Watson, A. J. Carmichael, and K. R. Seddon, "First Observation of Molecular Composition and Orientation at the Surface of a Room-Temperature Ionic Liquid," *Langmuir* **15,** 8429 (1999).

T. J. Gannon, M. Tassotto, and P. R. Watson, "Orientation of Molecules at a Liquid Surface Revealed by Ion Scattering and Recoiling," *Chem. Phys. Lett.* **300,** 163 (1999).

T. J. Gannon, and P. R. Watson, "An Inexpensive and Simple Power Supply for Ion Beam Lens Elements," *J. Vac. Sci. Tech. A* **15,** 2820 (1997).

R. F. Garrett, R. J. MacDonald, and D. J. O'Connor, "A Determination of the Ionization Probability for Aluminum Secondary Ion Emission," *Surf. Sci.* **138,** 432 (1984).

R. F. Garrett, R. J. MacDonald, and D. J. O'Connor, "Ion Neutralization in Secondary Ion Mass Spectrometry," *Nucl. Instrum. Meth. Phys. Res. B* **218,** 333 (1983).

J. E. Gayone, R. G. Pregliasco, E. A. Sánchez, and O. Grizzi, "Topographic and Crystallographic Characterization of a Grazing-Ion-Bombarded GaAs (110) Surface by Time-of-Flight Ion-Scattering-Spectrometry," *Phys. Rev. B* **56,** 4186 (1997).

J. E. Gayone, R. G. Pregliasco, E. A. Sánchez, and O. Grizzi, "Atomic-Structure Characterization of a H: GaAs(110) Surface by Time-of-Flight Ion-Scattering-Spectrometry," *Phys. Rev. B* **56,** 4194 (1997).

J. E. Gayone, E. A. Sánchez, and O. Grizzi, "Adsorption Kinetics and Surface Unrelaxation in H: GaAs(110) Studied by Time-of-Flight Scattering and Recoiling Spectrometry," *Surf. Sci.* **419,** 188 (1999).

J. E. Gayone, E. A. Sánchez, O. Grizzi, M. Passeggi, R. Vidal, and J. Ferrón, "Adsorption of Potassium on GaAs(110)," *Surf. Sci.* **454,** 137 (2000).

J. E. Gayone, E. A. Sánchez, R. G. Pregliasco, and O. Grizzi, "Investigation of the Dependence of the GaAs(110) Surface Derelaxation with Hydrogen Exposure," *Surf. Sci.* **377,** 597 (1997).

K. A. German, C. B. Weare, P. R. Varekamp, J. N. Andersen, and J. A. Yarmoff, "Site-Specific Neutralization of Low Energy Li$^+$ Scattered from Na/Al(100)," *Phys. Rev. Lett.* **70,** 3510 (1993).

K. A. German, C. B. Weare, P. R. Varekamp, J. A. Yarmoff, and J. N. Andersen, "Trajectory-Dependent Neutralization of Low Energy Li$^+$ Scattered from Clean and Na-Covered Al(100)," *J. Vac. Sci. Tech A* **11,** 2260 (1993).

K. A. German, C. B. Weare, and J. A. Yarmoff, "Erratum: Inner-Shell Promotions in Low Energy Li$^+$-Al Collisions at Clean and Alkali-Covered Al(100) Surfaces," *Phys. Rev. B* **53,** 10407 (1996).

K. A. German, C. B. Weare, and J. A. Yarmoff, "Inner-Shell Electron Promotion in Low Energy Li$^+$-Al(100) Collisions," *Phys. Rev. Lett.* **72,** 3899 (1994).

K. A. German, C. B. Weare, and J. A. Yarmoff, "Inner-Shell Promotions in Low Energy Li$^+$-Al Collisions at Clean and Alkali-Covered Al(100) Surfaces," *Phys. Rev. B* **50,** 14452 (1994).

R. Ghrayeb, M. Purushotham, M. Hou, and E. Bauer, "Estimate of Repulsive Interatomic Pair Potentials by Low-Energy Alkali-Metal-Ion Scattering and Computer Simulation," *Phys. Rev. B* **36,** 7364 (1987).

G. Gilarowski, J. Mendez, and H. Niehus, "Initial Growth of Cu on Ir(100)-(5 × 1)," *Surf. Sci.* **448,** 290 (2000).

G. Gilarowski, and H. Niehus, "Iridium on Cu(100): Surface Segregation and Alloying," *Phys. Stat. Sol. A* **173,** 159 (1999).

D. J. Godfrey, and D. P. Woodruff, "Elastic and Neutralization Effects in Structural Studies of Oxygen and Carbon Adsorption on Ni(100) Surfaces Studied by Low Energy Ion Scattering," *Surf. Sci.* **105,** 438 (1981).

D. J. Godfrey, and D. P. Woodruff, "Low Energy Ion Scattering Study of Oxygen Adsorption on Ni(100)," *Surf. Sci.* **89,** 76 (1979).

G. Gómez, J. Gayone, O. Grizzi, E. A. Sánchez, R. Pregliasco, M. Martiarena, E. García, and E. Goldberg, "Ion Fractions in 6 keV Ne, Ar and Na Scattering from GaAs(110) Surfaces," *Nucl. Instrum. Meth. Phys. Res. B* **125,** 268 (1997).

O. Grizzi, J. Gayone, G. Gómez, R. Pregliasco, and E. Sánchez, "Investigation of Hydrogen Covered Crystalline Surfaces by Low Energy Ion Scattering and Recoiling Spectrometry," *J. Nucl. Mat.* **248,** 428 (1997).

O. Grizzi, E. A. Sánchez, J. E. Gayone, L. Guillemot, V. A. Esaulov, and R. A. Baragiola, "Formation of Autoionizing Ne^{++} in Grazing Collisions with an Al(111) Surface," *Surf. Sci.* **469,** 71 (2000).

O. Grizzi, M. Shi, H. Bu, and J. W. Rabalais, "Time-of-Flight Scattering and Recoiling Spectrometry for Surface Analysis," *Rev. Sci. Instrum.* **61,** 740 (1990).

O. Grizzi, M. Shi, H. Bu, and J. W. Rabalais, "Time-of-Flight Scattering and Recoiling: I. The Structure of Oxygen on the W(211) Surface," *Phys. Rev. B* **40,** 10127 (1989).

O. Grizzi, M. Shi, H. Bu, and J. W. Rabalais, "Time-of-Flight Scattering and Recoiling: II. The Structure of Oxygen on the W(211) Surface," *Phys. Rev. B* **40,** 10147 (1989).

O. Grizzi, M. Shi, H. Bu, J. W. Rabalais, and R. A. Baragiola, "Ne** Autoionizing States and Ne$^+$ Charge Fractions Scattered from a Magnesium Surface," *Phys. Rev. B* **41,** 4789 (1990).

O. Grizzi, M. Shi, H. Bu, J. W. Rabalais, R. R. Rye, and P. Nordlander, "Determination of the Structure of Hydrogen on a W(211) Surface," *Phys. Rev. Lett.* **63,** 1408 (1989).

L. Guillemot, S. Lacombe, E. A. Sánchez, M. Maazouz, and V. E. Esaulov, "Inelastic Collisions of Ne Atoms and Ions with Si Surface and Effects of Initial Stages of Oxidation," *Surf. Sci.* **365,** 92 (1996).

L. Guillemot, S. Lacombe, V. Tuan, M. Maazouz, V. Esaulov, E. Sánchez, Y. Bandurin, A. Dashchenko, and V. Drobnich, "Dynamics of Excited State Production in the Scattering of Inert Gas Atoms and Ions of Mg and Al Surfaces," *Surf. Sci.* **365,** 353 (1996).

P. Häberle, and T. Gustafsson, "Medium-Energy Ion-Scattering Analysis of the C(2 × 2) Structure Induced by K on Au(110)," *Phys. Rev. B* **40,** 8218 (1989).

L. Hägg, A. Bárány, H. Cederquist, and U. Thumm, "Angular Differential Cross Sections in Slow Ion-C60 Interactions," *Phys. Scri. T* **80,** 205 (1999).

H. D. Hagstrum, "Studies of Adsorbate Electronic Structure Using Ion Neutralization and Photoemission Spectroscopies," *Electron and Ion Spectroscopy of Solids,* L. Fiermans, J. Vennik, and W. Dekeyser (Eds.), Plenum, New York, 1978.

H. D. Hagstrum, "The Determination of Energy-Level Shifts Which Accompany Chemisorption," *Surf. Sci.* **54,** 197 (1976).

H. D. Hagstrum, "Ion-Neutralization Spectroscopy," *Metals: Vol. 6,* Wiley & Sons, 1972.

H. D. Hagstrum, "Ion-Neutralization Spectroscopy," *J. Res. Natl. Bureau Standards A* **74,** 433 (1970).

H. D. Hagstrum, "Ion-Neutralization Spectroscopy of Solids and Solid Surfaces," *Phys. Rev.* **150,** 495 (1966).

H. D. Hagstrum, "Auger Ejection of Electrons from Metals by Ions," *Phys. Rev.* **104,** 672 (1956).

H. D. Hagstrum, "Auger Ejection of Electrons from Tungsten by Noble Gas Ions," *Phys. Rev.* **96,** 325 (1954).

H. D. Hagstrum, "Theory of Auger Ejection of Electrons from Metals by Ions," *Phys. Rev.* **96,** 336 (1954).

S. Han, S. C. Park, J. G. Lee, and H. Kang, "Ionic-to-Metallic Layer Transition in Cs Adsorption on Si(111)-(7 × 7): Charge-State Selective Detection of Adsorbate by Cs$^+$ Reactive Ion Scattering," *J. Chem. Phys.* **112,** 8660 (2000).

J. W. Hartman, M. H. Shapiro, T. A. Tombrello, and J. A. Yarmoff, "Dependence of Inner-Shell Vacancy Production upon Distance in Hard Li-Al Collisions," *Phys. Rev. B* **55,** 4811 (1997).

T. Hashizume, M. Katayama, D. Jeon, M. Aono, and T. Sakurai, "The Absolute Coverage of K on the Si(111)-3 × 1-K Surface," *Jpn. J. Appl. Phys.* **3,** L1263 (1993).

N. Hatke, M. Dirska, E. Luderer, A. Robin, M. Grether, A. Närmann, and W. Heiland, "Energy Loss and Resonant Coherent Excitation of Fast Highly Charged Ions on a Pt(110)-Surface," *Nucl. Instrum. Meth. Phys. Res. B* **135,** 307 (1998).

N. Hatke, S. Hustedt, J. Limburg, I. G. Hughes, R. Hoekstra, W. Heiland, and R. Morgenstern, "Energy Loss of Highly Charged Ions on an Al Surface," *Nucl. Instrum. Meth. Phys. Res. B* **115,** 165 (1996).

N. Hatke, A. Robin, M. Grether, A. Närmann, and W. Heiland, "Scattering of Multiply Charged Ions from Surfaces," *Int. J. Mass Spectrometry* **192,** 393 (1999).

W. Hayami, T. Aizawa, E. Asari, and R. Souda, "Three-Dimensional N-Atom Model for Computer Simulation of Impact-Collision ion Scattering Spectroscopy," *Surf. Sci.* **446,** 267 (2000).

W. Hayami, R. Souda, T. Aizawa, S. Otani, and Y. Ishizawa, "Structure Analysis of NbC(111)-O and NbC(111)-D Surfaces," *Surf. Sci.* **346,** 158 (1996).

W. Hayami, R. Souda, T. Aizawa, and T. Tanaka, "Structure Analysis of the HfB$_2$(0001) Surface by Impact-Collision Ion Scattering Spectroscopy," *Surf. Sci.* **415,** 433 (1998).

R. L. Headrick, P. Konarski, S. M. Yalisove, and W. R. Graham, "Medium-Energy Ion Scattering Study of the Initial Stage of Oxidation of Fe(001)," *Phys. Rev. B* **39,** 5713 (1989).

W. Heiland, "The Interaction of Molecular Ions with Surfaces," *Low Energy Ion-Surface Interactions,* Wiley & Sons, New York, 1994.

W. Heiland, "Charge Exchange Processes and Surface Chemistry," *Surf. Sci.* **251,** 942 (1991).

W. Heiland, "Inelastic-Surface Collisions," *Trends in Physics EPS-8 III,* 777(1991).

W. Heiland, "Interaction of Low-Energy Ions, Atoms and Molecules with Surfaces," *Inter of Charged Particles with Solids and Surfaces,* Plenum Press, New York, 1991.

W. Heiland, "Ion Scattering as a Tool for a Better Understanding of Chemical Reactions on Surfaces," in *Fundamental Aspects of Heterogeneous Catalysis Studies by Particle Beams,* H. H. Brongersma and R. A. van Santen (Eds.), Plenum Press, New York, 1991.

W. Heiland, "Low Energy Ion Beam Scattering for Surface Analysis," *Vacuum* **39,** 167 (1989).

W. Heiland, "Principles of Low Energy Ion Scattering," *Vacuum* **32,** 539 (1982).

W. Heiland, M. Aschoff, and S. Speller, "Low Energy Ion Scattering Analysis of Metal Alloy Surfaces," *Proc. of the Int. Symp. on Atomic Level Characterization, Microbeam Anal.* **141,** 251 (1998).

W. Heiland, U. Beitat, and E. Taglauer, "Scattering of Molecular and Atomic Hydrogen Ions from Single-Crystal Surfaces," *Phys. Rev. B* **19,** 1677 (1979).

W. Heiland, K. Brüning, and M. Aschoff, "Surface Analysis Using Low Energy Ion Beams," *Proc. of the 8th Int. Workshop on Ion Beam Surface Diagnostics,* Uzhgorod, Ukraine, 1998.

Heiland, W., H. Derks, and T. Bremer, "Slowing Down and Scattering of Ions in Solids," *Scan. Micro. Suppl.* **4,** 257 (1990).

W. Heiland, C. Höfner, N. Hatke, S. Hausmann, A. Närmann, H. Limburg, and R. Morgenstern, "Surface Channeling and Inelastic Effects," *Proc. Workshop on Ions Scattering Spectroscopy for Application on Surface Sci.,* Osaka, 1995.

W. Heiland, F. Iberl, and E. Taglauer, "Oxygen Adsorption on (110) Silver," *Surf. Sci.* **53,** 383 (1975).

W. Heiland, and A. Närmann, "Excitation and Loss Processes of Ions at Surfaces," *Nucl. Instrum. Meth. Phys. Res. B* **78,** 20 (1993).

W. Heiland, T. Schlathölter, and M. Vicanek, "Scattering of Small Molecules at Surfaces," *Phys. Status Sol.* **192,** 301 (1995).

W. Heiland, and E. Taglauer, "The Dissociation of Nitrogen Molecular Ions at Clean and Adsorbate Covered Surfaces," *Nucl. Inst. Meth. B* **194,** 667 (1982).

W. Heiland, and E. Taglauer, "The Backscattering of Low Energy Ions and Surface Structure," *Surf. Sci.* **68,** 96 (1977).

W. Herman, "Surface Structure Determination of $CeO_2(001)$ by Angle-Resolved Mass Spectrometry," *Phys. Rev. B* **59,** 14899 (1999).

W. Hetterich, H. Derks, and W. Heiland, "Determination of Pairwise Interaction Potential Parameters from a Double Scattering Experiment," *Appl. Phys. Lett.* **52,** 371 (1988).

W. Hetterich, H. Derks, and W. Heiland, "K^+ Ions Scattering from Ir(110): A Study of the Pairwise Interaction Potential," *Nucl. Instrum. Meth. Phys. Res. B* **33,** 401 (1988).

W. Hetterich, and W. Heiland, "The $(1 \times 3) \leftrightarrow (1 \times 1)$ Structure Transition of Ir(110)," *Surf. Sci.* **258,** 307 (1991).

W. Hetterich, and W. Heiland, "Low Energy Ion Scattering from the Ir(110) Surface: A Quest for the (1×2) Structure," *Surf. Sci.* **210,** 129 (1989).

W. Hetterich, W. Heiland, and H. Niehus, "Surface Crystallographic Data of the Ir(110) (1×3) Surface," *Surf. Sci.* **264,** L177 (1992).

W. Hetterich, C. Höfner, and W. Heiland, "An Ion Scattering Study of the Surface Structure and Thermal Vibrations on Ir(110)," *Surf. Sci.* **252,** 731 (1991).

W. Hetterich, C. Höfner, H. Niehus, and W. Heiland, "Ion Scattering Studies of the (1×1) and (1×3) Structure of the Ir(110) Surface," *Proc. of Symposium on Surface Sci. 3S91,* Austria, 1991.

W. Hetterich, U. Korte, G. Meyer-Ehmsen, and W. Heiland, "A Contribution to the $(1 \times n)$ Reconstructions of the Ir(110) Surface," *Surf. Sci.* **254,** L487 (1991).

R. Hoekstra, J. Manske, M. Dirska, G. Lubinski, M. Schleberger, and A. Närmann, "Circular Polarized Photon Emission Spectroscopy of keV Ion Scattering Off (un-)Magnetized Surfaces," *Nucl. Instrum. Meth. Phys. Res. B* **125,** 53 (1997).

C. Höfner, V. Bykov, and J. W. Rabalais, "Three-Dimensional Focusing Patterns of He^+ Ions Scattering From a Au{110} Surface," *Surf. Sci.* **393,** 184 (1997).

C. Höfner, W. Hetterich, W. Heiland, and H. Niehus, "Ion Scattering Studies of the Structure of the Ir(110) Surface," *Nucl. Instrum. Meth. Phys. Res. B* **76,** 328 (1992).

C. Höfner, and A. Närmann, "Stopping of Low Energy Ions at Metal Surfaces," *Appl. of Accelerators in Res. and Ind.,* Academic Press, New York, 1998.

C. Höfner, A. Närmann, A. Arnau, and W. Heiland, "Energy Loss of Fast Particles on Surfaces," *Symp. Surf. Sci.,* Kaprun, Austria, 1995.

C. Höfner, A. Närmann, and W. Heiland, "Energy Loss Spectra of H and He Scattered Off Clean and K-Covered Pd Surfaces," *Nucl. Instrum. Meth. Phys. Res. B* **93,** 113 (1994).

C. Höfner, A. Närmann, and W. Heiland, "Trajectory Calculations for H Scattering at Grazing Angles from Ni(110) and Ni(111)," *Nucl. Instrum. Meth. Phys. Res. B* **72,** 227 (1992).

C. Höfner, and J. W. Rabalais, "Deconstruction of the Au{110}-(1×2) Surface," *Phys. Rev. B* **58,** 9990 (1998).

C. Höfner, and J. W. Rabalais, "Surface and Subsurface Distortions of the Au{110}-(1 × 2) Structure," *Surf. Sci.* **400,** 189 (1998).

L. Houssiau, and P. Bertrand, "MARLOWE Simulations of He and Ne Ion Scattering on Cu₃Au(100) and Comparison with TOF-ISS Experiments," *Nucl. Instrum. Meth. Phys. Res. B* **125,** 328 (1997).

L. Houssiau, and P. Bertrand, "Direct Observation of the Rippling and the Order-Disorder Transition at the Cu₃Au (100) Surface by ToF-Ion Scattering," *Surf. Sci.* **352,** 978 (1996).

L. Houssiau, and P. Bertrand, "Observation of Surface Rippling on Cu3Au (100) Studied by ToF-ISS," *Nucl. Instrum. Meth. Phys. Res. B* **115,** 161 (1996).

L. Houssiau, and P. Bertrand, "Order-Disorder Phase Transition of the Cu₃Au (100) Surface Studied by ToF-Ion Scattering," *Nucl. Instrum. Meth. Phys. Res. B* **118,** 467 (1996).

L. Houssiau, and P. Bertrand, "Surface Structure Analysis by Time-of-Flight ISS: Influence of Primary Ion Nature and Energy for Scattering on Cu(110)," *Vacuum* **45,** 409 (1994).

L. Houssiau, and P. Bertrand, "2 keV-He and Ne Ion Scattering on a Cu(110) Surface: Experiment vs. Simulation," *Nucl. Instrum. Meth. Phys. Res. B* **90,** 247 (1994).

L. Houssiau, M. Graupe, R. Colorado, Jr., H. I. Kim, T. R. Lee, S. S. Perry, and J. W. Rabalais, "Characterization of the Surface Structure of CH₃ and CF₃ Terminated n-Alkanethiol Monolayers Self Assembled on Au{111}," *J. Chem. Phys.* **109,** 9134 (1998).

L. Houssiau, and J. W. Rabalais, "Scattering and Recoiling Imaging Spectrometry (SARIS) Study of Chlorine Chemisorption on Ni(110)," *Nucl. Instrum. Meth. Phys. Res. B* **157,** 274 (1999).

L. Houssiau, J. W. Rabalais, J. Wolfgang, and P. Nordlander, "Surface Structure and Electron Density Dependence of Scattered Ne⁺ Ion Fractions From Cd- and S-Terminated CdS(0001) Surfaces," *Phys. Rev. Lett.* **81,** 5153 (1998).

L. Houssiau, J. Wolfgang, P. Nordlander, and J. W. Rabalais, "Trajectory Dependence of Scattered Ne⁺ and Recoiled S⁺ Ion Fractions from the Cd- and S-Terminated CdS{0001} Surfaces," *J. Chem. Phys.* **110,** 8139 (1999).

L. Houssiau, J. Wolfgang, P. Nordlander, and J. W. Rabalais, "Surface Structure and Electron Density Dependence of Scattered Ne⁺ Ion Fractions from Cd- and S-Terminated CdS{0001} Surfaces," *Phy. Rev. Lett.* **81,** 5153 (1998).

C. C. Hsu, A. Bousetta, J. W. Rabalais, and P. Norlander, "Crystallographic Dependence of Recoiled O⁻ Ion Fractions from Ni{100}c(2 × 2)-O and NiO{100} Surfaces," *Phys. Rev. B* **47,** 2369 (1993).

C. C. Hsu, H. Bu, A. Bousetta, P. Norlander, and J. W. Rabalais, "Angular Dependence of Charge Transfer Probabilities Between O⁻ Ion Fractions from Ni{100}-c(2 × 2)-O Surface," *Phys. Rev. Lett.* **69,** 188 (1992).

C. C. Hsu, and J. W. Rabalais, "Structure Sensitivity of Scattered Ne+ Ion Fractions from a Ni{100} Surface," *Surf. Sci.* **256,** 77 (1991).

J. H. Huang, and R. S. Williams, "Surface-Structure Analysis of Au Overlayers on Si by Impact-Collision Ion-Scattering Spectroscopy: $\sqrt{3} \times \sqrt{3}$ and 6 × 6 Si(111)/Au," *Phys. Rev. B* **38,** 4022 (1988).

J. H. Huang, and R. S. Williams, "Surface Structure of Si(111)-5 × 1-Au Characterized by Impact Collision Ion Scattering Spectroscopy," *Surf. Sci.* **204,** 445 (1988).

J. H. Huang, and R. S. Williams, "The Atomic Structures of Si(111)-$\sqrt{3} \times \sqrt{3}$-and-6 × 6-Au Surfaces Studied by Low Energy Ion Scattering," *J. Vac. Sci. Tech. A* **6,** 689 (1988).

G. Hughes, J. Limburg, R. Hoekstra, R. Morgenstern, S. Hustedt, N. Hatke, and W. Heiland, "Negative Ion Production in Multicharged Ion-Surface Interactions," *Nucl. Instrum. Meth. Phys. Res. B* **98,** 458 (1995).

Th. M. Hupkens, "Low Energy Ion Scattering Study of Oxygen Adsorption on a Cu(100) Single Crystal Surface: I. Surface Characteristics," *Nucl. Instrum. Meth. Phys. Res. B* **9,** 277 (1985).

S. Hustedt, J. Freese, S. Mähl, W. Heiland, S. Schippers, J. Bleck-Neuhaus, M. Grether, R. Köhrbrück, and N. Stolterfoht, "Target Effects in the Interaction of Highly Charged Ions with a Al(110) Surface," *Phys. Rev. A* **50,** 4993 (1994).

S. Hustedt, J. Freese, S. Schippers, M. Grether, R. Köhrbrück, J. Bleck-Neuhaus, W. Heiland, and N. Stolterfoht, "Interaction of Highly Charged Ne Ions with Pt and Al Surfaces," *Proc. Symp. Surf. Sci.,* Les Arc 172 (1994).

S. Hustedt, N. Hatke, W. Heiland, J. Limburg, I. G. Hughes, R. Hoekstra, R. Morgenstern, "Scattering Potential Investigation in Highly Charged Ion-Surface Interaction," *Nucl. Instrum. Meth. Phys. Res. B* **98,** 454 (1995).

Y. Hwang, R. Souda, T. Aizawa, W. Hayami, S. Otani, and Y. Ishizawa, "Structure of Epitaxial MgO Layers on TiC(001) Studied by Time-of-Flight Impact-Collision Ion Scattering Spectroscopy," *Surf. Sci.* **380,** 45 (1997).

T. M. Ichinokawa, Y. Yokoyama, and K. Fukunaga, "Structure Analysis of Si(100)2 × n Surfaces by Ion Channeling and Blocking Spectroscopy," *Nucl. Instrum. Meth. Phys. Res. B* **33,** 611 (1988).

A. Ikeda, K. Sumitoimo, T. Nishioka, T. Yasue, T. Koshikawa, and Y. Kido, "Intermixing at Ge/Si (001) Interfaces Studied by Surface Energy Loss of Medium Energy Ion Scattering," *Surf. Sci.* **385,** 200 (1997).

U. Imke, K. J. Snowdon, and W. Heiland, "Theory of Charge Exchange in the Scattering of Molecular Ions from Simple Metals," *Phys. Rev. B* **34,** 41 (1986).

A. Ishii, "Theory of Auger Neutralization of He$^+$ Ions on NaCl Surface Due to the Local Electronic Structures," *Surf. Sci.* **192,** 172 (1987).

D. P. Jackson, "Approximate Calculation of Surface Debye Temperatures," *Surf. Sci.* **43,** 431 (1974).

D. P. Jackson, T. E. Jackmann, J. A. Davies, W. N. Unertl, and P. Norton, "Vibrational Properties of Au and Pt(110) Surfaces Deduced from Rutherford Backscattering Data," *Surf. Sci.* **126,** 226 (1983).

S. Jans, P. Wurz, R. Schletti, K. Brüning, K. Sekar, W. Heiland, J. Quinn, and R. E. Leuchtner, "Scattering of Atoms and Molecules Off a Barium Zirconate Surface," *Nucl. Instrum. Meth. B* **173,** 503 (2001).

Q. T. Jiang, P. Fenter, and T. Gustafsson, "Geometric Structure p(2 × 2)-S/Cu(001) Determined by Medium-Energy Ion Scattering," *Phys. Rev. B* **42** 9291 (1990).

Y. S. Jo, J. A. Schultz, T. R. Schuler, and J. W. Rabalais, "Scattering of CO$^+$ From Magnesium Surfaces: Molecular Ion Survival and Scattered Positive and Negative Ion Fractions," *J. Phys. Chem.* **89,** 2113 (1985).

K. Josek, Ch. Linsmeier, H. Knözinger, and E. Taglauer, "Ion Scattering Analysis of Alumina Supported Model Catalysts," *Nucl. Instrum. Meth. Phys. Res. B* **64,** 596 (1992).

J. I. Juaristi, and A. Arnau, "Interaction of Multiply Charged Ions with Metals," *Nucl. Instrum. Meth. Phys. Res. B* **115,** 173 (1996).

J. I. Juaristi, A. Arnau, P. M. Echenique, C. Auth, and H. Winter, "Charge State Dependence of the Energy Loss of Slow Ions in Metals," *Phys. Rev. Lett.* **82,** 1048 (1999).

J. I. Juaristi, A. Arnau, P. M. Echenique, C. Auth, and H. Winter, "Charge State Dependence of the Energy Loss of Slow Nitrogen Ions Reflected from an Aluminum Surface in Grazing Incidence," *Nucl. Instrum. Meth. Phys. Res. B* **157,** 87 (1999).

J. I. Juaristi, C. Auth, H. Winter, A. Arnau, K. Eder, D. Semrad, F. Aumayr, P. Bauer, and P. M. Echenique, "Unexpected Behavior of the Stopping of Slow Ions in Ionic Crystals," *Phys. Rev. Lett.* **84,** 2124 (2000).

I. Kamiya, M. Katayama, E. Nomura, and M. Aono, "Separation of Scattered Ions and Neutrals in CAICISS with an Acceleration Tube," *Surf. Sci.* **242,** 404 (1991).

H. Kang, K. D. Kim, and K. Y. Kim, "Molecular Identification of Surface Adsorbates. Reactive Scattering of Hyperthermal Cs^+ from a Ni(100) Surface Adsorbed with CO, C_6H_6, and H_2O," *J. Am. Chem. Soc.* **119,** 12002 (1997).

H. Kang, R. Shimizu, and T. Okutani, "ISS Measurement of Surface Composition of Au-Cu Alloys by Simultaneous Ion Bombardments with Ar^+ and He^+ Ions," *Surf. Sci.* **116,** L173 (1982).

H. Kang, M. C. Yang, K. D. Kim, and K. Y. Kim, "Reactive Scattering of Cs^+ from Chemisorbed Molecules on a Ni(100) Surface: Secondary Neutral Mass Spectrometry with Hyperthermal Ion Beam," *Int. J. Mass Spectrom.* **174,** 143 (1998).

D. S. Karpuzov, D. G. Armour, and I. N. Evdoimov, "Simulation of Ion Reflection from a Single Crystal at Grazing Angles," *Rad. Eff. Def. Sol.* **41,** 141 (1979).

M. Katayama, M. Aono, H. Oigawa, Y. Nannichi, H. Sugahara, and M. Oshima, "Surface Structure of InAs(001) Treated with $(NH_4)_2S_x$ Solution," *Jpn. J. Appl. Phys.* **30,** L786 (1991).

M. Katayama, B. V. King, E. Nomura, and M. Aono, "Structure Analysis of the CaF_2/Si(111) Interface in its Initial Stage of Formation by Coaxial Impact-Collision Ion Scattering Spectroscopy (CAICISS)," *Prog. Theoretical Phys.* **106,** 315 (1991).

M. Katayama, T. Nakayama, C. F. McConville, and M. Aono, "Surface and Interface Structural Control Using Coaxial Impact-Collision Ion Scattering Spectroscopy (CAICISS)," *Nucl. Instrum. Meth. Phys. Res. B* **99,** 598 (1995).

M. Katayama, E. Nomura, N. Kanekama, H. Soejima, and M. Aono, "Coaxial Impact-Collision Ion Scattering Spectroscopy (CAICISS): A Novel Method for Surface Structure Analysis," *Nucl. Instrum. Meth. Phys. Res. B* **33,** 857 (1988).

M. Katayama, E. Nomura, H. Soejima, S. Hayashi, and M. Aono, "Real-Time Monitoring of Molecular-Beam Epitaxy Processes with Coaxial Impact-Collision Ion Scattering Spectroscopy (CAICISS)," *Nucl. Instrum. Meth. Phys. Res. B* **45,** 408 (1990).

M. Katayama, R. S. Williams, M. Kato, E. Nomura, and M. Aono, "Structure Analysis of the Si(111)$\sqrt{3} \times \sqrt{3}$ R30°-Ag Surface," *Phys. Rev. Lett.* **66,** 2762 (1991).

K. Kato, T. Ide, S. Miura, A. Tamura, and T. Ichinokawa, "Si(100)2 × n Structures Induced by Ni Contamination," *Surf. Sci.* **194,** L87 (1988).

K. Kato, T. Ide, T. Nishimori, and T. Ichinokawa, "Formation and Atomic Configuration of Si(100)c(4 × 4) Structure," *Surf. Sci.* **207,** 177 (1988).

M. Kato, H. Ikegami, N. Inoue, T. Hasegawa, M. Katayama, and M. Aono, "A New Experimental Geometry of Elastic Recoil Detection Analysis (ERDA)," *Jpn. J. Appl. Phys.* **32,** 162 (1993).

M. Kato, M. Katayama, T. Chasse, and M. Ano, "Channeling and Backscattering of Low Energy Ions," *Nucl. Instrum. Meth. Phys. Res. B* **39**, 30 (1989).

M. Kato, V. Kempter, and R. Souda, "The Spectrum Line Width Observed in Electron Emission from the Autodetachment of the $He^{-*}(1s2s2;2S)$ Formed in Collisions of the He^+ Projectiles with Low Work Function Metallic Surfaces," *Nucl. Instrum. Meth. Phys. Res. B* **125**, 59 (1997).

M. Kato, D. J. O'Connor, K. Yamamoto, and R. Souda, "Neutralization of a Proton at Adsorbate-Covered Metal Surfaces," *Surf. Sci.* **363** 150 (1996).

M. Kato, R. S. Williams, and M. Aono, "Interatomic and Image Potentials in Low Energy Ion Scattering at Metal Surfaces," *Nucl. Instrum. Meth. Phys. Res. B* **33**, 462 (1988).

K. Kawamoto, K. Inari, T. Mori, and K. Oura, "A New Apparatus For Impact Collision Ion Scattering Spectroscopy," *Jpn. J. Appl. Phys.* **34**, 4917 (1995).

K. Kawamoto, T. Mori, S. Kujime, and K. Oura, "Observation of the Diffusion of Ag Atoms though an a-Si Layer on Si (111) by Low-Energy Ion Scattering," *Surf. Sci.* **363**, 156 (1996).

K. Kawamoto, and K. Oura, "Scattering Process of Low-Energy Ions from Binary Compound Surfaces at $180°$," *Jpn. J. Appl. Phys.* **34**, 4929 (1995).

H. Kawanowa, Y. Gotoh, S. Otani, and R. Souda, "Structure Analysis of the $WB_2(0001)$ Surface," *Surf. Sci.* **433**, 661 (1999).

H. Kawanowa, R. Souda, S. Otani, and Y. Gotoh, "Structure Analysis of a Graphitic Boron Layer at the $TaB_2(0001)$ Surface," *Phys. Rev. Lett.* **81**, 2264 (1998).

H. Khemliche, J. Limburg, R. Hoekstra, R. Morgenstern, N. Hatke, E. Luderer, and W. Heiland, "Energy Loss and Charge State Distribution of $N^{6.7+}$ Ion Al(110) Surface Collisions," *Nucl. Instrum. Meth. Phys. Res. B* **125**, 116 (1997).

Y. Kido, and T. Koshikawa, "Energy Straggling for Medium-Energy H^+ Beams Penetrating Cu, Ag, and Pt," *Phys. Rev. A* **44**, 1759 (1991).

Y. Kido, and T. Koshikawa, "Ion Scattering Analysis Programs for Studying Surface and Interface Structures," *J. Appl. Phys.* **67**, 187 (1990).

Y. Kido, T. Nishimura, Y. Furukawa, Y. Nakayama, T. Yasue, T. Koshikawa, P. C. Goppelt-Langer, S. Yamamoto, Z.Q. Ma, H. Naramoto, and T. Ueda, "Solid-Phase Epitaxial Growth of Ge on H^- Terminated and Oxidized Si(100) Surfaces," *Surf. Sci.* **327**, 225 (1995).

C. Kim, J. Ahn, V. Bykov, and J. W. Rabalais, "Element-, Velocity-, and Spatially-Resolved Images of Kr^+ Scattering and Recoiling from a CdS Surface," *Intern. J. Mass Spectrom. Ion Phys.* **174**, 305 (1998).

C. Kim, A. Al-Bayati, and J. W. Rabalais, "Time-Resolving, Position-Sensitive Detection System for Scattering and Recoiling Imaging Spectrometry," *Rev. Sci. Instrum.* **69**, 1289 (1998).

C. Kim, C. Höfner, A. Al-Bayati, and J. W. Rabalais, "Scattering and Recoiling Imaging Spectrometer (SARIS)," *Rev. Sci. Instrum.* **69**, 1676 (1998).

C. Kim, C. Höfner, V. Bykov, and J. W. Rabalais, "Element-, Time-, and Spatially-Resolved Images of Scattered and Recoiled Atoms," *Nucl. Instrum. Meth. Phys. Res. B* **125**, 315 (1997).

C. Kim, C. Höfner, and J. W. Rabalais, "Surface Structure Determination from Ion Scattering Images," *Surf. Sci.* **388**, L1085 (1997).

C. Kim, and J. W. Rabalais, "Focusing of He^+ Ions on Semi-Channel Planes in the Pt{111} Surface," *Surf. Sci.* **395**, 239 (1998).

C. Kim, and J. W. Rabalais, "Projections of Atoms in Terms of Interatomic Vectors," *Surf. Sci.* **385,** L938–L944 (1997).

S. Kim, Y. Kim, H. I. Kim, S. H. Lee, T. R. Lee, S. S. Perry, and J. W. Rabalais, "Chemisorption Site of Methanethiol on Pt{111}," *J. Chem. Phys.* **109,** 9574 (1998).

Y. Kim, S. S. Kim, E. Ada, Y. L. Yang, A. J. Jacobson, and J. W. Rabalais, "Scattered and Recoiled Ion Fractions from $LiTaO_3(100)$ Surfaces with Different Electrical Properties," *J. Chem. Phys.* **111,** 2720 (1999).

G. A. Kimmel, D. M. Goodstein, Z. H. Levine, and B. H. Cooper, "Local Adsorbate-Induced Effects on Dynamic Charge Transfer in Ion-Surface Interactions," *Phys. Rev. B* **43,** 9403 (1991).

B. King, M. Katayama, M. Aono, R. S. Daley, and R. S. Williams, "Analysis of CaF_2-Si(111) Using Coaxial Impact-Collision Ion Scattering Spectroscopy," *Vacuum* **41,** 938 (1990).

T. Kinoshita, Y. Tanaka, K. Sumitomo, F. Shoji, K. Oura, and I. Katayama, "Hydrogen-Induced Reconstruction of Si(111)-$\sqrt{3}$-Ag Surface Studied by TOF-ICISS," *Appl. Surf. Sci.* **60,** 183 (1992).

T. Kobayashi, G. Dorenbos, S. Shimoda, M. Iwaki, and M. Aono, "Neutralization Probability in Medium-Energy Ion Scattering," *Nucl. Instrum. Meth. Phys. Res. B* **118,** 584 (1996).

T. Kobayashi, C. F. McConville, G. Dorenbos, M. Iwaki, and M. Aono, "Depth Profile and Lattice Location Analysis of Sb Atoms in Si/Sb (d-doped)/Si(001) Structures Using Medium-Energy Ion Scattering Spectroscopy," *Appl. Phys. Lett.* **74,** 673 (1999).

T. Kobayashi, C. F. McConville, J. Nakamura, G. Dorenbos, H. Sone, T. Katayama, and M. Aono, "Study of Diffusion and Defects by Medium-Energy Coaxial Impact-Collision Ion Scattering Spectroscopy," *Defect and Diffusion Forum* **183,** 207 (2000).

B. J. Koeleman, S. T. de Zwart, A. L. Boers, B. Poelsema, and L. K. Verheij, "Information on Adsorbate Positions from Low-Energy Recoil Scattering: Adsorption of Hydrogen on Pt," *Phys. Rev. Lett.* **56,** 1152 (1986).

B. J. Koeleman, S. T. de Zwart, A. L. Boers, B. Poelsema, and L. K. Verheij, "Adsorption Study of Hydrogen on a Stepped Pt(997) Surface Using Low Energy Recoil Scattering," *Nucl. Instrum. Meth. Phys. Res. B* **218,** 225 (1983).

R. Köhrbrück, S. Hustedt, S. Schippers, W. Heiland, J. Bleck-Neuhaus, J. Kemmler, D. Lecler, and N. Stolterfoht, "Interaction of Ne (q = 8, 9, 10) with a Pt(110) Surface," *Nucl. Instrum. Meth. Phys. Res. B* **78,** 93 (1993).

R. Köhrbrück, K. Sommer, J. P. Biersack, J. Bleck-Neuhaus, S. Schippers, P. Roncin, D. Lecler, F. Fremont, and N. Stolterfoht, "Auger-Electron Emission from Slow, Highly Charged Ions Interacting with Solid Cu Targets," *Phys. Rev. A* **45,** 4653 (1992).

T. Koshikawa, T. Yasue, H. Tanaka, I. Sumita, and Y. Kido, "Surface Structure of Cu/Si (111) at High Temperature," *Surf. Sci.* **331,** 506 (1995).

T. Koshikawa, T. Yasue, H. Tanaka, I. Sumita, and Y. Kido, "Medium Energy Ion Scattering and STM Studies on Cu/Si(111)," *Nucl. Instrum. Meth. Phys. Res. B* **99,** 495 (1995).

E. W. Kuipers, and A. L. Boers, "The Quantum Efficiency of a Channeltron and a Magnetic Electron Multiplier for 0.5–5 keV H Atoms and Ions," *Nucl. Instrum. Meth. Phys. Res. B* **29,** 567 (1987).

R. Kumar, J. N. Chen, and J. W. Rabalais, "Ion Fractions and Collision Dynamics of 1–3 keV Ne^+, Ar^+, and Kr^+ Scattering From an Yttrium Surface," *Langmuir* **1,** 294 (1985).

R. Kumar, M. H. Mintz, and J. W. Rabalais, "Ion Survival Probabilities for 3 keV Ar^+ Scattering

From La, Yb, and Chemisorbed H_2, O_2, and H_2O on La Surfaces," *Surf. Sci.* **147,** 15 (1984).

J. Kuntze, J. Bömermann, T. Rauch, S. Speller, and W. Heiland, "Surface Reconstruction of the Ir(110)-Surface," *Surf. Sci.* **394,** 150 (1997).

J. Kuntze, S. Speller, W. Heiland, A. Atrei, G. Rovida, and U. Bardi, "Surface Structure and Composition of the Alloy $Au_3Pd(100)$ Determined by LEED and ISS," *Phys. Rev. B* **60,** 1535 (1999).

T. Kuroi, K. Umezawa, J. Yamane, F. Shoji, K. Oura, and T. Hanawa, "Ion Beam Analysis of the Concentration and Thermal Release of Hydrogen in Silicon Nitride Films Prepared by ECR Plasma CVD Method," *Jap. J. of Appl. Phys.* **27,** 1406 (1988).

P. Kürpick, and U. Thumm, "Hybridization of Ionic Levels at Metal Surfaces," *Phys. Rev. A* **58,** 2174 (1998).

P. Kürpick, and U. Thumm, "Basic Matrix Elements for Level Shifts and Widths of Hydrogenic Levels in Ion-Surface Interactions," *Phys. Rev. A* **54,** 1487 (1996).

P. Kürpick, U. Thumm, and U. Wille, "Ionization of Atoms Interacting with a Metal Surface Under the Influence of an External Electric Field," *Phys. Rev. A* **57,** 1920 (1998).

P. Kürpick, U. Thumm, and U. Wille, "Close-Coupling Calculations for Ion-Surface Interactions," *Nucl. Instrum. Meth. Phys. Res. B* **125,** 273 (1997).

P. Kürpick, U. Thumm, and U. Wille, "Resonance Formation of Hydrogenic Levels in Front of Metal Surfaces," *Phys. Rev. A* **56,** 543 (1997).

A. Kutana, I. L. Bolotin, and J. W. Rabalais, "Universal Expression for Blocking Cone Size Based on the ZBL Potential," *Surf. Sci.* **495,** 77–90 (2001).

S. Lacombe, L. Guillemot, M. Maazouz, N. Mandarino, E. Sanchez and V. Esaulov, "Scattering of Ne and He Atoms and Ions on a Na Surface," *Surf. Sci.* **410,** 70 (1998).

A. Liegl, and E. Taglauer, "Temperature Effects in Low-Energy Ion Scattering from Cu(115)," *Commun. Symp. on Surface Sci.,* La Plagne, 1992.

A. Liegl, and E. Taglauer, "Investigations on the Surface Structure of Cu(115) by Low Energy Ion Scattering Spectroscopy," *Proc. Symp. on Surface Sci.,* Obertraun 1991.

J. Limburg, C. Bos, T. Schlathölter, R. Hoekstra, R. Morgenstern, S. Hausmann, W. Heiland, and A. Närmann, "Energy Loss of keV He^{2+} Scattered Off an Al(110) Surface," *Surf. Sci.* **409,** 541 (1998).

J. Limburg, S. Schippers, I. Hughes, R. Hoekstra, R. Morgenstern, S. Hustedt, N. Hatke, and W. Heiland, "Probing Hollow Atom States Formed during Impact of Highly Charged Ions on Surfaces: N_6^{7+} and O^{7+} on Al(110) and Si(100)," *Nucl. Instrum. Meth. Phys. Res. B* **98,** 436 (1995).

J. Limburg, S. Schippers, I. Hughes, R. Hoekstra, R. Morgenstern, S. Hustedt, N. Hatke, and W. Heiland, "Velocity Dependence of KLL Auger Emission from Hollow Atoms during Collisions of Hydrogenic N^{6+} Ions on Surfaces," *Phys. Rev.* **51,** 3873 (1995).

Ch. Linsmeier, H. Knözinger, and E. Taglauer, "Ion Scattering and Auger Electron Spectroscopy Analysis of Alumina-Supported Rhodium Model Catalysts," *Surf. Sci.* **275,** 101 (1992).

Ch. Linsmeier, E. Taglauer, and H. Knözinger, "Ion Scattering Analysis of Rh Adlayers on Alumina Films," in *Fundamental Aspects of Heterogeneous Catalysis Studied by Particle Beams,* H. H. Brongersma and R. A. van Santen (Eds.), Plenum Press, New York, 1991.

N. Lorente, A. G. Borisov, D. Teillet-Billy, and J. P. Gauyacq, "Parallel Velocity Assisted Charge Transfer: F^- Ion Formation at Al (111) and Ag (110) Surfaces," *Surf. Sci.* **429,** 46 (1999).

N. Lorente, M. A. Cazalilla, J. P. Gauyacq, D. Teillet-Billy, and P. M. Echenique, "Auger Neutralization and De-Excitation of Helium at an Aluminum Surface: A Unified Treatment," *Surf. Sci.* **411,** L888 (1998).

N. Lorente, J. Merino, F. Flores, and M. Y. Gusev, "Negative Ion Formation on Al, Si and LiF," *Nucl. Instrum. Meth. Phys. Res. B* **125,** 277 (1997).

N. Lorente, and R. Monreal, "Self-Consistent LDA Calculation in Ion Neutralization at Metal Surfaces," *Surf. Sci.* **370,** 324 (1997).

N. Lorente, and R. Monreal, "Multielectron Neutralization Channels in Ion-Surface Scattering," *Phys. Rev. B* **53,** 9622 (1996).

N. Lorente, and R. Monreal, "Neutralization of Slow He^{2+} on Metal Surfaces: Theory for Auger and Cascade Electron Emission," *Surf. Sci.* **303,** 253 (1994).

N. Lorente, and R. Monreal, "A Theory for Auger Neutralization of He^+ Scattered Off Simple Metal Surfaces," *Nucl. Instrum. Meth. Phys. Res. B* **78,** 44 (1993).

N. Lorente, R. Monreal, and M. Alducin, "Local Theory of Auger Neutralization for Slow and Compact Ions Interacting with Metal Surfaces," *Phys. Rev. A* **49,** 4716 (1994).

N. Lorente, R. Monreal, M. Alducin, and P. Apell, "Radiative Neutralization of Small Clusters Impinging on Solid Surfaces," *Z. Phys. D* **41,** 143 (1997).

N. Lorente, R. Monreal, and M. Maravall, "Electron Emission in the Neutralization of Multiply-Charged Ions at Low Velocities on Metal Surfaces: The Effect of Secondary-Electron Cascades," *Nucl. Instrum. Meth. Phys. Res. B* **100,** 290 (1995).

N. Lorente, D. Teillet-Billy, and J. P. Gauyacq, "H_2^+ Scattered Off Al Surfaces: The Role of the Negative Ion Shape Resonance," *J. Chem. Phys.* **111,** 7075 (1999).

N. Lorente, D. Teillet-Billy, and J. P. Gauyacq, "The N_2^- Shape Resonance in Slow N_2^+ Collisions with Metallic Surfaces," *Surf. Sci.* **432,** 155 (1999).

N. Lorente, D. Teillet-Billy, and J. P. Gauyacq, "Theoretical Studies of Charge Transfer in Molecular Ion-Metal Surface Collisions," *Nucl. Instrum. Meth. B* **157,** 1 (1999).

N. Lorente, D. Teillet-Billy, and J. Gauyacq, "Charge Transfer as Responsible for H^{2+} Dissociation on Aluminum Surfaces," *Surf. Sci.* **402,** 197 (1998).

A. Losch, and H. Niehus, "NICASS—A New Method for Surface Analysis of Insulators," *Phys. Stat. Sol. A* **173,** 117 (1999).

A. Losch, and H. Niehus, "Structure Analysis of the KBr (100) Surface: An Investigation with a New Method for Surface Analysis of Insulators," *Surf. Sci.* **420,** 148 (1999).

C. M. Loxton, P. J. Martin, and R. J. MacDonald, "The Influence of Surface Oxygen Contamination on Elemental Analysis Using Photon Emission from Sputtered Atoms (Nb-V Alloys)," *J. Vac. Sci. Tech.* **20,** 388 (1982).

H. C. Lu, E. P. Gusev, E. Garfunkel, and T. Gustafsson, "An Ion Scattering Study of the Interaction of Oxygen with Si(111) Surface Roughening and Oxide Growth," *Surf. Sci.* **351,** 111 (1996).

K. M. Lui, Z. L. Fang, W. M. Lau, and J. W. Rabalais, "Computer Simulations of Hyperchanneling of Low Energy Ions in Semichannels of Pt(111)-(1 × 1)," *Nucl. Instrum. Meth. Phys. Res. B* **182,** 200–206 (2001).

K. M. Lui, Y. Kim, W. M. Lau, and J. W. Rabalais, "Adsorption Site Determination of Light

Elements on Heavy Substrates by Low Energy Ion Channeling," *J. Appl. Phys.* **86,** 5256 (1999).

K. M. Lui, Y. Kim, S. S. Kim, W. M. Lau, V. Bykov, and J. W. Rabalais, "How Do Hydrogen Atoms on Surfaces Affect the Trajectories of Heavier Scattered Atoms?" *J. Chem. Phys.* **111,** 11095 (1999).

K. M. Lui, Y. Kim, W. M. Lau, and J. W. Rabalais, "Quantitative Determination of Hydrogen Adsorption Site on the Pt(111)-(1 × 1) Surface by Low Energy Ion Channeling," *Appl. Phys. Lett.* **75,** 587 (1999).

S. B. Luitjens, A. J. Algra, and A. L. Boers, "Ion Fractions of Low Energy Ne (E_0 < 10 keV) Scattered from a Copper Single Crystal," *Surf. Sci.* **80,** 566 (1979).

S. B. Luitjens, A. J. Algra, E. P. Suurmeijer, and A. L. Boers, "Low Energy (5–16 keV) Noble Gas Particles Scattered from Step Structures at Copper Surfaces," *Le Vide* **201,** 1419 (1980).

S. B. Luitjens, A. J. Algra, E. P. Suurmeijer, and A. L. Boers, "The Measurement of Energy Spectra of Neutral Particles in Low Energy Ion Scattering," *Appl. Phys.* **21,** 205 (1980).

R. J. MacDonald, W. Heiland, E. Taglauer, and D. Menzel, "A Comparison of Surface Analysis Using Ion Scattering, Ion-Produced Photons, and Secondary Ion Emission," *Appl. Phys. Lett.* **33,** 576 (1978).

R. J. MacDonald, and P. J. Martin, "The Neutralization of Low Energy He^+ Scattered from Ni," *Surf. Sci.* **173,** 593 (1986).

R. J. MacDonald, and D. J. O'Connor, "Neutralization of He^+ and Ne^+ Scattered from Ag," *Surf. Sci.* **124,** 423 (1983).

R. J. MacDonald, D. J. O'Connor, and P. Higginbottom, "Neutralization Contributions in Low Energy Ion Scattering," *Nucl. Instrum. Meth. Phys. Res. B* **2,** 418 (1984).

R. J. MacDonald, D. J. O'Conor, and T. Wiklendt, "The Scattering of Low-Energy Atomic and Molecular Ions from Solid Surfaces," *Vacuum* **41,** 255 (1990).

R. J. MacDonald, D. J. O'Connor, J. Wilson, and Y. G. Shen, "Neutralization in Low Energy Ion Scattering," *Nucl. Instrum. Meth. Phys. Res. B* **33,** 446 (1988).

N. Mandarino, P. Zoccali, A. Oliva, M. Camarca, A. Bonanno, and F. Xu, "Near Threshold Behavior of the 2p Electron Excitation in Mg-Mg, Al-Al, and Si-Si Symmetric Collisions," *Phys. Rev. A* **48,** 2828 (1993).

G. Manicó, F. Ascione, N. Mandarino, A. Bonanno, P. Riccardi, P. Alfano, P. Zoccali, A. Oliva, M. Camarca, and F. Xu, "Double 2p Electron Excitation in Low-Energy Ne^+ Single Scattering from a Si Surface: An Energy Loss Study," *Surf. Sci.* **392,** L7 (1997).

J. Manske, M. Dirska, G. Lubinski, M. Schleberger, A. Närmann, and R. Hoekstra, "Surface Magnetism Studied by Polarized K-Light Emission After He^+ Scattering," *J. Magn. Mat.* **168,** 249 (1997).

L. Marchut, T. M. Buck, G. H. Wheatley, and C. J. McMahon, Jr., "Surface Structure Analysis Using Low Energy Ion Scattering: I. Clean Fe(001)," *Surf. Sci.* **141,** 549 (1984).

B. Maschhoff, J. Pan, R. A. Baragiola, and T. Madey, "Energy-Dependent Effects in Low-Energy Ion Scattering from TiO_2, Ti, and H_2O Ice," in *Fundamental Aspects of Heterogeneous Catalysis Studied Using Particle Beams,* Plenum, New York, 1991.

E. S. Mashkova, and V. B. Fleurov, "Small-Angle Ion Reflection from Single Crystals," *Rad. Eff. Def. Sol.* **80,** 227 (1984).

E. S. Mashkova, and V. A. Molchanov, *Medium-Energy Ion Reflection from Solids,* North Holland, Amsterdam, 1985.

E. S. Mashkova, V. A. Molchanov, and Y. G. Skripka, "Mechanism of the Scattering of Ions by Crystals," *Soviet Phys. Doklady* **16,** 440 (1971).

E. S. Mashkova, V. A. Molchanov, and Y. G. Skripka, "A Mechanism of Ion Scattering by Crystals," *Phys. Lett. A* **33,** 373 (1970).

F. Masson, S. Aduru, C. S. Sass, and J. W. Rabalais, "Scatter-Free Direct Recoil Spectra," *Chem. Phys. Lett.* **152,** 325 (1988).

F. Masson, H. Bu, M. Shi, and J. W. Rabalais, "Direct Determination of Crystal Surface Periodicity From Scattering and Recoiling Azimuthal Anisotropy," *Surf. Sci.* **249,** 313 (1991).

F. Masson, and J. W. Rabalais, "Surface Periodicity Exposed Through Shadowing and Blocking Effects," *Chem. Phys. Lett.* **179,** 63 (1991).

F. Masson, and J. W. Rabalais, "Surface Structural Analysis of the (1×2) and (1×3) Phases of Pt$\{110\}$ by Time-of-Flight Scattering and Recoiling Spectrometry (TOF-SARS)," *The Structure of Surfaces III,* **24,** 253, 1991.

F. Masson, and J. W. Rabalais, "Time-of-Flight Scattering and Recoiling Spectrometry (TOF-SARS) Analysis of Pt$\{110\}$: I. Quantitative Structured Study of the Clean (1×2) Surface," *Surf. Sci.* **253,** 245 (1991).

F. Masson, and J. W. Rabalais, "Time-of-Flight Scattering and Recoiling Spectrometry (TOF-SARS) Analysis of Pt$\{110\}$: II. The (1×2)-to-(1×3) Interconversion and Characterization of the (1×3) Phase," *Surf. Sci.* **253,** 258 (1991).

F. Masson, C. S. Sass, O. Grizzi, and J. W. Rabalais, "Application of Direct Recoil Spectrometry to Determination of the Saturation Coverage of Ethylene on Pt(111)," *Surf. Sci.* **221,** 299 (1989).

E. G. McRae, T. M. Buck, R. A. Malic, W. E. Wallace, and J. M. Sanchez, "Ordering and Layer Composition at the $Cu_3Au(110)$ Surface," *Surf. Sci.* **238,** L481 (1990).

B. Menner, G. Ohlendorf, F. Patorra, and V. Kempter, "Orientation and Alignment of Alkali p-States Excited in Low-Energy Collisions of Alkali Ions with Noble Gas Atoms," *Z. Phys. D* **17,** 237 (1990).

J. Merino, N. Lorente, F. Flores, and M. Y. Gusev, "Quantum Approach to Charge Transfer Between Low Energy H^+ Ions and Al Surfaces," *Nucl. Instrum. Meth. Phys. Res. B* **125,** 288 (1997).

J. Merino, N. Lorente, M. Y. Gusev, F. Flores, M. Maazouz, L. Guillemot, and V. A. Esaulov, "Charge Transfer of Slow H Atoms Interacting with Al: Dynamical Charge Evolution," *Phys. Rev. B* **57,** 1947 (1998).

J. Merino, N. Lorente, W. More, F. Flores, and M. Y. Gusev, "Charge Transfer of Slow Light Ions Interacting with Surfaces," *Nucl. Instrum. Meth. Phys. Res. B* **125,** 250 (1997).

J. Merino, N. Lorente, P. Pou, and F. Flores, "Charge Transfer of Slow H Atoms Interacting with Al: Atomic Levels and Linewidths," *Phys. Rev. B* **54,** 10959 (1996).

F. W. Meyer, C. C. Havener, S. H. Overbury, K. J. Snowdon, D. M. Zehner, W. Heiland, and H. Hemme, "Charge Exchange Processes Between Highly Charged Ions and Metal Surfaces," *Nucl. Instrum. Meth. Phys. Res. B* **23,** 234 (1987).

M. H. Mintz, U. Atzmony, and N. Shamir, "Initial Adsorption Kinetics of Oxygen on Polycrystalline Copper," *Surf. Sci.* **185,** 413 (1987).

M. H. Mintz, U. Atzmony, and N. Shamir, "Interrelations Between Planes Affecting Adsorption Kinetics on Polycrystalline Surfaces: Oxygen Adsorption on Copper," *Phys. Rev. Lett.* **59,** 90 (1987).

M. H. Mintz, J. A. Schultz, and J. W. Rabalais, "TOF Spectra of Direct Recoils: II. Interaction of H_2 With Clean and Oxidized Magnesium," *Surf. Sci.* **146,** 457 (1984).

J. Möller, "CO Adsorbiert auf Rh(001), Beobachtet in PES, ISS, und Sekundarelektronenspektren," *Cond. Matt.* **55,** 27 (1984).

J. Möller, H. Niehus, and W. Heiland, "Direct Measurement of Au(110) Surface Structural Parameters by Low Energy Ion Backscattering," *Surf. Sci.* **166,** L111 (1986).

J. Möller, K. J. Snowdon, W. Heiland, and H. Niehus, "Low Energy Ion Scattering from the Au(110) Surface," *Surf. Sci.* **178,** 475 (1986).

R. Monreal, and N. Lorente, "Dynamical Screening in Auger Processes Near Metal Surfaces," *Phys. Rev. B* **52,** 4760 (1995).

K. Morgenstern, H. Niehus, and G. Comsa, "Reconstruction of the Oxygen Covered $Cu_3Au(110)$-Surface Investigated by Low Energy Ion Backscattering and LEED," *Surf. Sci.* **338,** 1 (1995).

K. Morgenstern, M. Voetz, and H. Niehus, "Nitrogen Induced Reconstruction of the $Cu_3Au(110)$-Surface," *Phys. Rev. B* **54,** 17870 (1996).

Y. Muda, and T. Hanawa, "Theory of Ion Neutralization Near the Surface," *Surf. Sci.* **97,** 283 (1980).

H. Müller, H. Brenten, and V. Kempter, "Electron Emission in Slow Collisions of Inert Gas and Reactive Ions with W(110) Partially Covered by Alkali Atoms," *Nucl. Instrum. Meth. Phys. Res. B* **78,** 239 (1993).

H. Müller, D. Gador, H. Brenten, and V. Kempter, "Electron Emission in Low Energy Grazing Collisions of O^+ Ground State (^4S) and Excited State (^2D; ^2P) Ions with Alkalated W(110) Surfaces," *Surf. Sci.* **313,** 188 (1994).

H. Müller, D. Gador, H. Brenten, and V. Kempter, "Electron Emission in Low Energy Grazing Collisions of $O^{2+}X^2\Pi_g$ and $a^4\Pi_u$ Molecular Ions with W(110) Partially Covered by Alkali Atoms," *Surf. Sci.* **318,** 403 (1994).

H. Müller, D. Gador, and V. Kempter, "Dissociation and Electronic Transitions in Low Energy Grazing Collisions of CO_2^+ Ions with W(110) Surfaces," *Surf. Sci.* **33,** 313 (1995).

H. Müller, D. Gador, F. Wiegershaus, and V. Kempter, "Evidence for the Formation of $N^{-*}(^1D)$ Ions in Ion-Surface Collisions," *J. Phys. B* **29,** 715 (1996).

H. Müller, R. Hausmann, H. Brenten, and V. Kempter, "Electron Emission in Low-Energy Grazing Collisions of CO^+ Ions with W(110) Partially Covered by Cesium Atoms," *Chem. Phys.* **179,** 215 (1994).

H. Müller, R. Hausmann, H. Brenten, and V. Kempter, "Electron Emission in Low Energy Grazing Collisions of N^+ and N^{2+} with W(110) Partially Covered By Cs Atoms," *Surf. Sci.* **303,** 56 (1994).

H. Müller, R. Hausmann, H. Brenten, and V. Kempter, "Electron Emission in Low-Energy Grazing Collisions of C^+ Ions with W(110) Partially Covered by Alkali Atoms," *Surf. Sci.* **291,** 78 (1993).

H. Müller, R. Hausmann, H. Brenten, and V. Kempter, "Electrons from Intra- and Interatomic Auger Processes of H^+ and H^{2+} with W(110) Partially Covered by Alkali Atoms," *Surf. Sci.* **284,** 129 (1993).

H. Müller, R. Hausmann, H. Brenten, A. Niehaus, and V. Kempter, "Electron Emission in Collisions of Slow Rare Gas Ions with Partially Cesiated W(110)," *Z. Phys. D* **28,** 109 (1993).

H. Müller, and V. Kempter, "Modeling the Ion Impact Electron Spectra from Slow H$^+$ and O$^+$ Collisions with Clean and Alkalated W(110) Surfaces," *Surf. Sci.* **366,** 343 (1996).

D. R. Mullins, and S. H. Overbury, "Formation of Subsurface Carbon Induced by Oxygen Adsorption on Carbon-Covered W(001)," *Surf. Sci.* **210,** 501 (1989).

D. R. Mullins, and S. H. Overbury, "The Structure of Oxygen on W(001) by Low Energy Ion Scattering," *Surf. Sci.* **210,** 481 (1989).

D. R. Mullins, and S. H. Overbury, "The Structure of the Carburized W(001) Surface," *Surf. Sci.* **193,** 455 (1988).

D. R. Mullins, and S. H. Overbury, "The Structure and Composition of the NiAl(110) and NiAl(100) Surfaces," *Surf. Sci.* **199,** 141 (1988).

J. Murakami, T. Hashimoto, and I. Kusunoki, "Ion Scattering and Surface Recoiling from a Si Surface by Low Energy Ne$^+$ Bombardment," *Vacuum* **41,** 369 (1990).

M. Naitoh, F. Shoji, and K. Oura, "Hydrogen-Termination Effects of the Growth of Ag Thin Films on Si(111) Surfaces," *Surf. Sci.* **242,** 152 (1991).

A. Närmann, "Statistics of Energy-Loss and Charge Exchange of Penetrating Particles: Energy Loss Spectra and Applications," *Phys. Rev. A* **51,** 548 (1995).

A. Närmann, "Interrelation Between Charge Exchange and Energy Loss of Particles Interacting with Surfaces: A Brief Review and Recent Results," *Modern Phys. Lett. B* **5,** 561 (1991).

A. Närmann, H. Derks, and W. Heiland, "Channeling Effects in He Scattering from Ni(110)," *Surf. Sci.* **211,** 271 (1989).

A. Närmann, H. Derks, and W. Heiland, "Crystallographic Effects in Charge Exchange Processes: He Scattering from Ni(110)," *Surf. Sci.* **217,** 255 (1989).

A. Närmann, M. Dirska, J. Manske, and M. Schleberger, "Probing of Magnetic Surfaces with Inherent Monolayer Sensitivity by Ion Scattering," *Nucl. Instrum. Meth. Phys. Res. B* **136,** 1212 (1998).

A. Närmann, H. Franke, K. Schmidt, A. Arnau, and W. Heiland, "Scattering of H at Ni Surfaces: Charge Exchange and Energy Loss," *Nucl. Instrum. Meth. Phys. Res. B* **69,** 158 (1992).

A. Närmann, W. Heiland, R. Monreal, F. Flores, and P. M. Echenique, "Charge Exchange and Energy Loss of Particles Interacting with Surfaces," *Phys. Rev. B* **44,** 2003 (1991).

A. Närmann, W. Heiland, R. Monreal, F. Flores, and P. M. Echenique, "Experimental and Theoretical Investigations on the Interrelation of Charge Exchange Processes and Energy Loss," *Mat. Res. Soc. Proc.* **222,** 47 (1991).

A. Närmann, C. Höfner, T. Schlathölter, and W. Heiland, *Inelastic Phenomena of Low-Energy Particle-Surface Interactions,* Springer, Berlin, 1996.

A. Närmann, R. Monreal, P. M. Echenique, F. Flores, W. Heiland, and S. Schubert, "Charge Exchange and Energy Dissipation of Particles Interacting with Metal Surfaces," *Phys. Rev. Lett.* **64,** 1601 (1990).

A. Närmann, M. Schleberger, W. Heiland, C. Huber, and J. Kirschner, "Emission of Polarized Light by Slow Ions After Excitation Near a Magnetic Surface," *Surf. Sci.* **251,** 248 (1991).

A. Närmann, K. Schmidt, C. Höfner, W. Heiland, and A. Aarnau, "Energy Loss Spectra of H and H Scattered Off Metal Surfaces," *Nucl. Instrum. Meth. Phys. Res. B* **78,** 72 (1993).

A. Närmann, K. Schmidt, W. Heiland, R. Monreal, F. Flores, and P. M. Echenique, "Energy Loss of Light Ions and Neutrals from Surface Scattering," *Nucl. Instrum. Meth. Phys. Res. B* **48,** 378 (1990).

A. Närmann, and P. Sigmund, "Statistics of Energy Loss and Charge Exchange of Penetrating Particles: Higher Moments and Transients," *Phys. Rev. A* **49,** 4709 (1994).

A. Niehaus, "A Classical Model for Multiple-Electron Capture in Slow Collisions of Highly Charged Ions with Atoms," *J. Phys. B* **19,** 2925 (1986).

A. Niehof, and W. Heiland, "Experimental Realization of Low-Energy Surface Channeling," *Nucl. Instrum. Meth. Phys. Res. B* **48,** 306 (1990).

H. Niehus, "Characterization of Metal Alloy Systems by Scanning Tunneling Microscopy and Low Energy Ion Scattering," *Phys. Stat. Sol. B* **192,** 357 (1995).

H. Niehus, "Ion Scattering Spectroscopic Techniques," *Practical Surface Analysis II,* Wiley, New York, 1992.

H. Niehus, "Ion Scattering Spectroscopy and Scanning Tunneling Microscopy: A Powerful Combination for Surface Structure Analysis," *Appl. Phys. A* **53,** 388 (1991).

H. Niehus, "Surface Structure Determination Using Alkali and Noble Gas Ion Scattering," *J. Vac. Sci. Tech. A* **5,** 751 (1987).

H. Niehus, "Analysis of the Pt(110)-(1 × 2) Surface Reconstruction," *Surf. Sci.* **145,** 407 (1984).

H. Niehus, and C. Achete, "Oxygen Induced Mesoscopic Island Formation at Pd(110)," *Surf. Sci.* **369,** 9 (1996).

H. Niehus, and C. Achete, "Oxygen and Nitrogen at $Cu_3Au(100)$: A Surface Structure Investigation with NICISS and STM," *Proc. of Symp. on Surface Sci. 3S93,* Austria, 1993.

H. Niehus, and C. Achete, "Surface Structure Investigation of Nitrogen and Oxygen on $Cu_3Au(100)$," *Surf. Sci.* **289,** 19 (1993).

H. Niehus, and E. Bauer, "Quantitative Aspects of Ion Scattering Spectroscopy (ISS)," *Surf. Sci.* **47,** 222 (1975).

H. Niehus, Th. Baumann, M. Voetz, and K. Morgenstern, "Clean and Oxygen Covered $Cu_3Au(110)$: A Surface Structure Investigation with STM and NICISS," *Surf. Rev. Lett.* **3,** 1899 (1996).

H. Niehus, and G. Comsa, "Ion Scattering Spectroscopy in the Impact Collision Mode (ICISS): Surface Structure Information from Noble Gas and Alkali-Ion Scattering," *Nucl. Instrum. Meth. Phys. Res. B* **15,** 122 (1986).

H. Niehus, and G. Comsa, "Alkali Ion Impact Collision Scattering at Pt(111)," *Surf. Sci.* **152,** 93 (1985).

H. Niehus, and G. Comsa, "Real-Space Investigation of the Oxygen-Induced Ni(110)-(2 × 1) Phase," *Surf. Sci.* **151,** L171 (1985).

H. Niehus, and G. Comsa, "Determination of Surface Reconstruction with Impact-Collision Alkali Ion Scattering," *Surf.* 18 (1984).

H. Niehus, W. Heiland, and E. Taglauer, "Low Energy Ion Scattering at Surfaces," *Surf. Sci. Report* **17,** 213 (1993).

H. Niehus, C. Hiller, and G. Comsa, "Row Pairing Induced by Hydrogen Adsorption at Pd(110)," *Surf. Sci.* **151,** L171 (1985).

H. Niehus, K. Mann, B. N. Eldridge, and M. L. Yu, "Low-Energy Neutral/Ion Backscattering at As/Si(001)," *J. Vac. Sci. Tech. A* **6,** 625 (1988).

H. Niehus, and E. Preuss, "Low Energy Alkali Backscattering at Pt(111)," *Surf. Sci.* **119,** 349 (1982).

H. Niehus, W. Raunau, K. Besocke, R. Spitzl, and G. Comsa, "Surface Structure of NiAl(111)

Determined by Ion Scattering and Scanning Tunneling Microscopy," *Surf. Sci. Lett.* **225,** L8 (1990).

H. Niehus, and R. Spitzl, "Ion-Solid Interaction at Low Energies: Principles and Applications of Quantitative ISS," *Surf. Interface Anal.* **17,** 287 (1991).

H. Niehus, R. Spitzl, K. Besocke, and G. Comsa, "The N-Induced (2×3) Reconstruction of Cu(110): Evidence for Long Range, Highly Directional Interaction Between Cu-N Bonds," *Phys. Rev. B* **43,** 12619 (1991).

H. Niehus, M. Voetz, C. Achete, K. Morgenstern, and G. Comsa, "Surface Structure Investigation with Ion Scattering and Scanning Tunneling Microscopy at Oxygen and Nitrogen Covered Cu_3Au Surfaces," *Surface Sci. Principles and Current Applications,* Springer, Berlin, 1996.

D. Niemann, M. Rösler, M. Grether, and N. Stolterfoht, "Electron Emission from Slow Ne^{9+} Ions Impinging on an Al Surface," *Nucl. Instrum. Meth. Phys. Res. B* **135,** 460 (1998).

D. Niemann, M. Rösler, M. Grether, and N. Stolterfoht, "First Principle Calculation of Ion Induced Kinetic Electron Emission from Nearly Free-Electron Metals Below the Plasmon Threshold," *Nucl. Instrum. Meth. Phys. Res. B* **164,** 873 (2000).

D. J. O'Connor, "Compact Medium-Energy Electrostatic Analyzer," *J. Phys. E* **20,** 437 (1987).

D. J. O'Connor, "Structure Analysis of Oxygen on Ni(110) Using Low Energy Recoil Spectroscopy," *Surf. Sci.* **173,** 593 (1986).

D. J. O'Connor, B. V. King, R. J. MacDonald, Y. G. Shen, and X. Chen, "The Study of Surfaces Using Ion Beams," *Aust. J. Phys.* **43,** 601 (1990).

D. J. O'Connor, C. M. Loxton, R. J. MacDonald, and Y. G. Shen, "Oxygen Adsorption on Ni_3Al," *Proc. 7th Australian Conf. X-ray Anal. Assoc.,* NSW, 1988.

D. J. O'Connor, and R. J. MacDonald, "Inelastic Energy Loss of Low Energy Surface Scattered Ions," *Nucl. Instrum. Meth. Phys. Res. B* **170,** 495 (1980).

D. J. O'Connor, and R. J. MacDonald, "Short Communication: A Correction Factor to the Interatomic Potential Screening Function for Use in Computer Simulations," *Rad. Eff. Def. Sol.* **34,** 247 (1977).

D. J. O'Connor, R. J. MacDonald, W. Eckstein, and P. R. Higginbottom, "Surface Structure Analysis Using Low Energy Scattered and Recoiling Ions," *Nucl. Instrum. Meth. Phys. Res. B* **13,** 235 (1986).

D. J. O'Connor, Y. G. Shen, J. M. Wilson, and R. J. MacDonald, "The Role of the Electronic Structure in Charge Exchange Between Low Energy Ions and Surfaces," *Surf. Sci.* **197,** 277 (1988).

D. J. O'Connor, Y. G. Shen, J. M. Wilson, and R. J. MacDonald, "The Role of Charge Exchange in Surface Analysis," *Proc. 5th Australian Conf. on Nucl. Tech. Anal.,* NSW, 1987.

A. Oliva, A. Bonanno, M. Camarca, and F. Xu, "Ion-Induced Auger Emission From Si, Al, and Al_xMg_{1-x} Samples," *Nucl. Instrum. Meth. Phys. Res. B* **58,** 333 (1991).

J. Onsgaard, W. Heiland, and E. Taglauer, "Adsorption and Desorption of Oxygen and Carbon Monoxide on the Si(111) Surface Studied by Ion Scattering Spectroscopy," *Surf. Sci.* **99,** 112 (1980).

K. G. Orrman-Rossiter, D. R. Mitchell, S. E. Donnelly, C. J. Rossouw, S. R. Glanville, P. R. Miller, A. H. Al-Bayati, J. A. Van den Berg, and D. G. Armour, "Evidence for Competing Growth Phases in Ion-Beam-Deposited Epitaxial Silicon Films," *Phil. Mag. Lett.* **61,** 311 (1990).

C. Oshima, M. Aono, T. Tanaka, S. Kawai, S. Zaima, and Y. Shibata, "Clean TiC(001) Surface and Oxygen Chemisorption Studied by Work Function Measurement, Angle-Resolved X-ray Photoelectron Spectroscopy, Ultraviolet Photoelectron Spectroscopy, and Ion Scattering Spectroscopy," *Surf. Sci.* **102,** 312 (1981).

K. Oura, M. Naitoh, and F. Shoji, "Ion Beam Analysis of Hydrogen on Silicon Surfaces," *Microbeam Anal.* **2,** 139 (1993).

K. Oura, M. Naitoh, J. Yamane, and F. Shoji, "Hydrogen-Induced Reordering of the Si(111)-$\sqrt{3} \times \sqrt{3}$-Ag Surface," *Surf. Sci. Lett.* **230,** L151 (1990).

K. Oura, F. Shoji, and T. Hanawa, "Evidence for Solute Segregation on Cu-Mn Alloy Surfaces Studied by Low-Energy Ion Scattering," *Phys. Rev. B* **38,** 2188 (1988).

K. Oura, F. Shoji, and T. Hanawa, "Detection of Hydrogen on Solid Surfaces by Low-Energy Recoil Ion Spectroscopy," *Jap. J. of Appl. Phys.* **23,** L694 (1984).

K. Oura, K. Sumitomo, T. Kobayashi, T. Kinoshita, Y. Tanaka, F. Shoji, and I. Katayama, "Adsorption of H on Si(111)-$\sqrt{3} \times \sqrt{3}$-Ag: Evidence for Ag(111) Agglomerates Formation," *Surf. Sci.* **254,** L460 (1991).

K. Oura, Y. Tanaka, H. Morishita, and F. Shoji, "Low Energy Ion Scattering Study of Hydrogen-Induced Recording of Pb Monolayer Films on Si (111) Surfaces," *Nucl. Instrum. Meth. Phys. Res. B* **85,** 439 (1994).

K. Oura, M. Watamori, F. Shoji, and T. Hanawa, "Atomic Displacements of Si in the Si (111)-$(\sqrt{3} \times \sqrt{3})$R30°-Ag Surface Studied by High-Energy Ion Channeling," *Phys. Rev. B* **38,** 10146 (1988).

K. Oura, J. Yamane, K. Umezawa, M. Naitoh, F. Shoji, and T. Hanawa, "Hydrogen Adsorption on Si(100)-2 × 1 Surfaces Studied by Elastic Recoil Detection Analysis," *Phys. Rev. B* **41,** 1200 (1990).

S. H. Overbury, *Surface Imaging and Visualization,* CRC Press, Boca Raton, 1995.

S. H. Overbury, W. Heiland, D. M. Zehner, S. Datz, and R. S. Thoe, "Investigation of the Structure of Au(110) Using Angle Resolved Low Energy K^+ Ion Backscattering," *Surf. Sci.* **109,** 239 (1981).

S. H. Overbury, D. R. Mullins, M. T. Paffett, and B. E. Koel, "Surface Structure Determination of Sn Deposited on Pt(111) by Low Energy Alkali Ion Scattering," *Surf. Sci.* **254,** 45 (1991).

S. H. Overbury, D. R. Mullins, and J. F. Wendelken, "Surface Structure of Stepped NiAl(111) by Low Energy Li^+ Ion Scattering," *Surf. Sci.* **236,** 122 (1990).

S. H. Overbury, R. J. van den Oetelaar, and D. M. Zehner, "Surface Segregation In $Mo_{0.75}Re_{0.25}$ (001) Studied by Low Energy Alkali Ion Scattering," *Phys. Rev. B* **48,** 1718 (1993).

E. S. Parilis, L. M. Kishinevsky, N. Y. Turaev, B. E. Baklitzky, F. F. Umarov, V. Kh. Verleger, and I. S. Bitensky, *Atomic Collisions on Solids,* North-Holland, New York, 1993.

S. Park, H. Kang, and S. B. Lee, "Reaction Intermediate in Thermal Decomposition of 1, 3-Disilabutane to Silicon Carbide on Si(111): Comparative Study of Cs^+ Reactive Ion Scattering and Secondary Ion Mass Spectrometry," *Surf. Sci.* **450,** 117 (2000).

I. Paulini, W. Heiland, A. Arnau, F. Zarate, and P. Bauer, "Stopping Cross Sections of Protons and Deuterons in Lithiumniobate Near the Stopping Power Maximum," *Nucl. Instrum. Meth. Phys. Res. B* **118,** 39 (1996).

L. Pedemonte, M. Aschoff, K. Brüning, R. Tatarek, and W. Heiland, "Surface Structure and Surface Dynamics Analysis of the Ag(110)-Surface by Low Energy on Ion Scattering," *Nucl. Instrum. Meth. Phys. Res. B* **164,** 645 (2000).

X. D. Peng, and M. A. Barteau, "Characterization of the Crystallographic Structure of MgO Thin Films by ISS on the Basis of Trajectory-Dependent Neutralization," *Appl. Surf. Sci.* **44,** 87 (1990).

G. Piaszenski, M. Aschoff, S. Speller, and W. Heiland, "Structure Evaluation of an Alloy Single Crystal Surface, $Au_3Pd(113)$, by Low Energy Ion Scattering," *Nucl. Instrum. Meth. Phys. Res. B* **135,** 331 (1998).

E. Platzgummer, M. Borrell, C. Nagl, M. Schmid, and P. Varga, "Trajectory Dependent Neutralization of 1 keV He^+ Ions Scattered from Pb(111) and Pb Films on Cu(100)," *Surf. Sci.* **412,** 202 (1998).

E. Platzgummer, M. Sporn, R. Koller, S. Forsthuber, M. Schmid, W. Hofer, and P. Varga, "Temperature Dependent Segregation on Pt(111) and (100)," *Surf. Sci.* **419,** 236 (1999).

E. Platzgummer, M. Sporn, R. Koller, M. Schmid, W. Hofer, and P. Varga, "Temperature Dependent Segregation and (1×2) Missing-Row Reconstruction of $Pt_{25}Rh_{75}(110)$," *Surf. Sci.* **423,** 134 (1999).

E. Platzgummer, M. Sporn, R. Koller, M. Schmid, and P. Varga, "Temperature-Dependent Segregation Reversal and (1×3) Missing Row Structure of $Pt_{90}Co_{10}(110)$," *Surf. Sci.* **453,** 214 (2000).

B. Poelsema, L. K. Verheij, and A. L. Boers, "Study of Low Energy Noble Gas Ion Reflection from Monocrystalline Surfaces: Influence of Thermal Vibrations of the Surface Atoms," *Surf. Sci.* **64,** 554 (1977).

B. Poelsema, L. K. Verheij, and A. L. Boers, "Determination of the Surface Debye Temperature Using Quasi-Triple Collisions in Low Energy Ion Scattering," *Nucl. Instrum. Meth. Phys. Res. B* **132,** 623 (1976).

R. Poole, "Playing Three-Dimensional Pool," *Sci.* **246,** 995 (1989).

T. L. Porter, C. S. Chang, U. Knipping, and I. S. Tsong, "Impact-Collision Ion-Scattering-Spectrometry Study of Ni Layers Deposited on Si (111) at Room Temperature," *Phys. Rev. B* **36,** 9150 (1987).

T. L. Porter, C. S. Chang, and I. S. Tsong, "Si(111)-($\sqrt{3} \times \sqrt{3}$)Ag Surface Structure Studied by Impact-Collision Ion-Scattering Spectrometry," *Phys. Rev. Lett.* **60,** 1739 (1988).

T. L. Porter, D. M. Cornelison, C. S. Chang, and I. S. Tsong, "Impact Collision Ion Spectrometry Studies of the $NiSi_2(111)$ Surface," *J. Vac. Sci. Tech. A* **8,** 2497 (1990).

S. Priggemeyer, A. Brockmeyer, H. Dötsch, H. Koschmieder, D. J. O'Connor, and W. Heiland, "The Segregation of Pb on Yttrium-Iron-Garnet (YIG) Surfaces Studied by Ion Scattering Spectrometry (ISS)," *Appl. Surf. Sci.* **44,** 255 (1990).

S. Priggemeyer, H. Koschmieder, G. N. Van Wyk, and W. Heiland, "Analysis of Magneto-Optical Layers by Ion Scattering Spectrometry," *Fresenius J. Anal. Chem.* **341,** 343 (1991).

S. Pülm, A. Hitzke, J. Gunster, H. Müller, and V. Kempter, "Electron Emission from the Interaction of He Projectiles and UV Photons with LiF on W(110), *Rad. Eff. Def. Sol.* **128,** 151 (1994).

J. W. Rabalais, "An Imaging Spectrometry from Nature's Own Atomic Lenses," *Analytical Chem.* **170 A,** 207 (2001).

J. W. Rabalais, "Surface Structural Determination: Particle Scattering Methods," in *Encyclopedia of Chemical Physics and Physical Chemistry,* Vol. II, J. H. Moore and N. D. Spencer (Eds.), Institute of Physics Publishing, Bristol, UK, p. 1583–1610 (2001).

J. W. Rabalais, "Temporal and Spatial Resolution of Scattered and Recoiled Atoms for Surface Elemental and Structural Analysis," *Surf. Interface Anal.* **27,** 171 (1999).

J. W. Rabalais, "Direct Recoil Spectrometry," *CRC Crit. Rev. Sol. St. & Mat. Sci.* **14,** 319 (1998).

J. W. Rabalais, "Low Energy Ion Scattering," *Surf. Sci.* **299,** 219 (1994).

J. W. Rabalais, "Scattering and Recoiling Spectrometry," *Ency. Phys. Sci. Technol.,* **14,** 763 (1992).

J. W. Rabalais, "Surface Crystallography and Ion Scattering," *Chem. in Britain,* **28,** 37 (1992).

J. W. Rabalais, "Time-Of-Flight Scattering and Recoiling Spectrometry," *J. Vac. Sci. Technol. A* **9,** 1293 (1991).

J. W. Rabalais, "Time-Of-Flight Scattering and Recoiling Spectrometry (TOF-SARS) for Surface Structure Determinations," *Fund. Aspects Hetero. Catal. Studied by Particle Beams B: Physics,* **265,** 313, 1991.

J. W. Rabalais, "Scattering and Recoiling Spectrometry: An Ion's Eye View of Surface Structure," Sci. **250,** 521 (1990).

J. W. Rabalais, "Direct Recoil Spectrometry," *CRC Critical Rev. Sol. St. and Mat. Sci.* **14,** 319 (1988).

J. W. Rabalais, H. Bu, and C. D. Roux, "Impact-Parameter Dependence of Ar^+-Induced Kinetic Secondary Electron Emission From Ni{110}," *Phys. Rev. Lett.* **69,** 1391 (1992).

J. W. Rabalais, H. Bu, and C. D. Roux, "Surface and Adsorbate Structural Analysis From Time-of-Flight Scattering and Recoiling Spectrometry (TOF-SARS)," *Nucl. Instrum. Meth. B* **64,** (1992).

J. W. Rabalais, and J. N. Chen, "Inelastic Processes in Ion/Surface Collisions: Direct Recoil Ion Fractions as a Function of Kinetic Energy," *J. Chem. Phys.* **85,** 3615 (1986).

J. W. Rabalais, J. N. Chen, and R. Kumar, "Inelastic Processes in Ne^+ and Ar^+ Collisions with Mg and Y Surfaces Leading to Scattered Ion Fractions and Vacuum Ultraviolet Photon Emission," *Phys. Rev. Lett.* **55,** 1124 (1985).

J. W. Rabalais, J. N. Chen, R. Kumar, and M. Narayana, "Inelastic Processes in Ion/Surface Collisions: Scattered Ion Fractions and VUV Photon Emission for Ne^+ and Ar^+ Collisions With Mg and Y Surfaces," *J. Chem. Phys.* **83,** 6489 (1985).

J. W. Rabalais, J. N. Chen, R. Kumar, and M. Narayana, "Model for Scattered Ion Fractions Based on Equality in the Close Encounter," *Chem. Phys. Lett.* **120,** 406 (1985).

J. W. Rabalais, O. Grizzi, M. Shi, and H. Bu, "Surface Structure Determination from Scattering and Recoiling: W(211) and W(211)-p(1 × 2)-O," *Phys. Rev. Lett.* **63,** 51 (1989).

C. Rau, N. J. Zheng, M. Lu, and M. Rösler, "Ion-Induced Electron Emission from Magnetic and Nonmagnetic Surfaces," in *Ionization of Solids by Heavy Particles,* R. A. Baragiola (Ed.), Plenum Press, New York, 1993.

J. H. Rechtien, W. Mix, and K. J. Snowdon, "Evidence for Quantum Mechanical Interference Effects in Dissociative Scattering of H^{+2} and N^{+2} from Cu(111)," *Surf. Sci.* **259,** 26 (1991).

P. Riccardi, P. Barone, A. Bonanno, and A. Oliva, "Plasmon Excitation in Al by keV Ne and Ar Ions," *Nucl. Instrum. Meth. Phys. Res. B* **164,** 886 (2000).

P. Riccardi, P. Barone, A. Bonanno, A. Oliva, and R. A. Baragiola, "Angular Studies of Potential Electron Emission in the Interaction of Slow Ions with Al Surfaces," *Phys. Rev. Lett.* **84,** 378 (2000).

A. Robin, N. Hatke, A. Närmann, M. Grether, D. Plachke, J. Jensen, and W. Heiland, "Energy Loss of Fast N^{q+} Ions Scattered Off a Pt(110) Surface," *Nucl. Instrum. Meth. Phys. Res. B* **164,** 566 (2000).

M. T. Robinson, *Sputtering by Particle Bombardment,* Springer, Berlin, 1981.

W. D. Roos, J. du Plessis, G. N. van Wyk, E. Taglauer, and S. Wolf, "Surface Structure and Composition of NiAl(100) by Low-Energy Ion Scattering," *J. Vac. Sci. Tech. A* **14,** 1648 (1996).

M. Rösler, "Secondary Electron Emission from Simple Metals: Comparative Studies for Al, Mg, and Be," *Scan. Micro. Suppl.* **10,** 1025 (1996).

M. Rösler, "Theory of Particle-Induced Kinetic Electron Emission from Simple Metals: Comparative Studies of Different Excitation and Scattering Mechanisms for Al, Mg, and Be," *Appl. Phys. A* **61,** 595 (1995).

M. Rösler, "Theory of Ion-Induced Kinetic Electron Emission from Solids," in *Ionization of Solids by Heavy Particles,* Plenum Press, New York, 1993.

M. Rösler, "Contribution of Auger Processes to the Particle-Induced Electron Emission from Nearly-Free-Electron Metals," *Nucl. Instrum. Meth. Phys. Res. B* **58,** 309 (1991).

M. Rösler, and W. Brauer, "Theory of Electron Emission from Nearly-Free Electron Metals by Proton and Electron Bombardment," *Springer Tracts in Modern Physics* **122,** 1 (1991).

M. Rösler, and F. J. Garcia de Abajo, "Contribution of Charge Transfer Processes to Ion-Induced Electron Emission," *Phys. Rev. B* **54,** 17158 (1996).

C. Roux, H. Bu, and J. W. Rabalais, "Structure of the Ni-p(1×2)-H Surface From Time-of-Flight Scattering and Recoiling Spectrometry (TOF-SARS)," *J. Vac. Sci. Technol. A* **10,** 2143 (1992).

C. Roux, H. Bu, and J. W. Rabalais, "Structure of the Ni{110}-p(2×1)-OH Surface From Time-of-Flight Scattering and Recoiling Spectrometry," *Surf. Sci.* **279,** 1 (1992).

C. Roux, H. Bu, and J. W. Rabalais, "Sulfur Inhibition of the Oxygen Induced Reconstruction of the Ni{110} Surface," *Chem. Phys. Lett.* **200,** 60 (1992).

C. Roux, H. Bu, and J. W. Rabalais, "Structure of the Hydrogen Induced Ni{110}-p(1×2)-H Reconstructed Surface," *Surf. Sci.,* **259,** 253 (1991).

J. Ryu, T. Fuse, O. Kubo, T. Fujino, H. Tani, T. Harada, M. Katayama, and K. Oura, "Atomic-Hydrogen-Induced Self-Organization of Si(111)$\sqrt{3} \times \sqrt{3}$-In Surface Phase Studied by CAICISS and STM," *Surf. Sci.* **447,** 117 (2000).

J. Ryu, T. Fuse, O. Kubo, H. Tani, T. Fujino, T. Harada, A. A. Saranin, A. V. Zotov, M. Katayama, and K. Oura, "Adsorption of Atomic Hydrogen on the Si(001)4×3-In Surface Studied by Coaxial Impact Collision Ion Scattering Spectroscopy and Scanning Tunneling Microscopy," *J. Vac. Sci. Tech. B* **17,** 983 (1999).

J. Ryu, K. Kui, K. Noda, M. Katayama, and K. Oura, "CAICISS Studies of Atomic-Hydrogen-Induced Structural Changes of the Sb Terminated Si Surfaces," *Nucl. Instrum. Meth. Phys. Res. B* **136,** 1102 (1998).

J. Ryu, K. Kui, K. Noda, M. Katayama, and K. Oura, "The Effect of Hydrogen Termination on Ion Growth on Si(100) Surface," *Surf. Sci.* **401,** L425 (1998).

J. Ryu, K. Kui, K. Noda, M. Katayama, and K. Oura, "Atomic-Hydrogen-Induced Structural Change of the Si(100)-(2×1)-Sb Surface Studied by TOF-ICISS," *Appl. Surf. Sci.* **121,** 223 (1997).

J. Ryu, K. Kui, K. Noda, M. Katayama, and K. Oura, "Adsorption of Atomic Hydrogen on the Si(100)-(2x1)-Sb Surface," *Jpn. J. Appl. Phys.* **36,** 4435 (1997).

J. Ryu, K. Kui, Y. Tanaka, M. Katayama, K. Oura, and I. Katayama, "TOF-ICISS Observation of Pb Growth on the Si(111)-$\sqrt{3} \times \sqrt{3}$-Ag Surface," *Appl. Surf. Sci.* **113,** 393 (1997).

E. A. Sánchez, J. E. Gayone, M. L. Martiarena, and O. Grizzi, "Excitation of Volume Plasmons in Glancing Collisions of Protons with Al(111) Surfaces," *Phys. Rev. B* **61,** 14209 (2000).

E. A. Sánchez, L. Guillemot, and V. Esaulov, "Electron Transfer in the Interaction of Fluorine and Hydrogen with Pd(100): The Case of a Transition Metal Versus Jellium," *Phys. Rev. Lett.* **83,** 428 (1999).

C. S. Sass, and J. W. Rabalais, "Dissociative Scattering of 1.5–4.5 keV N_2^+ and N^+ on Gold and Graphite Surfaces," *J. Chem. Phys.* **89,** 3870 (1988).

K. Sato, S. Kono, T. Teruyama, K. Higashiyama, and T. Sagawa, "Ion Scattering Spectroscopic Study of Clean and Sn-Covered Ge(111) Surfaces," *Surf. Sci.* **158,** 644 (1985).

H. Schall, H. Brenten, K. H. Knorr, and V. Kempter, "Excitation in Low-Energy Li^+ Collisions with Cesiated W(110) Surfaces: Alignment and Orientation of the Collisionally Excited Li(2p) State," *Z. Phys. D* **16,** 161 (1990).

J. Scheer, K. Brüning, T. Fröhlich, P. Wurz, and W. Heiland, "Scattering of Small Molecules from a Diamond Surface," *Nucl. Instrum. Meth. Phys. Res. B* **157,** 208 (1999).

S. Schippers, S. Husted, W. Heiland, R. Köhrbrück, J. Kemmler, D. Lecler, and N. Stolterfoht, "Inner Shell Target Ionization by the Impact of N, O, and N on Pt(110)," in *Ionization of Solids by Heavy Particles,* R. A. Baragiola (Ed.), Plenum Press, New York, 1993.

S. Schippers, S. Oelschig, W. Heiland, L. Folkerts, R. Morgenstern, P. Eeken, I. F. Urazgil'din, and A. Niehaus, "Neutralization of He^{++} and He^+ on Pb Surfaces," *Surf. Sci.* **257,** 289 (1991).

T. Schlathölter, H. Franke, M. Vicanek, and W. Heiland, "The Interaction of Small Molecules with Pd and K Covered Pd Surfaces at Energies from 200 eV to 6 keV," *Surf. Sci.* **363,** 79 (1996).

T. Schlathölter, and W. Heiland, "Charge Exchange of Swift Molecules, H^{2+}, H_2, CO^{2+} and CO_2, at Pd(110) Surfaces," *Nucl. Instrum. Meth. Phys. Res. B* **100,** 352 (1995).

T. Schlathölter, and W. Heiland, "Low Energy Carbon Dioxide Scattering Form Pd(111) Surfaces," *Surf. Sci.* **331,** 311 (1995).

T. Schlathölter, and W. Heiland, "Scattering of Carbon Dioxide Molecules from Pd(111) Surfaces," *Surf. Sci.* **323,** 207 (1995).

T. Schlathölter, K. Schmidt, A. Närmann, and W. Heiland, "Charge Exchange and Dissociation of Hydrogen Molecules at Pd and Pd/K Surfaces," *Symp. Surface Sci.*, G. Betz and P. Varga, Kaprun, Austria **231** (1993).

T. Schlathölter, M. Vicanek, and W. Heiland, "Scattering of Fast N_2 from Pd(111): A Classical Trajectory Study," *Chem. Phys.* **106,** 4723 (1997).

T. Schlathölter, M. Vicanek, and W. Heiland, "Scattering of Fast N_2 from Pd(111): Orientation Influences on the Interaction Dynamics," *Rad. Eff. Def. Sol.* **141,** 175 (1997).

T. Schlathölter, M. Vicanek, and W. Heiland, "Interaction of Fast N_2 Molecules with Palladium Surfaces," *Surf. Sci.* **352,** 195 (1996).

T. Schlathölter, M. Vicanek, and W. Heiland, "Scattering of Homonuclear Diatomics from Pd(111)," *Surface Science Symposium,* Kaprun, Austria, 1995.

M. Schleberger, M. Dirska, J. Manske, and A. Närmann, "Measuring Hysteresis of Magnetic Surfaces by Ion Scattering," *Appl. Phys. Lett.* **71,** 3156 (1997).

M. Schleberger, S. Speller, C. Höfner, and W. Heiland, "Surface Channeling Experiments on Pb(110)," *Nucl. Instrum. Meth. Phys. Res. B* **90,** 274 (1994).

M. Schleberger, R. Tasker, A. Närmann, W. Heiland, C. Huber, and J. Kirschner, "Polarized

Light Emission After He$^+$ Scattering Off a Magnetized Fe(110) Surface," *Nucl. Instrum. Meth. Phys. Res. B* **58**, 384 (1991).

K. Schmidt, H. Franke, A. Närmann, and W. Heiland, "Scattering of Swift, Neutral, Molecular Hydrogen from Ni(110)," *J. Phys. Condens. Matter* **4**, 9869 (1992).

K. Schmidt, H. Franke, T. Schlathölter, C. Höfner, A. Närmann, and W. Heiland, "Scattering of Swift Molecules, H and CO, from Metal Surfaces," *Surf. Sci.* **301**, 326 (1994).

K. Schmidt, T. Schlathölter, and W. Heiland, "Dissociative Scattering of Hydrogen from Pd(110) and Pd(110) + K," *Chem. Phys. Lett.* **200**, 465 (1992).

H. K. Schmidt, J. A. Schultz, S. Tachi, S. Contarini, S. Aduru, J. W. Rabalais, and J. L. Margrave, "Direct Recoil Analysis of Light Elements (H,C,O,N) on Polyimide Surfaces (poly-isolindols quinazoline dione)," *J. Vac. Sci. Technol. A* **5**, 2961 (1987).

S. Schubert, U. Imke, and W. Heiland, "Negative Molecular State Formation and Dissociation of Molecules Scattered from Ni(110) and Ni(111) + K Surfaces," *Vacuum* **41**, 252 (1990).

S. Schubert, J. Neumann, U. Imke, K. J. Snowdon, P. Varga, and W. Heiland, "Neutralization and Dissociative Attachment in Molecular Ion-Surface Interaction: Scattering of N$_2^+$ and O^{2+} from Ni(111)," *Surf. Sci.* **171**, L375 (1986).

J. A. Schultz, C. R. Blakley, M. H. Mintz, and J. W. Rabalais, "Ion Fractions of Scattered Ne and Ar and Directly Recoiled, H, O, and Mg From Mg Surfaces," *Nucl. Instrum. Meth. B* **14**, 500 (1986).

J. A. Schultz, S. Contarini, Y. S. Jo, and J. W. Rabalais, "Methoxylation of Magnesium Studied by Direct Recoil Spectroscopy, SIMS, and XPS: Calibration of Relative Surface Hydrogen Concentration," *Surf. Sci.* **154**, 315 (1985).

J. A. Schultz, Y. S. Jo, and J. W. Rabalais, "Matrix Dependence of Secondary Ion Intensities From Mg(OH)$_2$ by Simulataneous Time-of-Flight SIMS and Direct Recoil Analysis," *Sol. State Comm.* **55**, 957 (1985).

J. A. Schultz, Y. S. Jo, S. Tachi, and J. W. Rabalais, "Simultaneous Direct Recoil and SIMS Analysis of H, C, and O on Si(100)," *Nucl. Instrum. Meth. B* **15**, 134 (1986).

J. A. Schultz, M. H. Mintz, T. R. Schuler, and J. W. Rabalais, "TOF Spectra of Direct Recoils: I. Chemisorption of O$_2$, H$_2$O, and CH$_3$OH on Magnesium," *Surf. Sci.* **146**, 438 (1984).

J. A. Schultz, E. Taglauer, P. Feulner, and D. Menzel, "Position Analysis of Light Adsorbates by Recoil Detection: H on Ru (001)," *Nucl. Instrum. Meth. Phys. Res. B* **64**, 588 (1992).

K. Sekar, J. Scheer, K. Brüning, and W. Heiland, "Interaction of Molecular Hydrogen with LiF(100) Surface," *Phys. Rev. B* **63**, 1 (2001).

K. Sekar, B. Sundaravel, I. H. Wilson, and W. Heiland, "Scanning Tunneling Microscopy and Ion Channeling Studies on Thin Co Films on Bromine Treated Si(100) Surfaces," *Appl. Surf. Sci.* **156**, 161 (2000).

S. P. Sharma, and T. M. Buck, " Double Scattering of Rare Gas Ions from Polycrystalline Gold and Copper," *J. Vac. Sci. Tech.* **12**, 468 (1975).

Y. G. Shen, D. J. O'Connor, and R. J. MacDonald, "The Composition and Structure Analysis of Ni$_3$Al(001) and Ni$_3$Al(110) Surfaces by Low Energy Ion Scattering Spectroscopy," *Proc 14th Australian Conf. on Cond. Matt. Physics,* Waga, 1990.

Y. G. Shen, D. J. O'Connor, and R. J. MacDonald, "Structure Analysis of O on Ni(001) and Ni$_3$Al(001) Surfaces by Low Energy Ion Beams," *Proc. 6th Australian Conf. on Nucl. Tech. Anal.,* Lucas Heights, 1989.

Y. G. Shen, D. J. O'Connor, and R. J. MacDonald, "Low Energy Ion Scattering from Clean

Si(110) Surface," *Proc. 5th Australian Conf. on Nucl. Tech. Anal.,* Lucas Heights, NSW, 1987.

M. Shi, H. Bu, and J. W. Rabalais, "Analysis of the Reconstructed Ir(110) Surface From Time-Of-Flight Scattering and Recoiling Spectrometry," *Phys. Rev. B* **42,** 2852 (1990).

M. Shi, O. Grizzi, and J. W. Rabalais, "Electron Exchange Processes of Scattered and Recoiled Particles With a LiCu Surface as a Function of Surface Composition," *Surf. Sci.* **235,** 67 (1990).

M. Shi, O. Grizzi, H. Bu, J. W. Rabalais, R. R. Rye, and P. Nordlander, "Time-Of-Flight Scattering and Recoiling Spectrometry: III. The Structure of Hydrogen on the W(211) Surface," *Phys. Rev. B* **40,** 10163 (1989).

M. Shi, O. Grizzi, and J. W. Rabalais, "Electron Exchange Processes of Scattered and Recoiled Particles with a LiCu Surface as a Function of Surface Composition," *Surf. Sci.* **235,** 67 (1990).

M. Shi, J. W. Rabalais, and V. A. Esaulov, "Neutralization and Ionization Processes in the Scattering of D^+, D_2^+, and D_3^+ on a Mg Surface," *Rad. Eff. Defects and Solids* **109,** 81 (1989).

M. Shi, Y. Wang, and J. W. Rabalais, "Structure of Hydrogen on the Si{100} Surface in the (2×1)-H Monohydride, (1×1)-H Dihydride, and c(4×4)-H Phases," *Phys. Rev. B* **48,** 1689 (1993).

T. H. Shin, S. J. Han, and H. Kang, "Efficient Hydration of Cs^+ Ions Scattered from Ice Films," *Nucl. Instrum. Meth. Phys. Res. B* **157,** 191 (1999).

F. Shoji, K. Kashihara, T. Hanawa, and K. Oura, "Neutralization and Inelastic Scattering Energy Loss of Low Energy He^+ Ions at InP(11)) Surfaces," *Nucl. Instrum. Meth. Phys. Res. B* **47,** 1 (1990).

F. Shoji, K. Kashihara, K. Sumitomo, and K. Oura, "Low-Energy Recoil-Ion Spectroscopy Studies of Hydrogen Adsorption on Si(100)-2 × 1 Surfaces," *Surf. Sci.* **242,** 422 (1991).

F. Shoji, K. Kusumura, and K. Oura, "A Si(100)-2 × 1: H Monohydride Surface Studied by Low-Energy Recoil-Ion Spectroscopy," *Surf. Sci.* **280,** L247 (1992).

F. Shoji, and K. Oura, "Low-Energy Recoil Ion Spectroscopy Studies of H on Si(111)-7 × 7," *Appl. Surf. Sci.* **60,** 66 (1991).

F. Shoji, K. Oura, and T. Hanawa, "A High Energy Ion Scattering/Channeling and Low Energy Ion Scattering Apparatus for Surface Analysis," *Vacuum* **42,** 189 (1991).

F. Shoji, K. Oura, and T. Hanawa, "High-Resolution Ion Scattering Spectrometry at Energies Ranging from 200 to 2000 eV," *Nucl. Instrum. Meth. Phys. Res. B* **36,** 23 (1989).

F. Shoji, A. Yamada, and K. Oura, "Inelastic Energy Loss of Recoiled Hydrogen Ions in Low-Energy He^+, Ne^+ and Ar^+ Collisions with Hydrogenated Silicon Surface," *Nucl. Instrum. Meth. Phys. Res. B* **115,** 196 (1996).

V. I. Shulga, "Two-Atomic Scattering of Ions in Their Reflection from Single Crystals," *Rad. Eff. Def. Sol.* **37,** 1 (1978).

V. I. Shulga, "Ion Beam Focusing by the Atomic Chains of a Crystal Lattice," *Rad. Eff. Def. Sol.* **26,** 61 (1975).

P. Sigmund, and A. Närmann, "Charge-Exchange and Energy Loss Statistics of Swift Penetrating Ions," *Laser and Particle Beams* **13,** 281 (1995).

J. H. Sizman, and C. Varelas, "Surface Channeling," *Nucl. Instrum. Meth. Phys. Res. B* **132,** 633 (1976).

K. J. Snowdon, C. C. Havener, F. W. Meyer, S. H. Overbury, D. M. Zehner, and W. Heiland, "Charge Exchange at Molecular-Orbital Pseudocrossings as an Important Mechanism for Nonkinetic-Electron Emission in Slow-Multicharged-Ion ($v < v_0$) Metal-Surface Scattering," *Phys. Rev. A* **38,** 2294 (1988).

K. J. Snowdon, D. J. O'Connor, and R. J. MacDonald, "Inelastic Energy Loss of Fast Ions for Direct Scattering and Skipping Trajectories on Metal Surfaces," *Rad. Eff. Def. Sol.* **109,** 33 (1989).

K. J. Snowdon, D. J. O'Connor, and R. J. MacDonald, "Observation of Skipping Motion in Small-Angle Ion-Surface Scattering," *Phys. Rev. Lett.* **61,** 1760 (1988).

K. J. Snowdon, D. J. O'Connor, and R. J. MacDonald, "Observation of Transient Adsorption or Skipping Motion in Small Angle Ion-Surface Scattering," *Appl. Phys. A* **47,** 83 (1988).

W. Soska, "Adsorption of Krypton on a Gold Surface Studied by the ICISS Method," *Surf. Sci.* **249,** 289 (1991).

R. Souda, "Bonding of Molecular Solids Probed by Low-Energy H^+ Scattering. Water and Oxygen on Pt(111)," *J. Phys. Chem. B* **105,** 5 (2001).

R. Souda, "Femtosecond Electron Dynamics on Solid Surfaces Probed by Ion Scattering and Stimulated Desorption of Secondary Ions," *Int. J. Mod. Phys. B* **14,** 1139 (2000).

R. Souda, "On the Mechanism of Ion Stimulated Desorption of O^+ from Transition-Metal Surfaces," *Int. J. Mass Spectrom. Ion Proc.* **202,** 199 (2000).

R. Souda, "Resonant Ion Stimulated Desorption of O^+ from the TiO_2(110) Surface: Effects of Oxygenation and Ion Bombardment," *Nucl. Instrum. Meth. Phys. Res. B* **160,** 453 (2000).

R. Souda, "Ion Stimulated Desorption of Secondary Ions from Ionic Compound Surfaces," *Phys. Rev. Lett.* **82,** 1570(1999).

R. Souda, "Resonant Ion Stimulated Desorption and Low Energy Proton Scattering Study of Interactions of Hydrogen and Oxygen with the $SrTiO_3$(100) Surface," *Phys. Rev. B* **60,** 6068 (1999).

R. Souda, T. Aizawa, W. Hayami, S. Otani, and Y. Ishizawa, "Neutralization of Low-Energy D^+ Scattering from Solid Surfaces," *Phys. Rev. B* **42,** 7761 (1990).

R. Souda, T. Aizawa, Y. Ishizawa, and C. Oshima, "Segregation of Ca Ions at the MgO(001) Surface Studied by Neutral Beam Incidence Ion Scattering Spectroscopy," *J. Vac. Sci. Tech. A* **8,** 3218 (1990).

R. Souda, T. Aizawa, C. Oshima, and Y. Ishizawa, "Electronic Excitation and Charge Exchange in Low-Energy He^+ Scattering from Solid Surfaces," *Nucl. Instrum. Meth. Phys. Res. B* **45,** 364 (1990).

R. Souda, T. Aizawa, C. Oshima, and Y. Ishizawa, "Inelastic Scattering of D, He, and Li Ions from KI," *Nucl. Instrum. Meth. Phys. Res. B* **45,** 369 (1990).

R. Souda, T. Aizawa, C. Oshima, S. Otani, and Y. Ishizawa, "Reionization Contributions in Low-Energy Scattering of He, Ne, and Ar Ions from Sn, Au, and $BaTiO_3$," *Surf. Sci.* **194,** L119 (1988).

R. Souda, T. Aizawa, S. Otani, and Y. Ishizawa, "Effects of Chemical Bonding on the Electronic Transition in Low Energy He^+ Scattering," *Surf. Sci.* **232,** 219 (1990).

R. Souda, T. Aizawa, S. Otani, and Y. Ishizawa, "Ion Neutralization Mediated by a Molecular State: He^+ on D-Covered TiC(111)," *Phys. Rev. B* **41,** 803 (1990).

R. Souda, T. Aizawa, S. Otani, Y. Ishizawa, and C. Oshima, "Oxygen Chemisorption on

Transition-Metal Carbide (100) Surfaces Studied by X-ray Photoelectron Spectroscopy and Low-Energy He$^+$ Scattering," *Surf. Sci.* **256,** 19 (1991).

R. Souda, and M. Aono, "Interactions of Low-Energy He$^+$, He^0, and He* with Solid Surfaces," *Nucl. Instrum. Meth. Phys. Res. B* **15,** 114 (1986).

R. Souda, M. Aono, C. Oshima, S. Otani, and Y. Ishizawa, "Trajectory-Dependent Neutralization Probability of Low-Energy He$^+$ Scattered from Solid Surfaces," *Nucl. Instrum. Meth. Phys. Res. B* **15,** 138 (1986).

R. Souda, M. Aono, C. Oshima, S. Otani, and Y. Ishizawa, "Mechanism of Electron Exchange Between Low Energy He$^+$ and Solid Surfaces," *Surf. Sci.* **150,** L59 (1985).

R. Souda, E. Asari, H. Kawanowa, S. Otani, and T. Tanaka, "Low Energy He$^+$ Ion Scattering from HfB$_2$(0001), TaB$_2$(0001), WB$_2$(0001), YB$_4$(001), and YB$_6$(001) Surfaces," *Surf. Sci.* **414,** 77 (1998).

R. Souda, E. Asari, H. Kawanowa, T. Suzuki, and S. Otani, "Capture and Loss of Valence Electrons During Low Energy H$^+$ and H$^-$ Scattering from LaB$_6$(100), Cs/Si(100), Graphite, and LiCl," *Surf. Sci.* **421,** 89 (1999).

R. Souda, E. Asari, H. Kawanowa, T. Suzuki, and S. Otani, "Interactions of SrF$_2$ and PrF$_3$ with TiC(111) and Si(111) Surfaces Studied by Low-Energy D$^+$ Scattering Spectroscopy," *Phys. Rev. B* **58,** 10054 (1998).

R. Souda, E. Asari, T. Suzuki, T. Tanaka, and T. Aizawa, "In Situ Observation of Charge Exchange and Surface Segregation of Hydrogen During Low Energy H$^+$ and H^{2+} Scattering from Semiconductor Surfaces," *Phys. Rev. Lett.* **81,** 465 (1998).

R. Souda, W. Hayami, T. Aizawa, and Y. Ishizawa, "Band Effects on Neutralization of Low-Energy D$^+$ Scattering from Ionic Crystals," *Phys. Rev. B* **43,** 10062 (1991).

R. Souda, W. Hayami, T. Aizawa, and Y. Ishizawa, "Charge Exchange in Low Energy D$^+$ Scattering from Mo(111)," *Surf. Sci.* **241,** 190 (1991).

R. Souda, W. Hayami, T. Aizawa, S. Otani, and Y. Ishizawa, "Auger-Type Electron Emission from Energetic He$^+$ Ions Interacting with the LaB$_6$(001) Surface," *Surf. Sci.* **363,** 133 (1996).

R. Souda, W. Hayami, T. Aizawa, S. Otani, and Y. Ishizawa, "Charge State of Potassium on Metal and Semiconductor Surfaces Studied by Low-Energy D$^+$ Scattering," *Phys. Rev. Lett.* **69,** 192 (1992).

R. Souda, W. Hayami, T. Aizawa, S. Otani, and Y. Ishizawa, "Chemical Effects in Low-Energy D$^+$ Scattering from Oxides," *Phys. Rev. B* **45,** 14358 (1992).

R. Souda, W. Hayami, T. Aizawa, S. Otani, and Y. Ishizawa, "Electron Emission from He$^+$ Interacting with a Cs Overlayer on W(110)," *Phys. Rev. B* **46,** 7318 (1992).

R. Souda, Y. Hwang, T. Aizawa, W. Hayami, K. Oyoshi, and S. Hishita, "Ca Segregation at the MgO(001) Surface Studied by Ion Scattering Spectroscopy," *Surf. Sci.* **287,** 136 (1997).

R. Souda, H. Kawanowa, S. Otani, and T. Aizawa, "Ion-Stimulated Desorption of O$^+$ from the Oxygenated TiC(111) Surface: The Role of Nonadiabatic Electronic Transitions and the Band Effect," *J. Chem. Phys.* **112,** 979 (2000).

R. Souda, C. Oshima, S. Otani, Y. Ishizawa, and M. Aono, "Structure Analysis of Oxygen Adsorbed TiC (111) by Impact Collision Ion Scattering Spectroscopy," *Surf. Sci.* **199,** 154 (1988).

R. Souda, T. Suzuki, H. Kawanowa, and E. Asari, "Low-Energy Reactive Ion Scattering as a Probe of Surface Femtochemical Reaction: H^+ and H^- Formation on Ionic Compound Surfaces," *J. Chem. Phys.* **110,** 2226 (1999).

R. Souda, T. Suzuki, H. Kawanowa, and E. Asari, "Effect of Surface Defects on Charge Exchange of Low-Energy Deuterium Ions Scattered from $SrCl_2$ and BaF_2," *Phys. Rev. B* **58,** 4143 (1998).

R. Souda, T. Suzuki, and K. Yamamoto, "Neutralization and Electronic Excitation During Low-Energy H and He Scattering from LiF," *Surf. Sci.* **397,** 63 (1998).

R. Souda, and K. Yamamoto, "Band Effect on Resonance Neutralization," *Nucl. Instrum. Meth. Phys. Res. B* **125,** 256 (1997).

R. Souda, K. Yamamoto, W. Hayami, and T. Aizawa, "Electronic Excitation During Low-Energy Ne Scattering from Mg and MgO(001)," *Surf. Sci.* **416,** 320 (1998).

R. Souda, K. Yamamoto, W. Hayami, T. Aizawa, and Y. Ishizawa, "Low-Energy He and Ne Scattering from Al(111): Reionization Versus Autoionization," *Surf. Sci.* **363,** 139 (1996).

R. Souda, K. Yamamoto, W. Hayami, T. Aizawa, and Y. Ishizawa, "Band Effect on Inelastic Rare Gas Collisions with Solid Surfaces," *Phys. Rev. Lett.* **75,** 3552 (1995).

R. Souda, K. Yamamoto, W. Hayami, T. Aizawa, and Y. Ishizawa, "Charge Exchange in Low-Energy D^+ Scattering from O^-, CO^-, and $CsCl^-$ Adsorbed Pt(111) Surfaces," *Nucl. Instrum. Meth. Phys. Res. B* **100,** 389 (1995).

R. Souda, K. Yamamoto, W. Hayami, T. Aizawa, and Y. Ishizawa, "Low-Energy H^+, He^+, N^+, O^+, and Ne^+ Scattering from Metal and Ionic-Compound Surfaces: Neutralization and Electronic Excitation," *Phys. Rev. B* **51,** 4463 (1995).

R. Souda, K. Yamamoto, W. Hayami, T. Aizawa, S. Otani, and Y. Ishizawa, "Capture and Loss of Valence Electrons in Low-Energy Proton Scattering from TiC(001)," *Surf. Sci.* **345,** 185 (1996).

S. Speller, M. Aschoff, J. Kuntze, A. Atrei, U. Bardi, E. Platzgummer, and W. Heiland, "The $Au_3Pd(001)$ Surface Studied by Ion Scattering and LEED," *Surf. Sci. Lett.* **6,** 829 (1999).

S. Speller, and W. Heiland, "Low Energy Ion Scattering and Scanning Tunneling Microscopy for Surface Structure Analysis," *Proc. XII Conference on Ion Interaction with Surfaces,* Russia, 1995.

S. Speller, S. Parascandola, and W. Heiland, "Structural Paramaters of Pd(110) and Pt(110)(1 × 2) in the Temperature Range of 300 K to 900 K Studied by Low Energy Ion Scattering," *Surf. Sci.* **383,** 131 (1997).

S. Speller, M. Schleberger, H. Franke, C. Müller, and W. Heiland, "Surface Melting and Roughening of Pb(110) Studied by Low Energy Ion Scattering," *Mod. Phys. Lett. B* **8,** 491 (1994).

S. Speller, M. Schleberger, and W. Heiland, "Structural Studies of the Pb(110) Surface with ISS and RHEED," *Surf. Sci.* **380,** 1 (1997).

S. Speller, M. Schleberger, and W. Heiland, "Ion Scattering Studies of the Pb(110) Surface from 160 to 590 K," *Surf. Sci.* **269,** 229 (1992).

S. Speller, M. Schleberger, A. Niehof, and W. Heiland, "Structural Effects on the Pb(110) Surface Between 160 and 580 K Observed by Low Energy Ion Scattering," *Phys. Rev. Lett.* **68,** 3452 (1992).

R. Spitzl, H. Niehus, and G. Comsa, "Structure Investigation of the Nitrogen Induced Cu(110)-(2 × 3) Phase with 180° Low Energy Impact Collision Ion Scattering Spectrometry," *Surf. Sci. Lett.* **250,** L355 (1991).

R. Spitzl, H. Niehus, and G. Comsa, "180° Low Energy Impact Collision Ion Scattering Spectroscopy," *Rev. Sci. Instrum.* **61,** 760 (1990).

R. Spitzl, H. Niehus, B. Poelsema, and G. Comsa, "H-Induced (1×2) Reconstruction of the Cu(110) Surface: Structure and Decomposition Kinetics," *Surf. Sci.* **239,** 243 (1990).

K. G. Standing, B. T. Chait, W. Ens, G. McIntosh, and R. Beavis, "Time-of-Flight Measurements of Secondary Organic Ions Produced by 1 keV to 16 keV Primary Ions," *Nucl. Instrum. Meth. Phys. Res. B* **198,** 33 (1982).

A. Steltenpohl, N. Memmel, E. Taglauer, Th. Fauster, and J. Onsgaard, "Pd, Au, and Codeposited Pd-Au Ultrathin Films on Ru(001)," *Surf. Sci.* **382,** 300 (1997).

N. Stolterfoht, D. Niemann, V. Hoffmann, M. Rösler, and R. Baragiola, "Plasmon Production by the Decay of Hollow Ne Atoms Near an Al Surface," *Phys. Rev. A* **61,** 52902 (2000).

P. Stracke, F. Wiegershaus, St. Krischok, V. Kempter, P. A. Zeijlmans van Emmichoven, A. Niehaus, and F. J. Garcia de Abajo, "Electron Emission in Slow Collisions of Protons with a LiF-Surface," *Nucl. Instrum. Meth. Phys. Res. B* **125,** 67 (1997).

P. Stracke, F. Wiegershaus, St. Krischok, H. Müller, and V. Kempter, "Electron Emission and Formation of Temporary Negative N^{-*} Ions in Slow Collisions of N^{+} with Insulator Surfaces (LiF; CsI)," *Nucl. Instrum. Meth. Phys. Res. B* **125,** 63 (1997).

K. Sumitomo, T. Kobayashi, F. Shoji, K. Oura, and I. Katayama, "Hydrogen-Mediated Epitaxy of Ag on Si(111) As Studied by Low-Energy Ion Scattering," *Phys. Rev. Lett.* **66,** 1193 (1991).

K. Sumitomo, K. Oura, I. Katayama, F. Shoji, and T. Hanawa, "A TOF-ISS/ERDA Apparatus for Solid Surface Analysis," *Nucl. Instrum. Meth. Phys. Res. B* **33,** 871 (1988).

K. Sumitomo, K. Tanaka, I. Katayama, F. Shoji, and K. Oura, "TOF-ICISS Study of Surface Damage Formed by Ar Ion Bombardment on Si(100)," *Surf. Sci.* **242,** 90 (1991).

M. M. Sung, J. Ahn, V. Bykov, J. W. Rabalais, D. D. Koleske, and A. E. Wickenden, "Composition and Structure of the GaN(0001)-(1×1) Surface," *Phys. Rev. B* **54,** 14652 (1996).

M. M. Sung, J. Ahn, V. Bykov, D. D. Koleske, A. E. Wickenden, and J. W. Rabalais, "Autocompensated Surface Structure of GaN Film on Sapphire," *ACS Symposium Series 681: Synthesis and Characterization of Advanced Materials,* **26,** 1998.

M. M. Sung, A. H. Al-Bayati, C. Kim, and J. W. Rabalais, "Mass and Charge Selection of Pulsed Ion Beams Using Sequential Deflection Pulses," *Rev. Sci. Inst.* **65,** 2953–2956 (1994).

M. M. Sung, V. Bykov, A. Al-Albayati, C. Kim, S. S. Todorov, and J. W. Rabalais, "From Scattering and Recoiling Spectrometry to Scattering and Recoiling Imaging," *Scan. Micro.* **9,** 321 (1995).

M. M. Sung, C. Kim, H. Bu, D. S. Karpuzov, and J. W. Rabalais, "Composition and Structure of the InP{100}-(1×1) and -(4×2) Structures," *Surf. Sci.* **322,** 116 (1995).

M. M. Sung, C. Kim, and J. W. Rabalais, "Method for Determining the Composition and Orientation of III-V {001} Semiconductor Surfaces," *Nucl. Instrum. Meth. Phys. Res. B* **118,** 522 (1996).

M. M. Sung, and J. W. Rabalais, "Comparison of Coaxial and Normal Incidence Ion Scattering," *Nucl. Instrum. Meth. Phys. Res. B* **108,** 389 (1996).

M. M. Sung, and J. W. Rabalais, "Composition and Reconstruction of the GaP{100}-(4×2)," *Surf. Sci.* **365,** 136 (1996).

M. M. Sung, and J. W. Rabalais, "Structure of the InAs{100}-(4 × 2) Surface," *Surf. Sci.* **356,** 161 (1996).

M. M. Sung, and J. W. Rabalais, "Orthogonal Alignment of the Missing-Rows and Dangling Bonds in {001} Surfaces of Different III-V Semiconductors With the Same Periodicity," *Surf. Sci.* **342,** L1137 (1995).

E. P. Suurmeijer, and A. L. Boers, "Low Energy Ion Reflection from Metal Surfaces," *Surf. Sci.* **43,** 309 (1973).

T. Suzuki, S. Hishita, K. Oyoshi, and R. Souda, "Initial Stage Growth Mechanism of Metal Adsorbates-Ti, Zr, Fe, Ge, and Ag-on MgO(001) Surface," *Surf. Sci.* **442,** 291 (1999).

T. Suzuki, S. Hishita, K. Oyoshi, and R. Souda, "Structure of α-Al$_2$O$_3$(0001) Surface and Ti Deposited on α-Al$_2$O$_3$(0001) Substrate: CAICISS and RHEED Study," *Surf. Sci.* **437,** 289 (1999).

T. Suzuki, S. Hishita, K. Oyoshi, and R. Souda, "Surface Segregation of Implanted Ions: Bi, Eu, and Ti at the MgO(100) Surface," *Appl. Surf. Sci.* **130,** 534 (1998).

T. Suzuki, and R. Souda, "Structure Analysis of CsCl Deposited on the MgO(001) Surface by Coaxial Impact Collision Atom Scattering Spectroscopy (CAICASS)," *Surf. Sci.* **442,** 283 (1999).

S. Tachi, J. A. Schultz, Y. S. Jo, and J. W. Rabalais, "Direct Recoil Time-Of-Flight Technique: In Situ Analysis of H, C, N, and O on Si(100) and Polymide Surfaces," *Proceedings of the 18th Conference on Solid State Devices and Materials,* Tokyo, 399 (1986).

E. Taglauer, "Ion Scattering Spectroscopy," *Surface Characterization,* Wiley VCH, Weinheim, 1997.

E. Taglauer, "Ion Scattering Studies of the Structure and Chemical Composition of Surfaces," *J. Surf. Anal.* **3,** 252 (1997).

E. Taglauer, "Low-Energy Ion Scattering and Rutherford Backscattering," in *Surface Analysis—The Principal Techniques,* J. C. Vickerman (Ed.), John Wiley & Sons, Chichester, 1997.

E. Taglauer, "Surface Chemical Composition," in *Handbook of Heterogeneous Catalysis,* G. Ertl, H. Knözinger, and J. Weitamp (Eds.), VCH Verlages, Weinheim, 1997.

E. Taglauer, "Probing Surfaces with Low-Energy Ions," *Nucl. Instrum. Meth. Phys. Res. B* **98,** 392 (1995).

E. Taglauer, "Probing Surfaces with Ions," *Surf. Sci.* **299,** 64 (1994).

E. Taglauer, "Ion Scattering Spectroscopy," *Ion Spectroscopies for Surface Analysis,* Plenum Press, New York, 1991.

E. Taglauer, "Low-Energy Ion Scattering Investigations of Catalysts II," in *Fundamental Aspects of Heterogeneous Catalysis Studied by Particle Beams,* H. H. Brongersma and R. A. van Santen (Eds.), Plenum Press, New York, 1991.

E. Taglauer, M. Beckschulte, R. Margraf, and D. Mehl, "Recent Developments in the Applications of Ion Scattering Spectroscopy," *Nucl. Instrum. Meth. Phys. Res. B* **35,** 404 (1988).

E. Taglauer, W. Englert, and W. Heiland, "Scattering of Low Energy Ions from Clean Surfaces: Comparison of Alkali and Rare Gas Ion Scattering," *Phys. Rev. Lett.* **45,** 740 (1980).

E. Taglauer, A. Kohl, W. Eckstein, and R. Beikler, "Aspects of Analyzing Supported Catalysts by Low-Energy Ion Scattering," *J. Mol. Catalysis A* **162,** 97 (2000).

E. Taglauer, and H. Knözinger, "Investigation of Catalyst Systems by Means of Low-Energy Ion

Scattering," *Surface Science: Principles and Applications,* Springer Proceedings in Physics **73,** 264 (1993).

E. Taglauer, and G. Staudenmaier, "Surface Analysis in Fusion Devices," *J. Vac. Sci. Tech. A* **5,** 1352 (1987).

E. Taglauer, St. Reiter, A. Liegl, and St. Schömann, "Ion Scattering and Scanning Tunneling Microscopy Studies of Stepped Cu Surfaces," *Nucl. Instrum. Meth. Res. B* **118,** 456 (1996).

E. Taglauer, A. Steltenpohl, R. Beikler, and L. Houssiau, "Scattered Ion Yields from Bimetallic Crystal Surfaces," *Nucl. Instrum. Meth. Phys. Res. B* **157,** 270 (1999).

Y. Tanaka, T. Kinoshita, K. Sumitomo, F. Shoji, K. Oura, and I. Katayama: "Ag Thin Film Growth on Hydrogen-Terminated Si(100) Surface Studied by TOF-ICISS," *Appl. Surf. Sci.* **60,** 195 (1992).

Y. Tanaka, H. Morishita, J. T. Ryu, I. Katayama, and K. Oura: "The Initial Stage of Pb Thin Film Growth on Si(111) Surface Studied by TOF-ICISS," *Nucl. Instrum. Meth. Phys. Res. B* **118,** 530 (1996).

Y. Tanaka, H. Morishita, J. T. Ryu, I. Katayama, and K. Oura: "Thin-Film Growth-Mode Analysis by Low Energy Ion Scattering," *Surf. Sci.* **363,** 161 (1996).

Y. Tanaka, H. Morishita, M. Watamori, K. Oura, and I. Katayama: "Structural Study of SrTiO3(100) Surfaces by Low Energy Ion Scattering," *Appl. Surf. Sci.* **82,** 528 (1994).

W. Tappe, A. Niehof, K. Schmidt, and W. Heiland, "Orientation Dependence of the Dissociation of H^{2+} at a Ni(110) Surface," *Europhys. Lett.* **15,** 406 (1991).

M. Tassotto, J. Gannon, and P. R. Watson, "Ion Scattering and Recoiling from Liquid Surfaces," *J. Chem. Phys.* **107,** 8899 (1997).

M. Tassotto, and P. R. Watson, "Detection Efficiency of a Channel Electron Multiplier for Low Incident Noble Gas Ions," *Rev. of Sci. Instrum.* **71,** 2704 (2000).

M. Tassotto, and P. R. Watson, "Simulation of Time-of-Flight Spectra in Direct Recoil Spectrometry for the Study of Recoil Depth Distributions and Multiple Scattering Contributions," *Surf. Sci.* **464,** 251 (2000).

M. Tassotto, and P. R. Watson, "A Correction Factor to the Screening Radius for Atomic Interactions Involving Hydrogen and its Application in Direct Recoil Spectrometry," *Surf. Sci.* **479,** 141 (2001).

S. V. Teplov, V. V. Zastavnjuk, V. Bykov, and J. W. Rabalais, "Computer Simulations of Ion Scattering and Recoiling Blocking Patterns for Surface Structure Analysis," *Surf. Sci.* **310,** 436 (1994).

S. V. Teplov, C. S. Chang, P. P. Kajarekar, T. L. Porter, and I. S. Tsong, "Energy Distribution of Ni^+ Ions Recoiled from a Ni(100) Surface," *Nucl. Instrum. Meth. Phys. Res. B* **35,** 151 (1988).

M. W. Thompson, and H. J. Pabst, "Trajectory Focusing in Surface Scattering and the Analysis of Surface Structure," *Rad. Eff. Def. Sol.* **37,** 105 (1978).

U. Thumm, "Interactions of Highly Charged Ions with Fullerenes," in *Trapping Highly Charged Ions—Fundamentals and Applications,* J. Gillaspy (Ed.), Nova Sci. Publishers, Huntington, 2001.

U. Thumm, "Soft Collisions of Highly Charged Ions with C60," *Comments At. Mol. Phys.* **34,** 119 (1999).

U. Thumm, "Hollow Ion Formation and Relaxation in Slow Bi^{46+}-C60 Collisions," *Phys. Rev. A* **55,** 479 (1997).

U. Thumm, "Electron Transfer and Emission in Slow Collisions of N^{5+} with C60," *J. Phys. B* **28,** 91 (1995).

U. Thumm, "Charge Exchange and Electron Emission in Slow Collisions of Highly Charged Ions with C60," *J. Phys. B* **27,** 3515 (1994).

U. Thumm, "Theory of Fast Ion-Surface Collisions: Emission of Conduction Band Electrons," *J. Phys. B* **25,** 421 (1992).

U. Thumm, A. Bárány, H. Cederquist, L. Hägg, and C. Setterlind, "Energy Gain in Collisions of Highly Charged Ions with C60," *Phys. Rev. A* **56,** 4799 (1997).

U. Thumm, T. Bastug, and B. Fricke, "Target-Electronic Structure Dependence in Highly Charged Ion-C60 Collisions," *Phys. Rev. A* **52,** 2955 (1995).

U. Thumm, J. Ducrée, P. Kürpick, and U. Wille, "Charge Transfer and Electron Emission in Ion–Surface Collisions," *Nucl. Instrum. Meth. B* **157,** 11 (1999).

U. Thumm, P. Kürpick, and U. Wille, "Size Quantization Effects in Atomic Level Broadening Near Thin Metallic Films," *Phys. Rev. B* **15,** 3067 (2000).

U. Thumm, and D. W. Norcross, "Angle-Differential and Momentum-Transfer Cross Sections for Low-Energy Electron-Cs Scattering," *Phys. Rev. A* **47,** 305 (1993).

U. Thumm, and D. W. Norcross, "Evidence of Very Narrow Shape Resonances in Low-Energy Electron-Cs Scattering," *Phys. Rev. Lett.* **67,** 3495 (1991).

B. Tilley, R. Souda, K. Yamamoto, W. Hayami, T. Aizawa, S. Otani, and Y. Ishizawa, "Dv-X Calculations of Charge Exchange Between D^+ and TiC, $TiC_xO_{(1-x)}$, and SrO Surfaces during Low Energy D^+ Scattering," *Jpn. J. Appl. Phys.* **36,** 3681(1997).

S. S. Todorov, H. Bu, K. H. Boyd, J. W. Rabalais, C. M. Gilmore, and J. A. Sprague, "Ion Beam Deposition of $^{107}Ag(111)$ Films on Ni(100)," *Surf. Sci.* **429,** 63 (1999).

I. S. Tsong, N. H. Tolk, T. M. Buck, J. S. Kraus, T. R. Pian, and R. Kelly, "Outer-Shell Processes in Ne^+ Collisions With a Ni(110) Surface," *Nucl. Instrum. Meth. Phys. Res. B* **194,** 655 (1982).

K. Umezawa, T. Kuroi, J. Yamane, F. Shoji, K. Oura, and T. Hanawa, "Quantitative Hydrogen Analysis by Simultaneous Detection of $^1H(^{19}F,\alpha\gamma)^{16}O$ At 6.46 MeV and ^{19}F-Erda," *Nucl. Instrum. Meth. Phys. Res. B* **33,** 634 (1988).

K. Umezawa, S. Nakanishi, and W. M. Gibson, "Surface Structure and Metal Epitaxy: Impact Collision Ion Scattering Spectroscopy Studies on Au-Ni(111)," *Surf. Sci.* **426,** 225 (1999).

K. Umezawa, S. Nakanishi, T. Yumura, W. M. Gibson, M. Watanabe, Y. Kido, S. Yamamoto, Y. Aoki, and H. Naramoto, "Surface Structure Analysis of Ni(111)-$\sqrt{3} \times \sqrt{3}$ R30°-Pb by Impact-Collision Ion-Scattering Spectroscopy," *Phys. Rev. B* **56,** 10585 (1997).

J. Umezawa, J. Yamane, T. Kuroi, F. Shoji, K. Oura, and T. Hanawa, "Nuclear Reaction Analysis and Elastic Recoil Detection Analysis of the Retention of Deuterium and Hydrogen Implanted into Si and GaAs Crystals," *Nucl. Instrum. Meth. Phys. Res. B* **33,** 638 (1988).

E. Van de Riet, J. M. Fluit, and A. Niehaus, "Determination of the Surface Vibrational Amplitude with LEIS," *Surf. Sci.* **231,** 368 (1990).

E. Van de Riet, and A. Niehaus, "Application of Low Energy Neutral Ionization Spectroscopy for Surface Structure Analysis," *Surf. Sci.* **243,** 43 (1991).

E. Van de Riet, J. B. Smeets, J. M. Fluit, and A. Niehaus, "The Structure of a Clean and Oxygen Covered Copper Surfaces Studied by Low Energy Ion Scattering," *Surf. Sci.* **214,** 111 (1989).

J. A. Van den Berg, L. K. Verheij, and D. G. Armour, "An Investigation of the Kinetics of

Structural Changes During the Early Oxidation Stages of a Ni(110) Surface Using Low Energy Ion Scattering (LEIS)," *Surf. Sci.* **84,** 408 (1979).

J. Van der Veen, "Ion Beam Crystallography of Surfaces and Interfaces," *Surf. Sci. Reports* **5,** 199 (1985).

W. F. Van der Weg, and D. J. Bierman, "Multiple Collisions of Ar and Cu Atoms on a Cu Surface," *Physics* **38,** 406 (1968).

J. N. Van Wunnik, R. Brako, K. Makoshi, and D. M. Newns, "Effect of Parallel Velocity on Charge Fraction in Ion-Surface Scattering," *Surf. Sci.* **126,** 618 (1983).

G. N. Van Wyk, W. Englert, and E. Taglauer, "An Ion Scattering Study of the Adsorption of Water on Stainless Steel," *Nucl. Instrum. Meth. Phys. Res. B* **35,** 504 (1988).

J. M. van Zoest, J. M. Fluit, T. J. Vink, and B.A. van Hassel, "Surface Structure Analysis of Oxidized Fe(100) by Low Energy Ion Scattering," *Surf. Sci.* **182,** 179 (1987).

I. Vaquila, I. L. Bolotin, T. Ito, B. N. Makarenko, and J. W. Rabalais, "Ion Fraction Map of He^+ Scattered From a Si(100)-(2×1) Surface," *Surf. Sci.* **496,** 187 (2002).

I. Vaquila, K. M. Lui, and J. W. Rabalais, "Surface Structure and Electron Density Dependence of Scattered Ne^+ Ion Fractions from the Si(100)-(2×1) Surface," *Surf. Sci.* **470,** 255 (2001).

P. Varga, W. Hofer, and H. Winter, "Ion-Neutralization of Multiple Charged Particles Near a Surface," *Scan. Elec. Micro.* **111,** 967 (1982).

P. Varga, W. Hofer, and H. Winter, "Auger Neutralization of Multiple Charged Noble Gas Ions at a Tungsten Surface," *Surf. Sci.* **117,** 142 (1982).

L. J. Varnerin, Jr., "Neutralization of Ions and Ionization of Atoms Near Metal Surfaces," *Phys. Rev.* **91,** 859 (1953).

L. K. Verheij, B. Poelsema, and A. L. Boers, "Neutralization and Ionization of Low Energy Helium Ions Scattering from a Copper Surface," *Rad. Eff. Def. Sol.* **34,** 163 (1977).

L. K. Verheij, B. Poelsema, and A. L. Boers, "Charge Exchange of Low Energy He^+ Ions and Atoms Scattered from a Copper Single Crystal," *Nucl. Instrum. Meth. Phys. Res. B* **132,** 565 (1976).

L. K. Verheij, B. Poelsema, and A. L. Boers, "The Scattering of Low Energy Helium Ions and Atoms Form a Copper Single Crystal: Elastic and Inelastic Effects," *Rad. Eff. Def. Sol.* **31,** 23 (1976).

L. K. Verheij, J. A. Van den Berg, and D. G. Armour, "Structure Analysis of an Oxidized Nickel(110) Surface Using Low Energy Ion Scattering," *Surf. Sci.* **84,** 408 (1979).

M. Vicanek, T. Schlathölter, and W. Heiland, "Molecule Scattering from Solid Surfaces: Orientation and Corrugation Effects," *Nucl. Instrum. Meth. Phys. Res. B* **125,** 194 (1997).

M. Vicanek, T. Schlathölter, and W. Heiland, "Scattering of Fast H_2 Molecules from Pd(110) Surfaces: Classical Trajectories Simulations," *Nucl. Instrum. Meth. Phys. Res. B* **115,** 206 (1996).

M. Voetz, H. Niehus, J. O'Connor, and G. Comsa, "Nitrogen Induced Ni(110)-(2×3) Reconstruction: A Structure Determination with Ion Scattering Spectroscopy and Scanning Tunneling Microscopy," *Surf. Sci.* **292,** 222 (1993).

T. Von Dem Hagen, M. Hou, and E. Bauer, "Structure Effects in Low Energy K^+ Ion Scattering from Single Crystal Surfaces," *Surf. Sci.* **117,** 134 (1982).

J. Vrijmoeth, A. G. Schins, and J. F. Van der Veen, "Structure Determination of the $CoSi_2(111)$ Surface Using Medium-Energy Ion Scattering," *Phys. Rev. B* **40,** 3121 (1989).

J. Vrijmoeth, J. F. Van der Veen, D. R. Heslinga, and T. M. Klapwijk, "Medium-Energy Ion-Scattering Study of a Possible Relation Between the Schottky-Barrier Height and the Defect Density at NiSi$_2$/Si(111) Interfaces," *Phys. Rev. B* **42,** 9598 (1990).

B. Walch, U. Thumm, M. Stöckli, and C. L. Cocke, "Angular Distributions of Projectiles following Electron Capture from C60 by 2.5 keV Ar $^{8+}$," *Phys. Rev. A* **58,** 1261 (1998).

J. H. Wang, and R. S. Williams, "Surface Structure Analysis of Au Overlayers on Si by Impact Collision Ion Scattering Spectroscopy: $\sqrt{3} \times \sqrt{3}$ and 6×6 Si(111)/Au," *Phys. Rev. B* **38,** 4022 (1988).

Y. Wang, H. Bu, T. E. Lytle, and J. W. Rabalais, "Structure of Ge Epilayers on Si{100}," *Surf. Sci.* **318,** 83 (1994).

Y. Wang, V. Bykov, and J. W. Rabalais, "Structure of a Si Epilayer on Ni{100}," *Surf. Sci.* **319,** 329 (1994).

Y. Wang, M. Shi, and J. W. Rabalais, "Coaxial Scattering as a Probe of Surface Structure: (1×2)-and (1×3)-Pt{110}," *Nucl. Inst. Meth. B* **62,** 505 (1992).

Y. Wang, M. Shi, and J. W. Rabalais, "Structure of the Si{100} Surface in the Clean (2×1), (2×1)-H Monohydride, (1×1)-H Dihydride, and c(4×4)-H Phases," *Phys. Rev. B* **48,** 1678 (1993).

Y. Wang, S. V. Teplov, and J. W. Rabalais, "Determination of the Structure of Subsurface Layers by Means of Coaxial Time-of-Flight Scattering and Recoiling Spectrometry (TOF-SARS)," *Nucl. Inst. Meth. Phys. Res. B* **90,** 237 (1994).

Y. Wang, S. V. Teplov, O. S. Zaporozchenko, V. Bykov, and J. W. Rabalais, "Coaxial Scattering Probe of the Surface and Subsurface Structure of the Si{100}-(2×1) and -(1×1)-H Phases," *Surf. Sci.* **296,** 213 (1993).

C. B. Weare, K. A. German, and J. A. Yarmoff, "Evidence for an Inhomogeneous to Homogeneous Transition in the Surface Local Electrostatic Potential of K-Covered Al(100)," *Phys. Rev. B* **52,** 2066 (1995).

C. B. Weare, and J. A. Yarmoff, "Resonant Neutralization of Li$^+$ Scattered from Alkali/Al(100) as a Probe of the Local Electrostatic Potential," *Surf. Sci.* **348,** 359 (1996).

C. B. Weare, and J. A. Yarmoff, "Resonant Neutralization of Li$^+$ Scattered from Cs/Al(100) as a Probe of the Surface Local Electrostatic Potential," *J. Vac. Sci. Tech. A* **13,** 1421 (1995).

C. B. Weare, J. A. Yarmoff, and Z. Sroubek, "The Effects of Charge Promotion on the Measured Charge State of Low-Energy Li$^+$ Scattered from Alkali/Al(100)," *Nucl. Instrum. Meth. Phys. Res. B* **119,** 352 (1996).

L. T. Weng, P. Bertrand, J. H. Stone-Masui, and W. E. Stone, "Desorption of Emulsifiers from Polystyrene Latexes Studied by Various Surface Techniques: A Comparison Between XPS, ISS, and Static SIMS," *Langmuir* **13,** 2943 (1997).

F. Wiegershaus, S. Krischok, D. Ochs, W. Maus-Friedrichs, and V. Kempter, "Electron Emission in Slow Collisions of He Projectiles (He*, He$^+$, He^{++}) with Li and LiF Surfaces," *Surf. Sci.* **345,** 91 (1996).

B. Willerding, W. Heiland, and K. J. Snowdon, "Neutralization of Fast Molecular Ions H^{+2} and N^{+2} at Surfaces," *Phys. Rev. Lett.* **53,** 2031 (1984).

R. Williams, M. Kato, R. S. Daly, and M. Aono, "Scattering Cross Sections for Ions Colliding Sequentially with Two Target Atoms," *Surf. Sci.* **225,** 355 (1990).

R. S. Williams, and J. A. Yarmoff, "Surface Structure Analysis from Low-Energy Ion Backscattering Angular Distributions," *Nucl. Instrum. Meth. Res. B* **218,** 235 (1983).

A. Wojciechowski, M. B. Medvedeva, K. Kh. Ferleger, K. Brüning, and W. Heiland, "Dissociative Scattering of H^{2+} on Metal Surfaces: Analysis of the High-Energy Parts of the Spectra of Scattered Molecule Constituents," *Nucl. Instrum. Meth. Phys. Res. B* **104,** 626 (2000).

I. Wojciechowski, M. B. Medvedeva, V. Kh. Ferleger, K. Brüning, and W. Heiland, "The Broadening of Energy Spectra of Atoms by a Solid Surface Under Molecular Ion Bombardment," *Nucl. Instrum. Meth. Phys. Res. B* **143,** 473 (1998).

D. P. Woodruff, "Trajectory and Collision Related Neutralization in Low Energy He^+ Ion Scattering," *Surf. Sci.* **116,** L219 (1982).

P. A. Wouters, P. A. Zeijlmans van Emmichoven, and A. Niehaus, "Angle Resolved Electron Spectra from Collisions of Doubly Charged Noble Gas Ions with a Metal Surface," *Surf. Sci.* **211,** 249 (1989).

P. Wurz, T. Fröhlich, K. Brüning, J. Scheer. W. Heiland, E. Hertzberg, and S. Fuselier, "Formation of Negative Ions by Scattering from a Diamond(111) Surface," *Proc. Week of Postdoc. Students,* Czech Republic, 1998.

C. Xu, and D. J. O'Connor, "Surface Relaxation Trend Study With Iron Surfaces," *Nucl. Instrum. Meth. Phys. Res. B* **53,** 315 (1991).

C. Xu, and D. J. O'Connor, "Fe(111) Surface Relaxation Analysis by In-and Out-of-Plane MEIS," *Nucl. Instrum. Meth. Phys. Res. B* **51,** 278 (1990).

C. Xu, and D. J. O'Connor, "Application of Out-of-Plane MEIS to Surface Structure Analysis," *Nucl. Instrum. Meth. Phys. Res. B* **42,** 251 (1989).

F. Xu, F. Ascione, N. Mandarino, P. Zoccali, P. Calaminici, A. Bonanno, A. Oliva, and N. Russo, "Ion Induced Atomic-Like LMM and L_2MM Auger Electron Emission from Al, Mg, Si, and Mg_xAl_{1-x}: Role of Symmetric and Asymmetric Collisions," *Phys. Rev. B* **48,** 9987 (1993).

F. Xu, R. A. Baragiola, A. Bonanno, P. Zoccali, M. Camarca, and A. Oliva, "Single to Triplet Conversion of Ne $2p^4$ Core Configuration at Metal Surfaces," *Phys. Rev. Lett.* **72,** 4041 (1994).

F. Xu, M. Camarca, A. Oliva, N. Mandarino, P. Zoccali, and A. Bonanno, "Al Double Core Electron Excitation in Asymmetric Collisions," *Surf. Sci.* **247,** 13 (1991).

F. Xu, N. Mandarino, A. Oliva, M. Camarca, A. Bonanno, and R. A. Baragiola, "Projectile L-Shell Electron Excitation in Slow Ne^+-Al Collisions," *Phys. Rev. A* **50,** 4048 (1994).

F. Xu, N. Mandarino, A. Oliva, P. Zoccali, M. Camarca, and A. Bonanno, "Al Target Electron Excitation in Asymmetric Collisions with Very Low Energy Ne^+ Projectiles," *Nucl. Instrum. Meth. Phys. Res. B* **90,** 564 (1994).

F. Xu, N. Mandarino, A. Oliva, P. Zoccali, M. Camarca, A. Bonanno, and R. A. Baragiola, "Projectile $L_{2,3}$-shell Electron Excitation in Very Low Energy Ne^+-Al Asymmetric Collisions," *Phys. Rev. A* **50,** 4040 (1994).

F. Xu, G. Manicò, F. Ascione, A. Bonanno, and A. Oliva, "Evidence of Charge Transfer from Na Adatoms to the Cs Substrate: A Collisionally Excited Autoionization Electron-Emission Study," *Phys. Rev. B* **54,** 10401 (1996).

F. Xu, G. Manicò, F. Ascione, A. Bonanno, A. Oliva, and R. A. Baragiola, "Inelastic Energy Loss in Low Energy Ne^+ Scattering from a Si Surface," *Phys. Rev. A* **57,** 1096 (1998).

F. Xu, P. Riccardi, A. Oliva, and A. Bonanno, "Ar L-shell and Metal M-shell Auger Electron Emission for 14 keV Ar^+ Ion Impact on Ca, Sc, Ti, V, Cr, Fe, Co, Ni, and Cu," *Nucl. Instrum. Meth. Phys. Res. B* **78,** 251 (1993).

Y. Yamamura, and W. Takeuchi, "Computer Studies of Trajectory Focusing Effects on the Total Reflection Coefficient," *Phys. Lett. A* **78,** 105 (1980).

Y. Yamamura, and W. Takeuchi, "Ion Focusing Effects on Total Reflection Coefficient Near the Semichannel Direction," *Rad. Eff. Def. Sol.* **49,** 251 (1980).

K. Yamashita, T. Yasue, T. Koshikawa, A. Ikeda, and Y. Kido, "High Depth Resolution Analysis of Cu/Si(111) $5' \times 5'$ Structure with Medium Energy Ion Scattering," *Nucl. Instrum. Meth. Res. B* **136,** 1086 (1998).

M. C. Yang, C. H. Hwang, and H. Kang, "Cs$^+$ Reactive Scattering from a Si(111) Surface Adsorbed with Water," *J. Chem. Phys.* **107,** 2611 (1997).

M. C. Yang, C. H. Hwang, J. K. Ku, and H. Kang, "Molecular Adsorbate Detection Using Hyperthermal Cs$^+$-Surface Scattering: Chemisorbed Water on Si(111)," *Surf. Sci.* **366,** L719 (1996).

M. C. Yang, W. Lee, and H. Kang, "Secondary Ion Mass Spectrometry without Secondary Ion Emission: Recombinative Scattering of Hyperthermal Cs$^+$ Ions from a Si(111) Surface Adsorbed with Water," *Chem. Phys.* **103,** 5149 (1995).

J. Yao, C. Kim, and J. W. Rabalais, "Design and Construction of a Triple-Axis UHV Goniometer," *Rev. Sci. Instrum.* **69,** 306 (1998).

J. Yao, P. B. Merrill, S. S. Perry, D. Marton, and J. W. Rabalais, "Thermal Stimulation of the Surface Termination of LaAlO$_3${100}," *J. Chem. Phys.* **108,** 1645 (1998).

J. A. Yarmoff, R. Blumenthal, and R. S. Williams, "Low Energy Ion Backscattering Angular Distributions of ^{6}Li$^+$ from Cu(001)," *Surf. Sci.* **165,** 1 (1986).

J. A. Yarmoff, D. M. Cyr, J. H. Huang, S. Kim, and R. S. Williams, "Impact Collision Ion Scattering Spectroscopy of Cu(110)-(2×1)-0 Using 5-keV ^{6}Li$^+$," *Phys. Rev. B* **33,** 3856 (1986).

J. A. Yarmoff, T. D. Liu, S. R. Qiu, and Z. Sroubek, "Mechanism of Electron Emission from Al(100) Bombarded by Slow Li$^+$ Ions," *Phys. Rev. Lett.* **80,** 2469 (1998).

J. A. Yarmoff, and C. B. Weare, "Resonant Charge Transfer During Scattering of Low Energy Li$^+$ from Cesiated Surfaces," *Nucl. Instrum. Meth. Phys. Res. B* **125,** 262 (1997).

J. A. Yarmoff, and R. S. Williams, "Apparatus for Low Energy Ion Scattering Spectroscopies: Imaging Angular Distributions and Collection of Angle-Resolved Energy Spectra," *Rev. Sci. Instrum.* **57,** 433 (1986).

J. A. Yarmoff, and R. S. Williams, "Quantitative Surface Structural Determination Using Impact Collision Ion Scattering Spectroscopy," *J. Vac. Tech. A* **4,** 1274 (1986).

J. A. Yarmoff, and R. S. Williams, "Computer Simulations of Ion Scattering Spectra for 2.4 keV Ne$^+$ Incident on Ni(001)," *Surf. Sci.* **127,** 461 (1983).

J. A. Yarmoff, and R. S. Williams, "Surface Crystallography from Low Energy Ion Backscattering Angular Distributions," *J. Phys. Chem.* **86,** 1927 (1982).

T. Yasue, and T. Koshikawa, "Effect of Hydrogen on Cu Formation on Si(111)," *Surf. Sci.* **377,** 923 (1997).

T. Yasue, C. Park, T. Koshikawa, and Y. Kido, "Structure and Concentration Analysis of Cu/Si(111) at Room Temperature with Medium Energy Ion Scattering," *Appl. Surf. Sci.* **70,** 428 (1993).

V. Y. Young, G. B. Hoflund, and A. C. Miller, "A Model for Analysis and Quantification of Ion Scattering Spectroscopy Data," *Surf. Sci.* **235,** 60 (1990).

M. L. Yu, and B. N. Eldridge, "Angular Anisotropy in Ion-Surface Charge Transfer," *Phys. Rev. B* **42,** 1000 (1990).

P. A. Zeijlmans van Emmichoven, A. Niehaus, P. Stracke, F. Wiegershaus, S. Krischok, V. Kempter, A. Arnau, F. J. Garcia de Abajo, and M. Penalba, "Electron Promotion in Collisions of Protons with a LiF Surface," *Phys. Rev. B* **59,** 10950 (1999).

B. Zhang, G. N. van Wyk, and E. Taglauer, "Analyzing Surface Compositions of Amorphous and Partly Crystallized CuTi Alloy Systems with AES and ISS," *Chinese J. of Atomic and Mol. Phys.* **9,** 2217 (1992).

F. Zhang, B. V. King, and D. J. O'Connor, "Low Energy Ion Scattering Investigation of the Order–Disorder Transition of the First Atomic Layer of the $Cu_3Au(100)$ Surface," *Phys. Rev. Lett.* **75,** 4646 (1995).

J. B. Zhou, H. C. Lu, T. Gustafsson, and P. Haberle, "Surface Structure of MgO(001): A Medium Energy Ion Scattering Study," *Surf. Sci.* **302,** 350 (1994).

J. F. Ziegler, J. P. Biersack, and U. Littmark, *The Stopping and Range of Ions in Solids,* Pergamon Press, New York, 1982.

P. Zoccali, A. Bonanno, M. Camarca, A. Oliva, and F. Xu, "Projectile and Target Autoionization Electron Emission in 700 eV Ne^+-Na/M (M = Cr, Cu, Mo, and Pt) Collisions," *Phys. Rev. B* **50,** 9767 (1994).

INDEX